*Isonitrile Chemistry*

# ORGANIC CHEMISTRY

## A SERIES OF MONOGRAPHS

ALFRED T. BLOMQUIST, *Editor*

*Department of Chemistry, Cornell University, Ithaca, New York*

1. Wolfgang Kirmse. CARBENE CHEMISTRY, 1964; 2nd Edition, *In preparation*
2. Brandes H. Smith. BRIDGED AROMATIC COMPOUNDS, 1964
3. Michael Hanack. CONFORMATION THEORY, 1965
4. Donald J. Cram. FUNDAMENTAL OF CARBANION CHEMISTRY, 1965
5. Kenneth B. Wiberg (Editor). OXIDATION IN ORGANIC CHEMISTRY, PART A, 1965; PART B, *In preparation*
6. R. F. Hudson. STRUCTURE AND MECHANISM IN ORGANO-PHOSPHORUS CHEMISTRY, 1965
7. A. William Johnson. YLID CHEMISTRY, 1966
8. Jan Hamer (Editor). 1,4-CYCLOADDITION REACTIONS, 1967
9. Henri Ulrich. CYCLOADDITION REACTIONS OF HETEROCUMULENES, 1967
10. M. P. Cava and M. J. Mitchell. CYCLOBUTADIENE AND RELATED COMPOUNDS, 1967
11. Reinhard W. Hoffman. DEHYDROBENZENE AND CYCLOALKYNES, 1967
12. Stanley R. Sandler and Wolf Karo. ORGANIC FUNCTIONAL GROUP PREPARATIONS, VOLUME I, 1968; VOLUME II, *In preparation*
13. Robert J. Cotter and Markus Matzner. RING-FORMING POLYMERIZATIONS, PART A, 1969; PART B, *In preparation*
14. R. H. DeWolfe. CARBOXYLIC ORTHO ACID DERIVATIVES, 1970
15. R. Foster. ORGANIC CHARGE-TRANSFER COMPLEXES, 1969
16. James P. Snyder (Editor). NONBENZENOID AROMATICS, I, 1969
17. C. H. Rochester. ACIDITY FUNCTIONS, 1970
18. Richard J. Sundberg. THE CHEMISTRY OF INDOLES, 1970
19. A. R. Katritzky and J. M. Lagowski. CHEMISTRY OF THE HETEROCYCLIC N-OXIDES, 1970
20. Ivar Ugi (Editor). ISONITRILE CHEMISTRY, 1971
21. G. Chiurdoglu (Editor). CONFORMATIONAL ANALYSIS, 1971

*In preparation*

Gottfried Schill. CATENANES, ROTAXANES, AND KNOTS

# Isonitrile Chemistry

*Edited by*
**Ivar Ugi**

*Department of Chemistry*
*University of Southern California*
*Los Angeles, California*

Academic Press    1971    New York and London

COPYRIGHT © 1971, BY ACADEMIC PRESS, INC.
ALL RIGHTS RESERVED
NO PART OF THIS BOOK MAY BE REPRODUCED IN ANY FORM,
BY PHOTOSTAT, MICROFILM, RETRIEVAL SYSTEM, OR ANY
OTHER MEANS, WITHOUT WRITTEN PERMISSION FROM
THE PUBLISHERS.

ACADEMIC PRESS, INC.
111 Fifth Avenue, New York, New York 10003

*United Kingdom Edition published by*
ACADEMIC PRESS, INC. (LONDON) LTD.
Berkeley Square House, London W1X 6BA

LIBRARY OF CONGRESS CATALOG CARD NUMBER: 73-84156

PRINTED IN THE UNITED STATES OF AMERICA

# Contents

|   |   |
|---|---|
| *List of Contributors* | ix |
| *Preface* | xi |

### 1. The Structure of Isonitriles
*J. A. Green II and P. T. Hoffmann*

|   |   |
|---|---|
| I. The History of the Structure of Isonitriles | 1 |
| II. Some Physicochemical Consequences of the Structure of the Isocyano Group | 4 |
| References | 6 |

### 2. Isonitrile Syntheses
*P. Hoffmann, G. Gokel, D. Marquarding, and I. Ugi*

|   |   |
|---|---|
| I. Introduction | 9 |
| II. The Dehydration of $N$-Monosubstituted Formamides and Related $\alpha$-Eliminations | 10 |
| III. The Classical Isonitrile Syntheses and Related Reactions | 17 |
| IV. Miscellaneous Reactions by Which Isonitriles Are Formed | 19 |
| References | 35 |

### 3. Kinetics of the Isonitrile–Nitrile Rearrangement
*Kenneth M. Maloney and B. S. Rabinovitch*

|   |   |
|---|---|
| I. Introduction | 41 |
| II. Thermal Rearrangement | 42 |
| III. External Excitation | 61 |
| References | 64 |

### 4. Simple $\alpha$-Additions
*T. Saegusa and Y. Ito*

|   |   |
|---|---|
| I. Introduction | 65 |
| II. $\alpha$-Addition Reactions of Isonitriles with Reactive Hydrogen Compounds | 67 |
| III. $\alpha$-Addition Reactions of Isonitriles with Reactive Halogen Compounds | 76 |
| IV. Oxidation of Isonitriles | 79 |

## Contents

|  |  |
|---|---|
| V. Reduction of Isonitriles | 80 |
| VI. Reactions with Carbenes and Nitrenes | 80 |
| VII. Reaction with Organometallic Compounds | 81 |
| VIII. Radical Reactions of Isonitriles | 84 |
| IX. Reactions Related to $\alpha$-Additions | 88 |
| References | 90 |

### 5. Cyclization Reactions
*H. J. Kabbe*

|  |  |
|---|---|
| I. Introduction | 93 |
| II. Four-Membered Rings | 95 |
| III. Five-Membered Rings | 97 |
| IV. Six-Membered Rings | 104 |
| V. Seven-Membered Rings | 106 |
| References | 106 |

### 6. The Reaction of Isonitriles with Boranes
*Joseph Casanova, Jr.*

|  |  |
|---|---|
| I. Isonitrile–Organoborane Reactions | 109 |
| II. Isonitrile–Organoaluminum and Cyanide–Organoaluminum Reactions | 124 |
| III. Cyanide–Organoborane Reactions | 125 |
| IV. Related Reactions: The Reaction of Carbon Monoxide and Other Lewis Bases with Organoboranes | 126 |
| References | 129 |

### 7. The Passerini Reaction and Related Reactions
*D. Marquarding, G. Gokel, P. Hoffmann, and I. Ugi*

|  |  |
|---|---|
| I. Introduction | 133 |
| II. $\alpha$-Acyloxycarbonamides | 133 |
| III. The Mechanism of the Passerini Reaction | 136 |
| IV. Tetrazoles | 139 |
| V. $\alpha$-Hydroxyamides | 140 |
| VI. $\alpha,\gamma$-Diketoamides | 142 |
| References | 143 |

### 8. Four-Component Condensations and Related Reactions
*G. Gokel, G. Lüdke, and I. Ugi*

|  |  |
|---|---|
| I. Introduction and General Remarks | 145 |
| II. Hydantoin-4-imides and 2-Thiohydantoin-4-imides | 149 |
| III. Acylated $\alpha$-Amino- and $\alpha$-Hydrazinocarbonamides | 155 |

|     |     |
| --- | --- |
| IV. Stereoselective Syntheses and the Reaction Mechanism of 4 C C | 161 |
| V. β-Lactams and Penicillanic Acid Derivatives | 181 |
| VI. Urethanes | 185 |
| VII. The Bucherer–Bergs Reaction | 186 |
| VIII. N-Alkyl-3-acyl-4-hydroquinoline-4-carbonamides | 186 |
| IX. Diacylimides, Amides, Thioamides, Selenoamides, and Amidines of α-Amino Acids | 189 |
| X. Tetrazoles | 193 |
| References | 197 |

## 9. Peptide Syntheses

*G. Gokel, P. Hoffmann, H. Kleimann, H. Klusacek, G. Lüdke, D. Marquarding, and I. Ugi*

|     |     |
| --- | --- |
| I. The General Concept | 201 |
| II. The Present Status of Classical Methods | 202 |
| III. Potential Advantages of the 4 C C Concept | 204 |
| IV. Isonitrile and Amine Components—Model Reactions | 204 |
| V. Tactics of the 4 C C Peptide Synthesis | 212 |
| References | 214 |

## 10. Coordinated Isonitriles

*Arnd Vogler*

|     |     |
| --- | --- |
| I. Introduction | 217 |
| II. Structure and Bonding in Isonitrile Complexes | 218 |
| III. Reactions of Coordinated Isonitriles | 222 |
| IV. The Synthesis of Metal Isonitrile Complexes | 231 |
| References | 232 |

## Addendum  Recent Developments in Isonitrile Chemistry  235

*G. W. Gokel*

|     |     |
| --- | --- |
| Author Index | 257 |
| Subject Index | 268 |

# List of Contributors

Numbers in parentheses indicate the pages on which the authors' contributions begin.

JOSEPH CASANOVA, JR. (*109*), *Department of Chemistry, California State College, Los Angeles, California*

G. GOKEL (*9, 133, 145, 201, 235*), *Department of Chemistry, University of Southern California, Los Angeles, California*

J. A. GREEN II (*1*), *Department of Chemistry, University of Southern California, Los Angeles, California*

P. T. HOFFMANN (*1, 9, 133, 201*), *Department of Chemistry, University of Southern California, Los Angeles, California; and Wissenschaftliches Hauptlaboratorium der Farbenfabriken Bayer, Leverkusen, Germany*

Y. ITO (*65*), *Department of Synthetic Chemistry, Kyoto University, Kyoto, Japan*

H. J. KABBE (*93*), *Wissenschaftliches Hauptlaboratorium der Farbenfabriken Bayer, Leverkusen, Germany*

H. KLEIMANN (*201*), *Wissenschaftliches Hauptlaboratorium der Farbenfabriken Bayer, Leverkusen, Germany*

H. KLUSACEK (*201*), *Department of Chemistry, University of Southern California, Los Angeles, California; and Wissenschaftliches Hauptlaboratorium der Farbenfabriken Bayer, Leverkusen, Germany*

G. LÜDKE (*145, 201*), *Department of Chemistry, University of Southern California, Los Angeles, California*

KENNETH M. MALONEY (*41*), *General Electric Lighting Research Laboratory, Cleveland, Ohio*

D. MARQUARDING (*9, 133, 201*), *Department of Chemistry, University of Southern California, Los Angeles, California; and Wissenschaftliches Hauptlaboratorium der Farbenfabriken Bayer, Leverkusen, Germany*

B. S. RABINOVITCH (*41*), *Fundamental Research Section, Battelle Memorial Institute, Pacific Northwest Laboratories, Richland, Washington*

T. SAEGUSA (*65*), *Department of Synthetic Chemistry, Kyoto University, Kyoto, Japan*

I. UGI (*9, 133, 145, 201*), *Department of Chemistry, University of Southern California, Los Angeles, California*

ARND VOGLER (*217*), *Lehrstuhl für Spez. PhysikChemie, Technische Universität, Berlin, Germany*

# *Preface*

After M. Passerini's papers appeared in the early thirties, the end of the classical era of isonitrile chemistry, very little was published in this field for almost three decades. During the past decade, however, a renaissance has occurred, numerous investigators have entered the field, the novel, intriguing results are evolving at an impressive rate.

Isonitriles are now easy to prepare and are useful intermediates for the synthesis of a wide variety of compounds. It can be predicted safely that in the near future isonitriles will no longer be a class of esoteric compounds, outside the mainstream of organic chemistry, but will be widely investigated and used in syntheses. Few areas of chemistry of broad interest can be covered in their entirety, comprehensively and in a unified manner. Isonitrile chemistry is one of these rarities.

The chemistry of isonitriles is not just the chemistry of one of the many functional classes of organic compounds; it is remarkably different from the rest of organic chemistry because the isonitriles are the only class of stable organic compounds containing formally divalent carbon. This divalent carbon accounts for the wide variety of reactions, particularly the multicomponent reactions. In fact, all reactions that lead to isonitriles and all subsequent transformations are transitions of the isonitrile carbon from the formally divalent state to the tetravalent state and vice versa, a transition which is unique within the organic chemistry of stable compounds.

This work should prove useful to anyone requiring information on the chemistry of isonitriles. It provides an introduction to as well as a comprehensive coverage of isonitrile chemistry, from its beginnings, around the middle of the last century, to 1970. The most recent developments in the field are covered in an Addendum written with the generous help of an impressive number of isonitrile chemists who responded to my request to point out recent publications and to submit recent unpublished results.

This work is comprised of ten chapters, which correspond to the major aspects of isonitrile chemistry, and an Addendum. An attempt has been made to organize the book in the following manner: Chapter 1 deals with general properties, Chapter 2 reviews isonitrile syntheses, Chapters 3 to 9 cover the major reactions, and Chapter 10 is devoted to the coordination chemistry of isonitriles. The Addendum is a compilation of recent advances in the field. In a few years there will probably be other major areas such as reactions of isonitriles with organometallic reagents, radicals, and reactions via catalytically active complexes. These three fields can be anticipated on the basis of the most recent advances, but there will surely be others.

A variety of industrial uses for isonitriles can also be foreseen because of their biocidal properties as well as their utility in building up and/or crosslinking macromolecular systems by multicomponent reactions of polyfunctional reactants. The increasingly important technological aspects of the isonitriles are covered only where the applications involve their specific chemical properties.

Isonitrile chemistry, by virtue of the unique valency status of the isonitrile carbon, contrives many intriguing problems for the physical chemist. It offers novel synthetic approaches (particularly because of the ability to participate in multicomponent reactions) to a wide variety of nitrogen-containing organic compounds, most notably the peptide and related derivatives of the $\alpha$-amino acids. The reported biosynthesis of some isonitriles as well as the pronounced effects of some isonitriles on living organisms provide a link to biology and biochemistry. The coordination properties of isonitriles are not only of interest to coordination chemists but also to those engaged in homogeneous catalysis.

I gratefully acknowledge the fact that the present volume is the product of the common effort of a great number of active isonitrile chemists, not only of those who contributed as authors, but also of many colleagues who participated in helpful discussions, made stimulating suggestions, and pointed out to the authors pertinent published and unpublished work.

I am further indebted to the Western Research Application Center (WESRAC) for helping to scan the literature for recent publications.

*Isonitrile Chemistry*

# Chapter 1

# The Structure of Isonitriles

*J. A. Green II and P. T. Hoffmann*

I. The History of the Structure of Isonitriles . . . . . . . 1
II. Some Physicochemical Consequences of the Structure of the Isocyano Group . 4
    References. . . . . . . . . . . . 6

## I. THE HISTORY OF THE STRUCTURE OF ISONITRILES

The history of isonitriles actually began several years before they were identified as a discrete class of compounds. Several chemists, trying to prepare alkyl cyanides from alkyl iodides and silver cyanide, isolated considerable amounts of substances whose "horrifying" odor often led to termination of the preparation.

In 1859, eight years before Gautier's work first appeared, Lieke[42] reacted allyl iodide and silver cyanide and obtained in reasonable yield a liquid with a "penetrating" odor, which he believed to be allyl cyanide. He tried to transform the presumed allyl cyanide into crotonic acid by acidic hydrolysis, but was surprised to obtain only formic acid. Study of this "anomalous" hydrolysis reaction was discontinued because of "continuing complaints in the neighborhood about the vile odor." Lieke carried out all his experiments outdoors because "opening a vessel containing the nitrile [*sic*] is sufficient to taint the air in a room for days."

Several years later, Meyer[48] described methyl- and ethyl "cyanide," which he had obtained by alkylation of silver cyanide without realizing that he had isolated the isonitriles. It was not until the fundamental work by Gautier[12-20] that these unpleasant smelling compounds were known to be "isomers of the ordinary nitriles."

At the same time, Hofmann[31-34] synthesized several isonitriles, among them phenyl isocyanide,* by reacting amines with chloroform and potassium

---

* In accordance with generally accepted usage, the term *isonitriles* is used for the general class of compounds, whereas the term *isocyanide* is used for specific designations (e.g., ethyl isocyanide or alkyl isocyanide).

hydroxide. Gautier and Hofmann started a lengthy series of studies which lasted the next few decades and which dealt with the peculiar bonding relationships in the new class of compounds.

Gautier saw isonitriles as "true homologs of hydrocyanic acid,"[16] since, like the acid, "they have the greatest of deleterious effects on an organism,"[17] and by hydrolysis are converted into formic acid and "substituted ammonia." Somewhat later, he observed that methyl and ethyl isonitrile, whose "detestable odors were at the same time reminiscent of artichokes and phosphorus,"[17] were perhaps not poisonous, since no ill effects resulted when he dropped them into the eyes and mouth of a dog.*[14]

On the basis of his hydrolysis results, Gautier[16] developed the first structural formula for ethyl isonitrile (I):

$$C_2H_5 \begin{Bmatrix} C \\ N \end{Bmatrix}$$

(I)

In contrast to the isomeric propionitrile (II), the "lone" carbon in isonitrile (I)

$$C_2H_5C{\}}N$$

(II)

is attached to the ethyl radical via the nitrogen atom. Since the terminal carbon may be di- as well as tetravalent, he finally suggested two structural formulas, III and IV, which were discussed further by Nef[49-54] 25 years later.

$$C_2H_5-N=C \qquad C_2H_5-N\equiv C$$

(III) (IV)

Because of the inordinately large number of observed α-addition reactions of the isonitrile carbon, Nef settled on the structural formula (V) which emphasizes the unsaturated, formally divalent character of the terminal carbon.[49]

$$C_2H_5-N=C=$$

(V)

---

* With a few exceptions, isonitriles exhibit no appreciable toxicity to mammals. As has been found in the toxicological laboratories of Farbenfabriken Bayer A.G., Elberfeld, Germany, oral and subcutaneous doses of 500–5000 mg/kg of most of the isonitriles can be tolerated by mice, yet there are exceptions like 1,4-diisocyanobutane which is extremely toxic ($LD_{50,mice} < 10$ mg/kg).

In 1930, a third, polar structure (VI) was proposed by Lindemann and Wiegrebe[43] in analogy to the structure of carbon monoxide as postulated by Langmuir[40] and which best complied with the new octet rule. In support of their proposed structure, they cited parachor measurements as evidence of the triple bond. Indeed, parachor results predict no significant contribution from resonance with a double bond structure, such as V.[4,25]

$$R-\overset{\oplus}{N}\equiv\overset{\ominus}{C}$$

(VI)

In the same year, Hammick and co-workers[29] found the partial dipole moment of the isonitrile–NC group to be opposite to that of the nitrile–CN group. Further dipole measurements with 4-substituted phenyl isocyanides[29,55] were consistent with the dipolar structure (VI) and support the linear C—N—C linkage which such a structure implies.[63] Sidgwick summarized these and other early experiments in an excellent review of structural studies of isonitriles.[62]

Soon thereafter, Brockway[5] presented electron diffraction data, later corroborated by Gordy and Pauling,[26] which supported a predominantly triple-bonded structure.[56]

The early normal coordinate analyses of isonitrile vibrational spectra by Lechner,[41] and later by Badger and Bauer,[1] yielded only limited information, indicating an almost exclusively triple-bonded structure, although not ruling out double-bond character entirely. As early as 1931, Dadieu[7] had proposed that the Raman band between 1960 and 2400 cm$^{-1}$ in isonitrile spectra was evidence for a triple bond.

Finally, two decades after the proposal of Lindemann and Wiegrebe, extensive microwave studies provided perhaps the most conclusive evidence for structure VI.[6,37] These results prove the linearity of the C—N—C bond system beyond doubt. Microwave dimensions for methyl isocyanide and the isomeric acetonitrile are given in Table I.

TABLE I

MOLECULAR DIMENSIONS OF CH$_3$NC AND CH$_3$CN[37]

|  | $d_{CH}$(Å) | $d_{CC}$(Å) | $d_{C-N}$(Å) | $d_{N\equiv C(C\equiv N)}$(Å) | <HCH |
|---|---|---|---|---|---|
| CH$_3$NC | 1.094 | — | 1.427 | 1.167 | 109°46′ |
| CH$_3$CN | 1.092 | 1.460 | — | 1.158 | 109°8′ |

Thus, the early evidence fully established the triple-bond representation (VI); the equivalent structural representation (VII) is now generally being

$$R-N{\equiv}C$$
(VII)

used. The unique system of bonding orbitals of the isocyano group leads to a number of consequences in the physicochemical properties of isonitriles, which of course may serve as latter-day confirmation of the assigned structure.

## II. SOME PHYSICOCHEMICAL CONSEQUENCES OF THE STRUCTURE OF THE ISOCYANO GROUP

The strengths of the C≡N bonds in isonitriles and nitriles are approximately equal, as is indicated by the similar C≡N stretch frequencies at ca. 2150 and 2250 cm$^{-1}$, respectively.[30] Force constants have been reported,[8,38,41,44] the most recent and probably most accurate being those reported by Duncan[8]: $k_{NC} = 16.7$ mdyne/Å and $k_{CN} = 18.1$ mdyne/Å. In addition, heat of formation calculations[65] based on thermodynamic data[11] yield similar values for the isocyano and cyano groups, i.e., $\Delta H_f = 88$–$98$ kcal/mole.

In a comparative study of the structures of the cyano and isocyano groups, Bak and co-workers[2] have calculated electron densities using the nuclear positions and dipole moments ($\mu = 3.92$ D for CH$_3$CN and $\mu = 3.83$ D for CH$_3$NC).[21] The centers of negative charge ($t_{6-}$) and positive charge ($t_{6+}$) are at quite similar positions with regard to the nitrogen nucleus, suggesting similar electron distributions between carbon and nitrogen, as shown in Figs. 1 and 2.

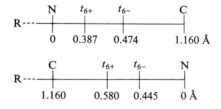

Fig. 1. *Charge distribution in isocyano and cyano groups.*[2]

R—N⌒C   or   R—N⌒C

R—C⌒N   or   R—C⌒N

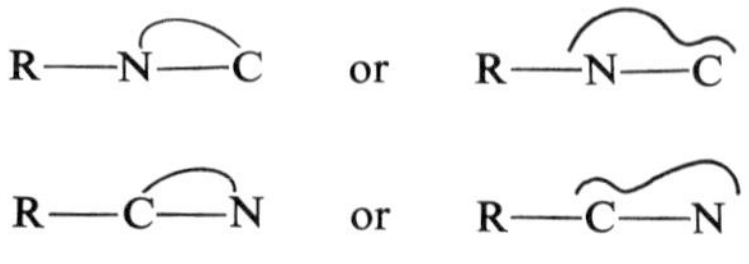

Fig. 2. *Possible π-electron density curves for isocyano and cyano groups.*[2]

## 1. The Structure of Isonitriles 5

The molar bond refraction of the isocyano group is observed to be greater than that of the cyano group attached to the same residue.[22,68] Lippincott et al.[45] have proposed on the basis of quantum-mechanical calculations that the polarizability of the isocyano group is greater, as the bond refraction data would indicate.[3] Since the bonding electron distributions are about the same,[2] Gillis suggests that the higher bond refraction of the isocyano group results from looser binding of the lone pair electrons on carbon,[22] presumably due to the lower electronegativity of carbon.

If Gillis' suggestion[22] is correct, then isonitriles rather than nitriles might be expected to be the stronger Lewis bases. Purcell[57] has made the same prediction by MO calculations. The prediction is borne out by IR hydrogen-bonding studies of isonitriles with alcohols,[35,59,60] hydrogen-bonding solvents[24] (e.g., chloroform), and hydrocarbons[10] (e.g., $C_6H_5C\equiv CH$).

These studies call attention to an interesting property of the infrared spectra of isonitriles. Horrocks and Mann[35] have found that the —NC stretch frequency *increases* with increasing solvent polarity, even to values higher than gas phase, in contrast to other multiple bonded systems, such as carbonyl compounds. For example, for $t$-butyl isocyanide, $\nu_{NC}$ increases from 2131.3 cm$^{-1}$ in cyclohexane to 2137.8 cm$^{-1}$ in acetonitrile, with the gas-phase value at ca. 2134 cm$^{-1}$. Apparently, polar solvents enhance the contribution of the polar triple-bonded structure (VI) and thus increase the frequency.

These authors[35] have observed a similar frequency shift for $p$-tolyl isocyanide, which they interpret as evidence for the absence of appreciable contribution from resonance structures $Ar^{\ominus}=N^{\oplus}=C$: in this system. Ugi and Meyr[67] observed a slight frequency decrease when strongly electron-withdrawing $p$-substituents were introduced into phenyl isocyanide. The effect is small, however, and the question of its interpretation is open.

Very thorough infrared studies have been conducted, notably by Gordy and Williams,[27,28] and more recently by Thompson and Williams.[66] All the fundamental vibrations of methyl isocyanide have been assigned. The results of normal coordinate analyses have already been correlated with the structural concept above. In addition, Williams[69] has calculated a set of thermodynamic parameters from infrared and microwave spectra.

Nuclear quadrupole coupling in isonitriles is very low,[36] indicating a near-zero electric field gradient about nitrogen.[9] This allows measurement of $^{14}N$—$^1H$ spin–spin coupling constants in isonitriles, whereas such measurements are not possible for most other organic nitrogen compounds. Several groups have taken advantage of this specific property of the isocyanide group to obtain NMR values containing $^{14}N$ terms.[39,46,47,58] Further, Spiesecke[64] has used $^{13}C$ NMR to obtain accurate structural assignments.

The near-UV spectra of aliphatic isonitriles[67] are of unusually low intensity, absorption taking place mainly below 2000 Å ($\epsilon = 0.3 \pm 0.1$ liter mole$^{-1}$ cm$^{-1}$

for cyclohexyl isocyanide at 2537 Å[61]). No detailed studies of the spectra and the species involved have been made, and the photochemistry of isonitriles has been almost totally neglected. Shaw and Pritchard[61] studied the light-induced (2537 Å) vapor-phase rearrangement of $CH_3NC$ to $CH_3CN$, but they did not elaborate regarding the possible nature of excited states, other than to allude to a suspected triplet intermediate.

The mass spectra of aliphatic isonitriles are quite similar to those of the corresponding nitriles. $\beta$-Bond cleavage predominates, presumably through a cyclic transition state, although $\alpha$-bond cleavage occurs to a greater extent in isonitriles than in nitriles,[23] reflecting the weaker R—N bond. In aromatic isonitriles, the main mode of fragmentation is expulsion of hydrogen cyanide,[70] in analogy to the behavior of the nitriles.

## REFERENCES

1. Badger, R. M., and Bauer, S. H., *J. Amer. Chem. Soc.* **59**, 303 (1937).
2. Bak, B., Hansen-Nygaard, L., and Restrup-Andersen, J., *J. Mol. Spectrosc.* **2**, 54 (1958).
3. Barrow, G. M., "Physical Chemistry," 2nd ed., p. 433. McGraw-Hill, New York, 1966.
4. Bhagwat, W. V., and Shukla, R. P., *J. Indian Chem. Soc.* **28**, 106 (1951).
5. Brockway, L. O., *J. Amer. Chem. Soc.* **58**, 2516 (1936).
6. Costain, C. C., *J. Chem. Phys.* **29**, 864 (1958).
7. Dadieu, A., *Chem. Ber.* **64B**, 358 (1931).
8. Duncan, J. L., *Spectrochim. Acta* **20**, 1197 (1964).
9. Emsley, J. W., Feeney, J., and Suttcliffe, L. H., "High Resolution Nuclear Magnetic Resonance Spectroscopy," Vol. 2, p. 1040. Pergamon Press, Oxford, 1966.
10. Ferstandig, L. L., *J. Amer. Chem. Soc.* **84**, 3533 (1962).
11. Franklin, J. L., *Ind. Eng. Chem.* **41**, 1070 (1949).
12. Gautier, A., *Ann. Chim. (Paris)* [4] **17**, 103 (1869).
13. Gautier, A., *Ann. Chim. (Paris)* [4] **17**, 203 (1869).
14. Gautier, A., *Ann. Chim. (Paris)* [4] **17**, 218 (1869).
15. Gautier, A., *Justus Liebigs Ann. Chem.* **142**, 289 (1867).
16. Gautier, A., *Justus Liebigs Ann. Chem.* **146**, 119 (1869).
17. Gautier, A., *Justus Liebigs Ann. Chem.* **146**, 124 (1868).
18. Gautier, A., *Justus Liebigs Ann. Chem.* **149**, 29 (1869).
19. Gautier, A., *Justus Liebigs Ann. Chem.* **149**, 155 (1869).
20. Gautier, A., *Justus Liebigs Ann. Chem.* **151**, 239 (1869).
21. Ghosh, S. N., Trambarulo, R., and Gordy, W., *J. Chem. Phys.* **21**, 308 (1963).
22. Gillis, R. G., *J. Org. Chem.* **27**, 4103 (1962).
23. Gillis, R. G., and Occolowitz, J. L., *J. Org. Chem.* **28**, 2924 (1963).
24. Gillis, R. G., and Occolowitz, J. L., *Spectrochim. Acta* **19**, 873 (1963).
25. Glasstone, S., and Lewis, D., "Elements of Physical Chemistry," p. 267. Van Nostrand, Princeton, New Jersey, 1960.
26. Gordy, W., and Pauling, L., *J. Amer. Chem. Soc.* **64**, 2952 (1942).
27. Gordy, W., and Williams, R. L., *J. Chem. Phys.* **3**, 664 (1935).
28. Gordy, W., and Williams, R. L., *J. Chem. Phys.* **4**, 85 (1936).
29. Hammick, D. L., New, R. G. A., Sidgwick, N. V., and Sutton, L. E., *J. Chem. Soc.* p. 1876 (1930).

30. Herzberg, G., "Molecular Spectra and Molecular Structure," Vol. 2, p. 332. Van Nostrand, Princeton, New Jersey, 1959.
31. Hofmann, A. W., *Justus Liebigs Ann. Chem.* **144**, 114 (1867).
32. Hofmann, A. W., *Justus Liebigs Ann. Chem.* **146**, 107 (1868).
33. Hofmann, A. W., *Chem. Ber.* **3**, 766 (1870).
34. Hofmann, A. W., *C. R. Acad. Sci.* **65**, 484 (1867).
35. Horrocks, W. D., and Mann, R. H., *Spectrochim. Acta* **19**, 1375 (1963).
36. Kemp, M. K., *J. Phys. Chem.* **71**, 765 (1967).
37. Kessler, M., Ring, H., Trambarulo, R., and Gordy, W., *Phys. Rev.* **79**, 54 (1950).
38. Krishnamachari, S. L. N. G., *Indian J. Phys.* **28**, 463 (1954).
39. Kuntz, I. D., Jr., Schleyer, P. von R., and Allerhand, A., *J. Chem. Phys.* **35**, 1533 (1961).
40. Langmuir, I., *J. Amer. Chem. Soc.* **41**, 868 and 1543 (1919).
41. Lechner, F., *Wien Ber.* **141**, 291 (1932).
42. Lieke, W., *Justus Liebigs Ann. Chem.* **112**, 316 (1859).
43. Lindemann, H., and Wiegrebe, L., *Chem. Ber.* **63**, 1650 (1930).
44. Linnett, J. W., *J. Chem. Phys.* **8**, 91 (1940).
45. Lippincott, E. R., Nagarajan, G., and Stutman, J. M., *J. Phys. Chem.* **70**, 78 (1966).
46. Lowenstein, A., and Margalit, Y., *J. Phys. Chem.* **69**, 4152 (1965).
47. McFarlane, M. W., *J. Chem. Soc., A* p. 1660 (1967).
48. Meyer, E., *J. Prakt. Chem.* [*1*] **67**, 147 (1866).
49. Nef, I. U., *Justus Liebigs Ann. Chem.* **270**, 267 (1892).
50. Nef, I. U., *Justus Liebigs Ann. Chem.* **280**, 291 (1894).
51. Nef, I. U., *Justus Liebigs Ann. Chem.* **287**, 265 (1895).
52. Nef, I. U., *Justus Liebigs Ann. Chem.* **298**, 202 (1897).
53. Nef, I. U., *Justus Liebigs Ann. Chem.* **298**, 368 (1897).
54. Nef, I. U., *Justus Liebigs Ann. Chem.* **309**, 126 (1899).
55. New, R. G. A., and Sutton, L. E., *J. Chem. Soc.* p. 1415 (1932).
56. Pauling, L., "The Nature of the Chemical Bond," 3rd ed., p. 270. Cornell Univ. Press, Ithaca, New York, 1960.
57. Purcell, K. F., *J. Amer. Chem. Soc.* **89**, 247 (1967).
58. Ray, J. D., Piette, L. H., and Hollis, D. P., *J. Chem. Phys.* **29**, 1022 (1958).
59. Schleyer, P. von R., and Allerhand, A., *J. Amer. Chem. Soc.* **84**, 1322 (1962).
60. Schleyer, P. von R., and Allerhand, A., *J. Amer. Chem. Soc.* **85**, 866 (1963).
61. Shaw, D. H., and Pritchard, H. O., *J. Phys. Chem.* **70**, 1230 (1966).
62. Sidgwick, N. V., *Chem. Rev.* **9**, 77 (1931).
63. Smyth, C. P., "Dielectric Behavior and Structure," pp. 249, 283, 317, and 321. McGraw-Hill, New York, 1955.
64. Spiesecke, H., *Z. Naturforsch. A* **23**, 467 (1968).
65. Szwarc, M., and Taylor, J. W., *Trans. Faraday Soc.* **47**, 1293 (1951).
66. Thompson, H. W., and Williams, R. L., *Trans. Faraday Soc.* **48**, 502 (1952).
67. Ugi, I., and Meyr, R., *Chem. Ber.* **93**, 239 (1960).
68. Vogel, A. I., Cresswell, I., Jeffery, G. H., and Leicester, J., *J. Chem. Soc.* p. 514 (1952).
69. Williams, R. L., *J. Chem. Phys.* **25**, 656 (1956).
70. Zeeh, B., *Org. Mass Spectrom.* **1**, 315 (1968).

# Chapter 2

# Isonitrile Syntheses

P. Hoffmann, G. Gokel, D. Marquarding, and I. Ugi

I. Introduction . . . . . . . . . . . . 9
II. The Dehydration of N-Monosubstituted Formamides and Related α-Eliminations 10
   A. The Phosgene Method . . . . . . . . . . 11
   B. Other Dehydrating Agents . . . . . . . . . 14
   C. Related α-Eliminations . . . . . . . . . . 16
III. The Classical Isonitrile Syntheses and Related Reactions . . . . 17
   A. The Alkylation of Cyanides . . . . . . . . . 17
   B. Dichlorocarbene Reactions . . . . . . . . . 18
IV. Miscellaneous Reactions by Which Isonitriles Are Formed . . . . 19
   A. Redox Reactions . . . . . . . . . . . 19
   B. Reactions Related to the Beckmann Rearrangement . . . . 20
   C. Formation of Isonitriles from Cyclic Precursors . . . . . 22
V. Table of the Known Isonitriles . . . . . . . . . 24
   References . . . . . . . . . . . . 35

## I. INTRODUCTION

A hundred years have passed since Gautier[34-41] and Hofmann[72-76] discovered the isonitriles. For the first ninety or so years thereafter, relatively little effort was expended in the study of these exceptionally reactive molecules. Early investigations were stimulated by interest in the fundamental questions of isomerism (R—CN vs R—NC) and whether carbon could exist in the divalent state (see Chapter 1).

There can be little doubt that the study of these compounds was delayed by the lack of enthusiasm over the odor. It should also be pointed out, however, that although some delay was due to the odor, the deterrent is greatly outweighed by the fact that isonitriles can be detected* even in trace amounts by this odor, and most of the routes leading to the formation of isonitriles were discovered by the "scent" of the reaction mixture.

* The characteristic IR absorption at 2120–2180 $cm^{-1}$ is useful for the quantitative and qualitative analysis of isonitriles. Quantitative IR analysis can be performed by comparison of the isonitrile peak with the chloroform band (2393 $cm^{-1}$) in chloroform solutions.[8] The reaction of isonitriles with oxalic acid or polysulfides[168] can also be used for assays. The complexing properties of isonitriles provide a basis for color tests.[25,139]

The principal reason for the relatively small volume of publications on the subject is that convenient and generally applicable methods for the preparation of isonitriles have become available only in the past decade, when isonitrile syntheses by dehydration of N-monosubstituted formamides (1) were developed.

## II. THE DEHYDRATION OF N-MONOSUBSTITUTED FORMAMIDES AND RELATED α-ELIMINATIONS

Considering their hydrolysis products (1), Gautier suggested that isonitriles might be regarded as derivatives of formic acid and primary amines,[37,38] and that it should therefore be possible to prepare them by a dehydration reaction. However, Gautier did not succeed in preparing isonitriles by eliminating water from the formates of primary amines by agents like phosphorus pentoxide. If Gautier had used the formamides instead and had avoided acidic reaction conditions by which isonitriles are destroyed, he would have been successful in these attempts.

$$R-NH_2 + HCO_2H \underset{+H_2O}{\overset{-H_2O}{\rightleftharpoons}} R-NH-CHO \underset{+H_2O}{\overset{-H_2O}{\rightleftharpoons}} R-NC \quad (1)$$

A wide variety of acylating agents (phosgene (see Section II, A), cyanuric chloride,[176] thionyl chloride,[166] benzenesulfonyl chloride and toluenesulfonyl chloride (see Section II, B), phosphorus tribromide,[166] phosphorus trichlorides,[166] phosphorus oxychloride (see Section II, B), phosphorus pentachloride,[166] phosphorus pentoxide,[166] and triphenylphosphine dibromide[6]) dehydrates N-monosubstituted formamides in the presence of bases (trialkylamines and dialkylarylamines, pyridine, quinoline, sodium hydroxide, potassium carbonate, and potassium t-butoxide). Particularly suitable combinations for the preparation of isonitriles are phosgene/triethylamine or sodium hydroxide, benzenesulfonyl chloride or toluenesulfonyl chloride/pyridine or quinoline, and phosphorus oxychloride/pyridine or potassium t-butoxide.

The combination of dicyclohexylcarbodiimide and pyridine hydrochloride is also effective in dehydrating cyclohexylformamide in DMF solution.[70]

Presumably the dehydration of N-monosubstituted formamides by acylating agents, e.g., phosgene, proceeds by a sequence of several steps (2):

$$R-NH-C\underset{H}{\overset{O}{\diagdown}} \xrightarrow[2R'_3N]{COCl_2} \left[R-\overset{\oplus}{NH}\cdots C\underset{H}{\overset{O-CO-Cl}{\diagdown}} Cl^-\right] \xrightarrow{-HCl} R-N=C\underset{H}{\overset{O-CO-Cl}{\diagdown}} \xrightarrow{-CO_2} \left[R-N=C\underset{H}{\overset{Cl}{\diagdown}}\right] \longrightarrow$$

$$R-NC + 2R'_3\overset{\oplus}{N}HCl^{\ominus} + CO_2 \quad (2)$$

It is not known whether the final step of (2) is an α-elimination of a proton and a chlorocarbonate or a chloride anion.

The elimination of hydrogen sulfide from N-monosubstituted thioformamides with cyanogen bromide or picryl chloride in the presence of bases proceeds in a similar manner.[7]

The N-monosubstituted formamides, which are the starting materials for isonitriles, are easily prepared from the corresponding primary amines.

N-Alkylformamides are generally readily available from primary aliphatic amines and the calculated quantity of commerical grade 70–95% formic acid, by refluxing in chlorobenzene, toluene, or xylene and removing the water formed with the aid of a water separator. N-Arylformamides are obtained by heating primary arylamines for 2–15 hr at 70–150°C with two to ten times their weight of 85–100% formic acid. Formic acid in combination with acetic anhydride or distilled formic acetic anhydride are excellent reagents for the preparation of the formamides, in particular from α-amino acid derivatives.

The formamides used as starting materials must be free from both formic acid and disubstituted formamides. Formamides which are sparingly soluble in the solvent chosen for dehydration are suspended in that solvent by thoroughly stirring in a homogenizer before the reaction is carried out.

## A. The Phosgene Method

### 1. General Procedure

Of all dehydrating agents the combination of phosgene* and bases (aqueous base solutions of pH = 7.5–9.5[3] (see also Section II, A,2,c,ii), tertiary amines,†[162,164] like trimethylamine, triethylamine, tri-n-butylamine, N,N-dimethylcyclohexylamine, N,N-diethylaniline, pyridine, and quinoline) has been found to be the most convenient, economical, and versatile. Most of the isonitriles recorded in this chapter were obtained by the phosgene method (cf. Section V).

In a preferred variation on the phosgene method, a fast current of phosgene is, without external cooling, led into a vigorously stirred solution or suspension of an N-alkylformamide or N-arylformamide in triethylamine/methylene chloride until the refluxing caused by the heat of the reaction ceases. The isonitrile is isolated by introducing ammonia‡ into the reaction mixture,

---

\* The hydrogen chloride content of the phosgene should be low.

† The tertiary amines used must be free from water and from amines which can undergo acylation.

‡ One recent report[96] recommends the use of liquid ammonia as base and solvent in the phosgene dehydration of N-monosubstituted formamides. Because ammonia reacts much faster with phosgene than with formamides, this technique is synthetically ineffective if not dangerous.

filtering to remove the precipitated ammonium chloride, and concentrating the filtrate *in vacuo*. The crude isonitrile is purified by distillation, recrystallization, reprecipitation, or chromatography, whichever is applicable. A particularly suitable method for the purification of crystalline isonitriles which are thermally unstable is by treating them (in a homogenizer*) with solvents which dissolve the impurities from the suspended isonitrile, followed by suction filtration.

The preparation of low boiling isonitriles (bp < 100°C at 760 mm Hg) or those isonitriles which are sensitive to ammonia necessitates a slightly different procedure. General modifications of the phosgene method are presented below to accommodate these special difficulties.

## 2. Examples[164]

*a. Ethyl Isocyanoacetate.* The solution of 105 g (1.06 mole) of phosgene (caution!!) in 900 ml of methylene chloride is added to a refluxing solution of 131 g (1.00 mole) of ethyl *N*-formylglycinate in 320 ml of triethylamine and 500 ml of methylene chloride. After concentrating *in vacuo*, 200 ml of benzene is added. The filtered solution is concentrated once more, and the residue is distilled *in vacuo*; bp 76–78°C/4 mmHg, yield: 87 g (77%).

*b. t-Butyl Isocyanide.* (i) 1.00 kg (10.1 moles) of phosgene (caution!!) is delivered through a wide tube into a stirred solution of 1.01 kg (10.0 moles) of *N-t*-butylformamide in 1.30 kg of trimethylamine and 7.0 liters of *o*-dichlorobenzene in a flask with a reflux condenser charged with a freezing mixture of ice and salt (20°C). Water is added, the layers are separated, and the nonaqueous layer is dried over anhydrous potassium carbonate or magnesium sulfate and fractionated; bp 90–92°C/750 mmHg, yield: 681 g (82%).

(ii) 1.00 kg (10.1 moles) of phosgene is added to 1.01 kg (10.0 moles) of *t*-butylformamide in 5.4 liters of tri-*n*-butylamine and 2.5 liters of 1,2,4-trichlorobenzene at 10–20°C. After adding 50 g of ammonia, the reaction product is distilled at 120–150 mmHg (bath temperature 80–85°C) into a receiver cooled by dry ice and purified by fractionation, yield: 648 g (78%).

(iii) If the product of preparation b,ii is isolated by steam distillation, the yield is 413 g (50%).

(iv) If the tri-*n*-butylamine of b,ii is replaced by 3.50 kg of *N*,*N*-diethylaniline, the yield of *t*-butyl isocyanide is 582 g (70%); on replacement by 3.00 kg of quinoline, the yield is 222 g (24%).

*c. Cyclohexyl Isocyanide.* (i) 1.27 kg (10.0 moles) of *N*-cyclohexylformamide, 3.20 liters of triethylamine, and 4.50 liters of methylene chloride are stirred, and phosgene (caution!!) is introduced through a wide tube rapidly enough (300–400 g/hr) to cause vigorous refluxing. When the solution ceases boiling

* For example, Ultra-Turrax from Janke & Kunkel K.G., Staufen i. Br., Germany.

(after 1.04 kg of phosgene has been added), the introduction of phosgene is immediately stopped and the solution is cooled to 20°C. About 400 g of gaseous ammonia is added to the solution over a period of 1–2 hr, and the mixture is filtered and then concentrated *in vacuo*. The residue is distilled; bp 67–72°C/14 mmHg, yield: 995 g (88%).

(ii) The solution of 2.54 kg (20 moles) of cyclohexylformamide in 6 liters of methylene chloride and 3 liters of water is stirred vigorously. At 5–10°C (pH of aqueous phase = 7.5–8.5) 2.70 kg (27.3 moles) of phosgene is introduced during 1 hr. The pH of the reaction mixture is kept constant by adding 45% aqueous sodium hydroxide ($\approx$ 5 kg). The nonaqueous layer is separated, dried over anhydrous potassium carbonate, and evaporated. The residue is distilled *in vacuo*; yield 2.00 kg (92%).

*d. 4,4′-Diisocyanodiphenylmethane.* 254 g (1.00 mole) of finely ground 4,4′-diformylaminodiphenylmethane is suspended in 650 ml of triethylamine and 1.00 liter of methylene chloride and thoroughly homogenized. The suspension is stirred and 200 g (2.02 moles) of phosgene (caution!!) is introduced into the refluxing reaction mixture. After saturating with ammonia at 20°C, the precipitated ammonia chloride is filtered off. The filtrate is concentrated *in vacuo* and the residue (219 g) is stirred at 0°C in a homogenizer with 150 ml of ether and 8 ml of isopropanol, and the product is filtered off; mp 131–133°C, yield: 181 g (83%).

3. SPECIAL CASES

Isonitriles with functional groups which are reactive toward phosgene are prepared by first protecting these groups and then deprotecting after phosgenation. Only such protective groups can be used which can be removed under neutral or basic conditions, e.g., *O*-acetyl- or trimethylsilyl for alcohols and methyl- or trimethylsilyl for carboxyls.[71]

Optically active α-isocyano esters (II) (preferably R′ = CH$_3$ or (CH$_3$)$_3$Si[71]) are needed for four-component peptide syntheses (see Chapter 9); they are prepared by phosgenating the corresponding *N*-formyl α-amino esters (I) in the presence of *N*-methyl morpholine* between −60 and −20°C (3):

$$\underset{(I)}{\text{CHO—NH—}\overset{R}{\underset{*}{\text{CH}}}\text{—CO}_2\text{R}'} \xrightarrow[\text{CH}_3\text{—N}\diagup\diagdown\text{O}]{\text{COCl}_2} \underset{(II)}{\text{CN—}\overset{R}{\underset{*}{\text{CH}}}\text{—CO}_2\text{R}'} \longrightarrow \longrightarrow$$

$$\underset{(III)}{\text{CN—}\overset{R}{\underset{*}{\text{CH}}}\text{—CO—NH—}\overset{R''}{\underset{*}{\text{CH}}}\text{—CO—}\cdots} \quad (3)$$

* When other tertiary amines are used, appreciable racemization is observed.

From the α-isocyano esters (II), the α-isocyano acids, or their sodium salts, respectively, are obtained. These can be used for the preparation of activated α-isocyano esters, which react with C-terminally protected amino acids or peptides to form "isocyano peptides" (III); some of the "isocyano peptide esters" can be obtained by phosgenation of formyl peptide esters.[71,145]

A variant of the phosgene method is the "one-step synthesis" of ester isonitriles (4). Hydroxyalkylformamides or hydroxyarylformamides (IV) react with phosgene to form isocyanodialkyl or isocyanodiaryl carbonates (V).

$$2CHO-NH-A-OH + 3COCl_2 + 6R_3N \rightarrow$$
$$(IV)$$
$$CN-A-O-CO-O-A-NC + 2CO_2 + 6R_3N \cdot HCl \quad (4)$$
$$(V)$$

If hydroxyalkylformamides or hydroxyarylformamides are reacted with acylating halides (acyl chlorides, chlorophosphates) before treatment with phosgene, the reaction (5) yields the isocyano derivatives of the corresponding esters (VI).[164]

$$Acyl-Cl + HO-A-NH \cdot CHO + COCl_2 \rightarrow Acyl-O-A-NC + CO_2 + 3R_3N \cdot HCl$$
$$(VI) \quad (5)$$

An examination of Section V reveals that the phosgene method can be made to yield almost any isonitrile, even the "isocyanoamides."[15–18]* It can be used to synthesize mono- and polyisonitriles of nearly any type provided that structural elements are avoided which destabilize the isonitrile.†

### B. Other Dehydrating Agents

The dehydration of $N$-monosubstituted formamides to isonitriles can be effected with aryl sulfochlorides or phosphorus oxychloride, but these reagents require more laborious procedures than those involving phosgene. Moreover, the phosgene method generally gives the highest yield. In laboratories, however, where safety provisions are inadequate for the safe handling of the highly toxic phosgene, phosphorus oxychloride or some sulfochloride is the reagent of choice.

---

* The simplest member of the isocyanamide family is isodiazomethane[110]; with regard to its chemical reactivity it can be considered "isocyanamide" ($NH_2$—NC).

† For example, no heterocyclic isonitriles with the isocyano group as an α-substituent of a heteroatom has ever been obtained. 2,4-dinitrophenyl isocyanide and 1,4-dichloro-2,5-diisocyanobenzene cannot be obtained either.

Benzenesulfonyl chloride[51,55,56] and toluenesulfonyl chloride[22,62,78,92] in pyridine (6) are particularly suitable for the preparation of small quantities of isonitriles.

$$R\text{—NHCHO} + Ar\text{—SO}_2Cl + 2Py \rightarrow R\text{—NC} + Py \cdot ArSO_3H + Py \cdot HCl \quad (6)$$

Quinoline[22,78] is a useful base and solvent for the preparation of low-boiling, aliphatic isocyanides, since its high boiling point facilitates the isolation of the reaction product by distillation.

The reaction of a formamide with a dehydrating halide in the presence of a base was first carried out by Hagedorn and Tönjes[51,55,56] during an attempt to elucidate the structure of xanthocillin (VII, R=H). $O,O'$-Dimethylxanthocillin (VII, R=CH$_3$—) was formed from "$O,O'$-dimethylxanthocillin dihydrate" by benzenesulfonyl chloride in pyridine.

The antibiotic xanthocillin was discovered by Rothe[142] in 1948 in cultures of *Penicillium notatum* Westling and *Penicillium chrysogenum*. It is the only known naturally occurring isonitrile. It originates from tyrosine[45] by oxidative dimerization. The isocyano group is presumably formed from a formylamino group by the dehydrating action of a high energy phosphate, such as ATP.

As noted above, formamides are dehydrated by phosphorus oxychloride[88,136,165,169] and pyridine (7):

$$2R\text{—NHCHO} + POCl_3 + 4Py \rightarrow 2R\text{—NC} + Py \cdot HPO_3 + 3Py \cdot HCl \quad (7)$$

This method gives 58–95 % yields of aliphatic isocyanides but only 7–54 % of aromatic isocyanides.

The elimination of water from $N$-arylformamides by phosphorus oxychloride proceeds satisfactorily, however, if potassium $t$-butoxide is used as the base (8). The reason for this is probably that the anions of $N$-arylformamides[65] are more readily $O$-acylated by phosphorus oxychloride than the starting material itself.

$$2[R\text{—N}\cdots\text{CH}\cdots\text{O}]^-K^+ + POCl_3 + 2(CH_3)_3COK \rightarrow$$
$$2RNC + KPO_3 + 3KCl + 2(CH_3)_3COH \quad (8)$$

It is possible in this fashion to prepare aromatic mono- and diisocyanides in 56–88 % yield.

Phosphorus oxychloride also dehydrates $N$-formylhydrazones[47] to form isocyanamide derivatives (see footnote, p. 14) and is useful for the preparation of aliphatic $\beta$-oxo-, $\beta$-hydroxy-, and $\beta$-chloroisocyanides; $\alpha,\beta$-unsaturated isocyanides are prepared by dehydrochlorination of $\beta$-chloroisocyanides [see also Matteson and Bailey[100,101] and (10)], as done in the last steps of the recent synthesis (9) of $O,O'$-dimethylxanthocillin (VII, R = CH$_3$).[47–53]

$$\text{CH}_3\text{O}-\underset{\underset{\text{OHC—NH}}{|}}{\text{C}_6\text{H}_4-\text{CO—CH}_2} \xrightarrow[\text{2. I}_2]{\text{1. C}_2\text{H}_5\text{ONa}}$$

$$\text{CH}_3\text{O}-\text{C}_6\text{H}_4-\underset{\underset{\text{OCH—NH}}{|}}{\text{CO—CH}}-\underset{\underset{\text{NH—CHO}}{|}}{\text{CH—CO}}-\text{C}_6\text{H}_4-\text{OCH}_3 \xrightarrow{\text{NaBH}_4}$$

$$\text{CH}_3\text{O}-\text{C}_6\text{H}_4-\underset{\underset{\text{OHC—NH}}{|}}{\overset{\overset{\text{OH}}{|}}{\text{CH}}-\text{CH}}-\underset{\underset{\text{NH—CHO}}{|}}{\overset{\overset{\text{OH}}{|}}{\text{CH}-\text{CH}}}-\text{C}_6\text{H}_4-\text{OCH}_3 \xrightarrow[\text{Py}]{\text{POCl}_3}$$

$$\text{CH}_3\text{O}-\text{C}_6\text{H}_4-\underset{\underset{\text{CN}}{|}}{\overset{\overset{\text{Cl}}{|}}{\text{CH}}-\text{CH}}-\underset{\underset{\text{NC}}{|}}{\overset{\overset{\text{Cl}}{|}}{\text{CH}-\text{CH}}}-\text{C}_6\text{H}_4-\text{OCH}_3 \xrightarrow[\text{Py}]{\text{KOH}}$$

$$\text{CH}_3\text{O}-\text{C}_6\text{H}_4-\text{CH}=\underset{\underset{\text{CN}}{|}}{\text{C}}-\underset{\underset{\text{NC}}{|}}{\text{C}}=\text{CH}-\text{C}_6\text{H}_4-\text{OCH}_3 \quad (9)$$
$$\text{(VII)}$$

Hydrogen cyanide is eliminated from VIII by the potassium *t*-butoxide/phosphorus oxychloride synthesis of 1-cyclohexenyl isocyanide (10).[170]

$$\underset{\text{(VIII)}}{\underset{\underset{\text{NH—CHO}}{|}}{\overset{\overset{\text{CN}}{|}}{\text{C}_6\text{H}_{10}}}} \xrightarrow{(\text{CH}_3)_3\text{COK}} \text{C}_6\text{H}_{10}-\overset{\ominus}{\text{N}}\text{···CH···O} \xrightarrow[\text{POCl}_3]{(\text{CH}_3)_3\text{COK}} \underset{\text{(10)}}{\text{C}_6\text{H}_9-\text{NC}}$$

## C. Related α-Eliminations

α-Eliminations of small molecules from formimino ethers and *N*-hydroxy-formamidines provide a somewhat less general route to isonitriles. The base-catalyzed α-elimination (11) of ethanol from IX yields phenyl isocyanide.[138]

$$\text{C}_6\text{H}_5-\text{N}=\text{CH}-\text{OC}_2\text{H}_5 \rightarrow \text{C}_6\text{H}_5-\text{NC} + \text{C}_2\text{H}_5\text{OH} \quad (11)$$

The adducts X[79] of *N*-methylene arylamines and nitrosoarenes decompose on heating to form isonitriles,[20,31] presumably by an α-elimination via a cyclic mechanism (12).

$$Ar\text{---}N\text{=}CH_2 + ON\text{---}Ar' \rightarrow Ar\text{---}N\text{=}CH\text{---}\underset{\underset{OH}{|}}{N}\text{---}Ar' \quad (12)$$

$$[Ar\text{---}N\text{=}C\overset{H}{\underset{H\leftarrow O}{\text{---}N}}\text{---}Ar'] \rightarrow Ar\text{---}NC + Ar'NHOH$$

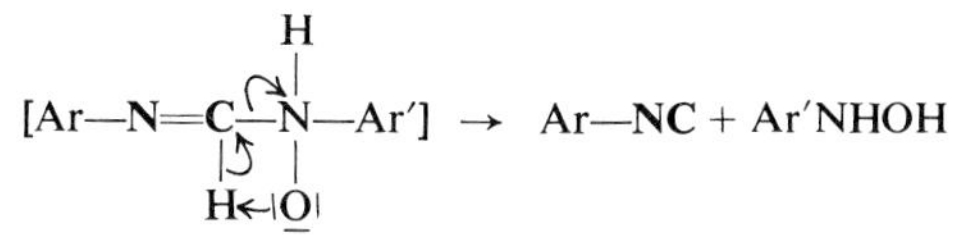

Ar = $C_6H_5$—, 4-Br—$C_6H_4$—, 4-Cl—$C_6H_4$—, 4-$CH_3$—$C_6H_4$
Ar' = $C_6H_5$—, 4-Cl—$C_6H_4$—

## III. THE CLASSICAL ISONITRILE SYNTHESES AND RELATED REACTIONS

### A. The Alkylation of Cyanides

Hydrogen cyanide is tautomeric with the parent compound of the isonitriles (13), i.e., hydrogen isocyanide.[19,27,63,85,104,107,140] It reacts with diazomethane to form a mixture of acetonitrile and methyl isocyanide (14).[134]

$$H\text{---}C\text{≡}N \rightleftarrows \overset{\ominus}{C\text{⋯}N} \rightleftarrows H\text{---}N\text{⇌}C \quad (13)$$

$$CH_2N_2 + HCN \rightarrow CH_3\text{---}CN + CH_3\text{---}NC \quad (14)$$

Hydrogen cyanide adds to ethylene, under the influence of a silent discharge, yielding ethyl isocyanide (15).[4,33]

$$C_2H_4 + HCN \xrightarrow{\Delta E_{el}} C_2H_5\text{---}NC \quad (15)$$

Nucleophilic attack of the ambident cyanide ion[42,43] on alkylating agents like alkyl halides, alkali monoalkyl sulfates and dialkyl sulfates leads to nitriles as the major products (16). Only small amounts of isonitriles are formed, generally less than 25% (16).[46,83,105,118,171]

$$R\text{---}X + CN^\ominus \rightarrow R\text{---}CN + R\text{---}NC \quad (16)$$

Appreciable yields of isonitriles can only be obtained by alkylation procedures which involve the intermediate formation of a transition metal–isonitrile complex.

Lieke,[90] Meyer,[106] and Gautier[34] prepared the first isonitriles by alkylating silver cyanide with various alkyl iodides to form isonitrile complexes (see Chapter 10). Treatment of these complexes with potassium cyanide liberates the free isonitrile (17) in yields ranging downward from $55\%$[58,80] of the silver cyanide consumed.*

$$R-I + [AgCN] \longrightarrow [R-NC \cdot AgI] \xrightarrow{KCN} R-NC \quad (17)$$

Similarly, the reaction of cuprous cyanide with an alkyl iodide yields a complex which may be decomposed into the corresponding isonitrile.[46,59] A much lower yield $(0-10\%)$[46] of the isonitrile is obtained if either Zn, Cd, or Ni is substituted for Cu(I) or Ag.

Olefins from which tertiary carbonium ions can be generated add hydrogen cyanide in the presence of cuprous halides. At 100°C, the olefin, hydrogen cyanide, and the cuprous halide (in the molar ratio 4:4:1) react to form 0.88–1.6 moles of $C_4$—$C_8$ $t$-alkyl isocyanide per mole of cuprous halide (18). Diolefins yield ditertiary diisocyanides. The cuprous halide–isonitrile complexes are the intermediates in the synthesis of isonitriles from olefins.[128]

$$(CH_3)_2C{=}CH_2 \xrightarrow[CuX]{HCN} [t\text{-}C_4H_9-NC \cdot CuX] \xrightarrow{NaCN} t\text{-}C_4H_9NC \quad (18)$$

Alkyl isocyanides can also be obtained by treating the alkylation products of alkali metal or silver hexacyanoferrates (II) or hexacyanocobaltates (III) with the hydroxides[67] or cyanides[46,59] of the group 1 metals, or even by simple heating.[168]

When ethanolic solutions of hydrogen hexacyanoferrate (II) and hydrogen cyanide are heated to 120°C, up to $40\%$ of ethyl isocyanide is formed.[60,61] The partial "esterification" of the hydrogen hexacyanoferrate (II) (19) is followed by the replacement of isonitrile ligands by hydrogen cyanide. Hydrogen hexacyanocobaltate (III) behaves in a similar fashion.[66–69]

$$H_4[Fe(CN)_6] \xrightarrow{nC_2H_5OH} H_{4-n}[(C_2H_5-NC)_nFe(CN)_{6-n}] \xrightarrow{nHCN} nC_2H_5-NC \quad (19)$$

## B. Dichlorocarbene Reactions

The "carbylamine reaction" (20), i.e., the reaction of primary amines with chloroform and strong bases, such as ethanolic potassium hydroxide solution,[13,72–74,129,132] solid alkali hydroxides,[9,24,91,95,146,157] or potassium $t$-butoxide,[136,155] has been recommended for the qualitative detection of primary amines[76] and was considered for a long time to be the most useful

---

* Similar procedures are useful for the preparation of Si-, Ge-, and Sn-isocyanides (see Section V).
† Bromoform[28] and iodoform[13] are less suitable.

method for the preparation of isonitriles.[29,32,81,143,146,149] For the carbylamine reaction, yields up to 20% with ethanolic potassium hydroxide solution as the base, up to 45% with solid alkali,* and up to 55% with potassium *t*-butoxide have been claimed.

$$R-NH_2 \xrightarrow[\text{[CCl}_2]]{\text{CHCl}_3,\text{ Base}} \left[ R-\overset{\oplus}{N}H_2-\overset{\ominus}{C}Cl_2 \xrightarrow{\beta\text{-Elim.}} R-N=CHCl \right] \xrightarrow{\alpha\text{-Elim.}} R-NC \quad (20)$$

In 1897, Nef[108,119] interpreted the Hofmann carbylamine reaction as the addition of dichlorocarbene to primary amines, followed by $\beta$-elimination of one molecule of hydrogen chloride and $\alpha$-elimination of another.[175] A similar mechanism is followed in the formation of isonitriles (15–43% yield) by the thermal decomposition of sodium trichloroacetate in the presence of arylamines, such as aniline, *p*-toluidine, or *p*-anisidine (21)[86] and possibly

$$3Ar-NH_2 + CCl_3-CO_2Na \rightarrow Ar-NC + 2Ar-NH_2 \cdot HCl + NaCl + CO_2 \quad (21)$$

also in the reaction of primary amines and carbon tetrachloride with sodium[158] or copper.[7]

Carbodiimides (e.g., diisopropyl carbodiimide) are cleaved by dichlorocarbene to form isonitriles and isonitrile dichlorides (22).[150]

$$i\text{-}C_3H_7-N=C=N-i\text{-}C_3H_7 + C_6H_5-Hg-CCl_2Br \rightarrow$$
$$i\text{-}C_3H_7-NC + i\text{-}C_3H_7-N=CCl_2 + C_6H_5-Hg-Br \quad (22)$$

## IV. MISCELLANEOUS REACTIONS BY WHICH ISONITRILES ARE FORMED

### A. Redox Reactions

Isonitriles can also be prepared by a variety of redox reactions. The scope of these reactions is generally not as wide as the majority of the $\alpha$-eliminations (Section III) because drastic conditions are required.

Cyanates[99,136] and isocyanates[112,113] are reduced by heating with phosphorus(III) compounds (23). The reduction of isothiocyanates (24) succeeds

$$R-O-C\equiv N \quad \text{or} \quad R-N=C=O \rightarrow R-NC \quad (23)$$

with a variety of agents, such as triethylphosphine,[76] copper,[172] triphenyltin hydride,[93] or during complex formation,[98] and also by photolysis.[148]

* Even higher yields (up to 85%) were reported in the literature, but they are probably due to the presence of unreacted amine in the distillate.[96]

$$R-N=C=S \xrightarrow{\text{Red.}} R-NC \qquad (24)$$

Tertiary phosphines dehalogenate isocyanide dichlorides (25).[97]

$$R-N=CCl_2 + R'_3P \rightarrow R-NC + R'_3PCl_2 \qquad (25)$$

Aliphatic isocyanide dichlorides are also reduced by iodide ion. The intermediate isocyanide diiodide is unstable and dissociates spontaneously into isonitrile and iodine (26).[135]

$$R-N=CCl_2 \xrightarrow{KI} [R-N=CI_2] \longrightarrow R-NC + I_2 \qquad (26)$$

The reduction of XI by magnesium provides an access to trifluoromethyl isocyanide (27).[94]

$$CF_3-NH-CF_2Br \xrightarrow{Mg} CF_3-NC \qquad (27)$$
$$(XI)$$

During the oxidation of XII with sodium chlorite, hypochlorite or mercuric oxide, isonitriles are formed, presumably via a diazo intermediate (28).[89]

$$R-NH-CS-NH-NH_2 \rightarrow [R-N=C=N_2] \rightarrow R-NC \qquad (28)$$
$$(XII)$$
$$R = c\text{-}C_6H_{11}-, C_6H_5-$$

Brackmann and Smit[14] believe that the copper(II) chloride catalyzed oxidation of *n*-butylamine/methanol with oxygen leads to *n*-butyl isocyanide.

## B. Reactions Related to the Beckmann Rearrangement

The *O*-tosyl oximes of 3,5-disubstituted 4-hydroxybenzaldehyde and of *p*-dimethylaminobenzaldehyde eliminate *p*-toluenesulfonic acid to give mixtures of the corresponding nitriles and isonitriles (40–92% yield of mixtures). The isonitrile is formed from the *syn*-isomer (XIII) by an abnormal Beckmann rearrangement (29).[109,156] A similar interpretation can be given

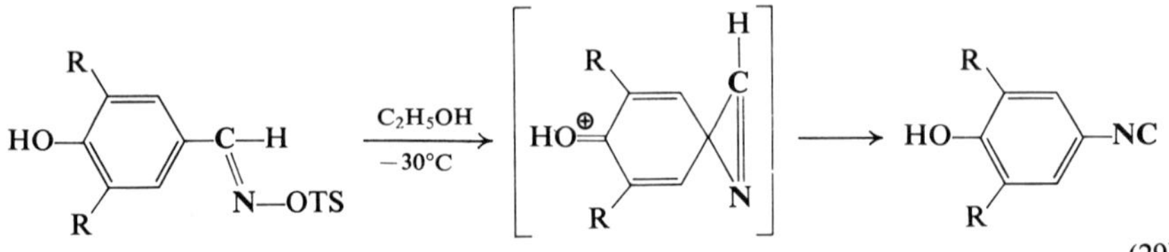

(XIII)

R = Br, Cl, CH$_3$, C(CH$_3$)$_3$

(29)

for the boron trifluoride–mercuric oxide catalyzed reaction of the *syn*-oximes (XIV) of benzaldehyde, *p*-methylbenzaldehyde, and *p*-nitrobenzaldehyde with methylketene diethylacetal (30).[111,114] Similar treatment of the *anti*-oximes yields only the nitriles (60–80%). Aryl cyanates also transform aryl aldoximes

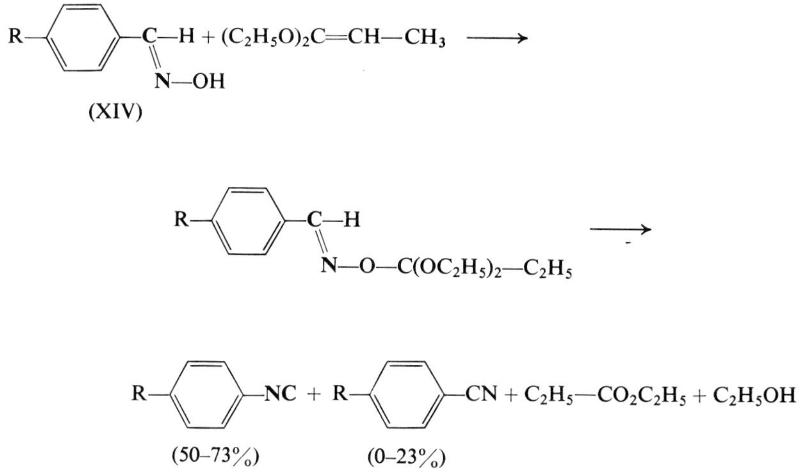

$$R-\underset{(XIV)}{\underset{N-OH}{C_6H_4-C(-H)}} + (C_2H_5O)_2C=CH-CH_3 \longrightarrow$$

$$R-C_6H_4-\underset{N-O-C(OC_2H_5)_2-C_2H_5}{C-H} \longrightarrow$$

$$R-C_6H_4-NC + R-C_6H_4-CN + C_2H_5-CO_2C_2H_5 + C_2H_5OH \quad (30)$$
$$(50-73\%) \qquad (0-23\%)$$

into isonitriles.[44]

Werner and Piquet[174] attempted a Beckmann rearrangement with benzenesulfonyl chloride and alkali on γ-benzil monoxime (XV). They obtained phenyl isocyanide, benzoic acid, and a sulfur compound which was hydrolyzed by base to yield benzoic acid and phenyl isocyanide (31).

The reaction was interpreted as a Beckmann rearrangement of the γ-benzil monoxime *O*-benzenesulfonate to phenylglyoxyl anilide *O*-benzenesulfonate

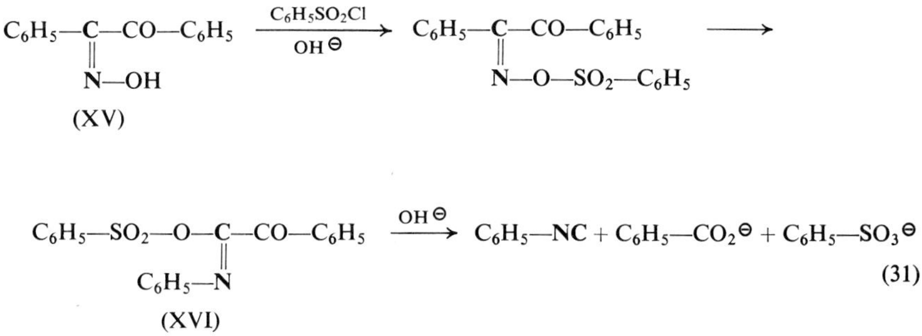

$$\underset{(XV)}{\underset{N-OH}{C_6H_5-C-CO-C_6H_5}} \xrightarrow[OH^\ominus]{C_6H_5SO_2Cl} \underset{N-O-SO_2-C_6H_5}{C_6H_5-C-CO-C_6H_5} \longrightarrow$$

$$\underset{(XVI)}{\underset{C_6H_5-N}{C_6H_5-SO_2-O-C-CO-C_6H_5}} \xrightarrow{OH^\ominus} C_6H_5-NC + C_6H_5-CO_2^\ominus + C_6H_5-SO_3^\ominus \quad (31)$$

(XVI), followed by alkali-induced fragmentation of the latter. This interpretation is in accordance with the α-fragmentation of α-keto imidochlorides (XVII) by alkali[116-118] or by amines (32).[161,163] An α-fragmentation also

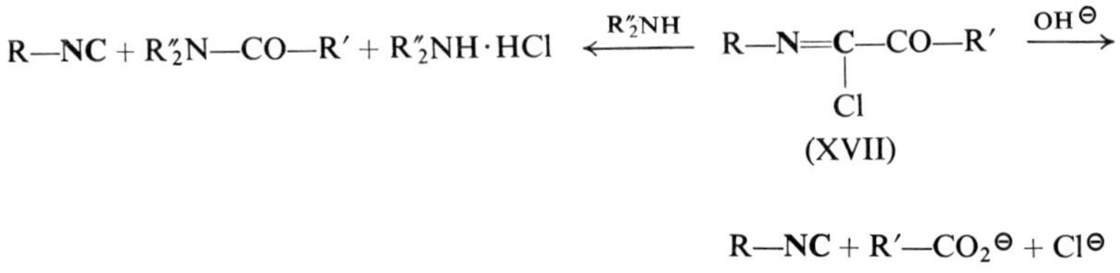

$$R-NC + R''_2N-CO-R' + R''_2NH \cdot HCl \xleftarrow{R''_2NH} R-N=C(Cl)-CO-R' \xrightarrow{OH^\ominus}$$

(XVII)

$$R-NC + R'-CO_2^\ominus + Cl^\ominus \quad (32)$$

occurs during the anomalous Beckmann rearrangement of benzoin oxime (XVIII) (33).[173]

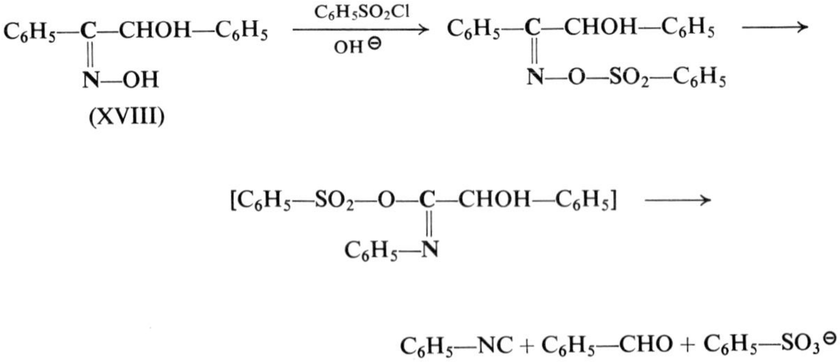

$$C_6H_5-C(=N-OH)-CHOH-C_6H_5 \xrightarrow[OH^\ominus]{C_6H_5SO_2Cl} C_6H_5-C(=N-O-SO_2-C_6H_5)-CHOH-C_6H_5 \longrightarrow$$

(XVIII)

$$[C_6H_5-SO_2-O-C(=N-C_6H_5)-CHOH-C_6H_5] \longrightarrow$$

$$C_6H_5-NC + C_6H_5-CHO + C_6H_5-SO_3^\ominus \quad (33)$$

## C. Formation of Isonitriles from Cyclic Precursors

Quinazoline 3-oxide is converted by acetic anhydride into 2-isocyanobenzonitrile (34).[64]

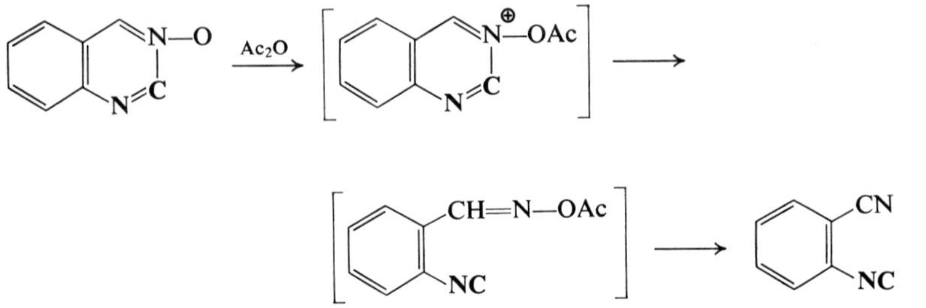

(34)

Ploquin[137] has reported the preparation of phenyl isocyanide from 2-methylpyridine by reaction with dichlorocarbene.

The formation of isonitriles by the anomalous Hofmann degradation of β-arylglycidamides (XIX) (25) presumably occurs by a complex mechanism with a rearrangement of the carbon skeleton. It should be investigated by $^{14}$C labeling techniques.

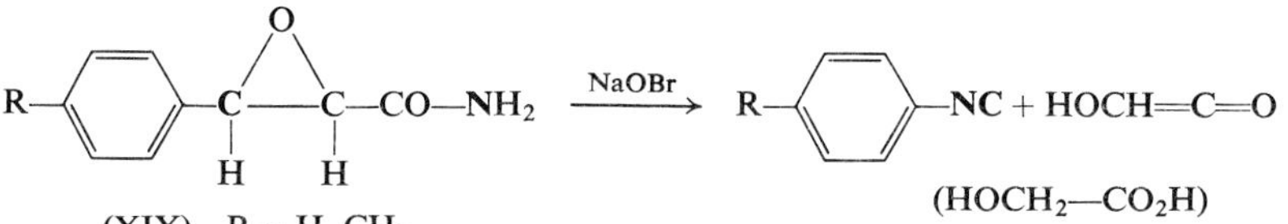

$$R\text{-}C_6H_4\text{-}\underset{H}{\overset{}{C}}\text{-}\underset{H}{\overset{O}{\overset{|}{C}}}\text{-}CO\text{-}NH_2 \xrightarrow{\text{NaOBr}} R\text{-}C_6H_4\text{-}NC + HOCH\text{=}C\text{=}O \quad (35)$$

(XIX)  R = H, CH$_3$              (HOCH$_2$—CO$_2$H)

α-Lactams (XX) isomerize and decompose on heating (36).[152-154]

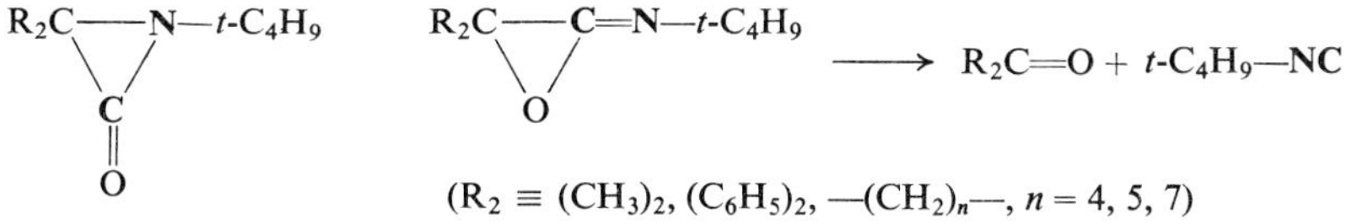

$$R_2C\text{-}\underset{\underset{O}{\overset{\|}{C}}}{\overset{}{}}\text{-}N\text{-}t\text{-}C_4H_9 \quad\quad R_2C\text{-}\underset{O}{\overset{}{}}\text{-}C\text{=}N\text{-}t\text{-}C_4H_9 \longrightarrow R_2C\text{=}O + t\text{-}C_4H_9\text{-}NC \quad (36)$$

$(R_2 \equiv (CH_3)_2, (C_6H_5)_2, \text{-}(CH_2)_n\text{-}, n = 4, 5, 7)$

p-Tolyl isocyanide is one of the products of peracid oxidation[137] of diphenyl-ketene p-tolylimine (XXI).[82] Mechanistically (36) and (37) are closely related.

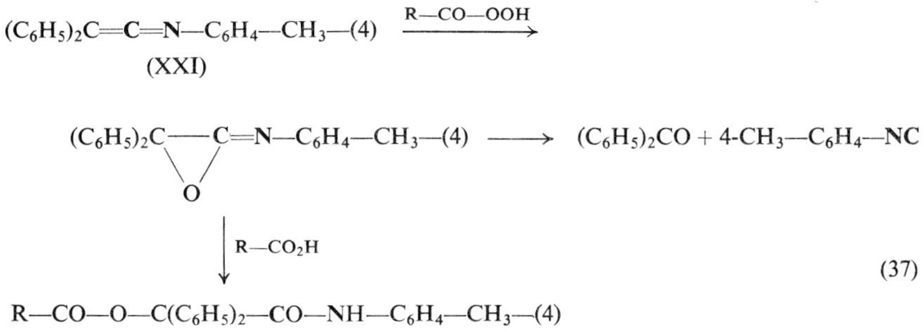

$$(C_6H_5)_2C\text{=}C\text{=}N\text{-}C_6H_4\text{-}CH_3\text{-}(4) \xrightarrow{R\text{-}CO\text{-}OOH}$$
(XXI)

$$(C_6H_5)_2C\text{-}\underset{O}{\overset{}{}}\text{-}C\text{=}N\text{-}C_6H_4\text{-}CH_3\text{-}(4) \longrightarrow (C_6H_5)_2CO + 4\text{-}CH_3\text{-}C_6H_4\text{-}NC$$

$$\downarrow R\text{-}CO_2H \quad\quad\quad (37)$$

$$R\text{-}CO\text{-}O\text{-}C(C_6H_5)_2\text{-}CO\text{-}NH\text{-}C_6H_4\text{-}CH_3\text{-}(4)$$

The formation of small quantities of cyclohexyl isocyanide from N-cyclohexyldichloroacetamide and potassium t-butoxide[160] and of 4-chlorophenyl isocyanide from N-(4-chlorophenyl)dichloromethylsulfonamide (38)[30] occur presumably by similar mechanisms via three-membered cyclic intermediates.

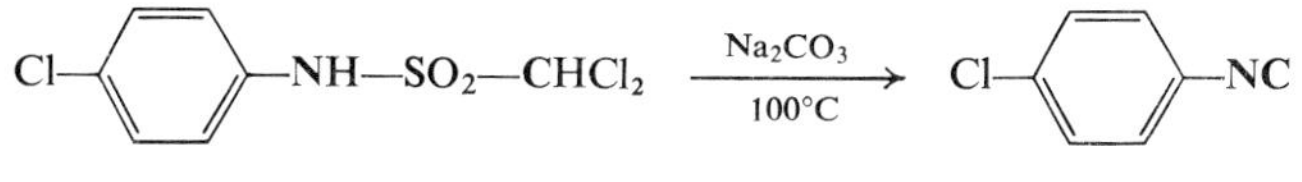

$$Cl\text{-}C_6H_4\text{-}NH\text{-}SO_2\text{-}CHCl_2 \xrightarrow[100°C]{Na_2CO_3} Cl\text{-}C_6H_4\text{-}NC \quad (38)$$

TABLE I

THE KNOWN ISONITRILES*

| Compound | Yield (%) | Reference |
|---|---|---|
| $C_2$ | | |
| 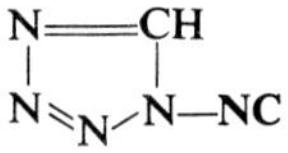 ![N=CH / N≡N–N–NC triazole structure] | — | 57 |
| $CF_3$—NC | 37.5 | 101 |
| $CH_3$—NC | 67 | 6, 10, 22, 23, 24, 35, 46, 74, 78, 80, 83, 91, 134, 164, 168, 171 |
| $C_3$ | | |
| $CH_2$—$(NC)_2$ | — | 122, 124 |
| $CH_2$=$CH$—$NC$ | 49 | 100, 101 |
| $C_2H_5$—NC | 65 | 4, 6, 10, 22, 23, 33, 35, 46, 60, 74, 80, 99, 112, 113, 115, 164, 171 |
| $(CH_3)_2N$—NC | — | 17 |
| $C_4$ | | |
| CN—$(CH_2)_2$—NC | 64 | 29, 124, 159, 164 |
| $CH_2$=$CH$—$CH_2$—NC | 62 | 46, 90, 164 |
| $n$-$C_3H_7$—NC | 95 | 46, 171 |
| $i$-$C_3H_7$—NC | 75 | 23, 32, 39, 40, 150, 164, 166 |
| $(CH_3)_2Si(NC)_2$ | 54 | 103 |
| $(CH_3)_3Ge$—NC | — | 151 |
| $(CH_3)_3Si$—NC | 80 | 11, 12, 103 |
| $(CH_3)_3Sn$—NC | — | 151 |
| $C_5$ | | |
| $c$-$C_4H_7$—NC | 24 | 23 |
| $C_2H_5O$—CO—$CH_2$—NC | 77 | 164 |
| $CH_3$—NH—CO—O—$(CH_2)_2$—NC | 73 | 164 |
| $n$-$C_4H_9$—NC | 75 | 28, 32, 37, 46, 78, 92, 95, 99, 164–166, 176 |
| $i$-$C_4H_9$—NC | 95 | 22, 46, 78, 171 |
| $s$-$C_4H_9$—NC | 50 | 22, 23 |
| $t$-$C_4H_9$—NC | 82 | 10, 23, 25, 78, 95, 121, 128, 152–154, 164–166 |
| $(C_2H_5)_2N$—NC | 67 | 15, 17 |
| $C_6$ | | |

TABLE I (*Continued*)

| Compound | Yield (%) | Reference |
|---|---|---|
| 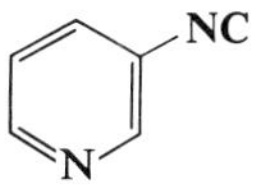 3-pyridyl—NC | 71 | 125 |
| 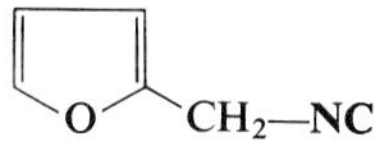 2-furyl—CH$_2$—NC | 77 | 125, 164 |
| CN—(CH$_2$)$_4$—NC | 58 | 29, 124, 159 |
| C$_2$H$_5$O—CO—(CH$_2$)$_2$—NC | 64 | 164 |
| 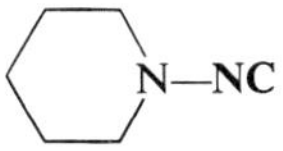 piperidinyl N—NC | 36 | 17 |
| n-C$_5$H$_{11}$—NC | — | 74 |
| (CH$_3$)$_2$(C$_2$H$_5$)C—NC | — | 128 |
| (CH$_3$)$_2$CH—(CH$_2$)$_2$—NC | 91 | 46, 74, 171 |
| (CH$_3$)$_2$N—(CH$_2$)$_3$—NC | 66 | 157, 162, 164 |
| (CH$_3$)$_3$Si—Si(CH$_3$)$_2$—NC | — | 26 |
| C$_7$ | | |
| C$_6$Cl$_5$—NC | 64 | 164 |
| 2,4,6-Br$_3$—C$_6$H$_2$—NC | 86 | 164 |
| 2,6-Br$_2$—C$_6$H$_3$—NC | 93 | 164 |
| 2,6-Cl$_2$—C$_6$H$_3$—NC | 97 | 164 |
| 3,4-Cl$_2$—C$_6$H$_3$—NC | 42 | 164 |
| 3,5-Br$_2$—4-OH—C$_6$H$_2$—NC | 92 | 109 |
| 3,5-Cl$_2$—4-OH—C$_6$H$_2$—NC | 50 | 109 |
| 4-Br—C$_6$H$_4$—NC | — | 79 |
| 2-Cl—C$_6$H$_4$—NC | 43 | 166 |
| 4-Cl—C$_6$H$_4$—NC | 54 | 23, 24, 30, 79, 166 |
| 3-NO$_2$—C$_6$H$_4$—NC | 93 | 13, 164, 166 |
| 4-NO$_2$—C$_6$H$_4$—NC | 68 | 13, 114, 162, 164, 166 |
| C$_6$H$_5$—NC | 76 | 1, 6, 23, 25, 73, 74, 86, 91, 95, 111, 114–116, 162, 164–166, 173, 174 |
| 4-OH—C$_6$H$_4$—NC | 90 | 109 |
| 4-H$_2$N—C$_6$H$_4$—NC | 30 | 126 |
| 4-NH$_2$—SO$_2$—C$_6$H$_4$—NC | 5 | 126, 146 |
| OC-[O(CH$_2$)$_2$NC]$_2$ | 86 | 164 |
| $\Delta^1$-c-C$_6$H$_9$—NC | 74 | 170 |

TABLE I (*Continued*)

| Compound | Yield (%) | Reference |
|---|---|---|
| c-C$_6$H$_{11}$—NC | 88 | 6, 113, 141, 162, 164–166, 169, 176 |
| CH$_3$O—CO—CH—i-C$_3$H$_7$—NC | 76 | 162, 164 |
| t-C$_4$H$_9$O—CO—CH$_2$—NC | 77 | 162, 164 |
| 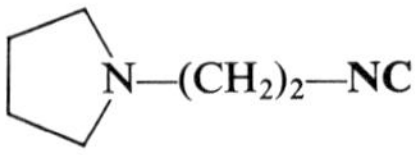 pyrrolidine-N—(CH$_2$)$_2$—NC | 64 | 164 |
| 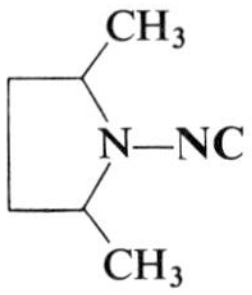 2,5-dimethylpyrrolidine-N—NC | 60 | 17 |
| CH$_3$—NH—CO—O—CH$_2$—C(CH$_3$)$_2$—NC | 73 | 164 |
| n-C$_6$H$_{13}$—NC | 66 | 32, 92, 113 |
| (CH$_3$)(C$_2$H$_5$)$_2$C—NC | — | 128 |
| (CH$_3$)$_2$(n-C$_3$H$_7$)C—NC | — | 128 |
| (CH$_3$)$_2$(i-C$_3$H$_7$)C—NC | — | 128 |
| (C$_2$H$_5$)$_2$N—(CH$_2$)$_2$—NC | 15 | 157 |
| (i-C$_3$H$_7$)$_2$N—NC | 82 | 17 |
| (C$_2$H$_5$)$_3$Si—NC | 64 | 11 |
| C$_8$ | | |
| C$_6$Cl$_4$(NC)$_2$-(1,3) | 61 | 159, 164 |
| C$_6$Cl$_4$(NC)$_2$-(1,4) | 84 | 159, 164 |
| 2-CF$_3$—4-Cl—C$_6$H$_3$—NC | 80 | 164 |
| 2,4,6-Cl$_3$—C$_6$H$_2$—CH$_2$—NC | 49 | 164 |
| 2-CH$_3$O—3,5,6-Cl$_3$—C$_6$H—NC | 65 | 164 |
| C$_6$H$_4$—(NC)$_2$-(1,2) | 32 | 159, 164 |
| C$_6$H$_4$—(NC)$_2$-(1,3) | 83 | 124, 159, 164 |
| C$_6$H$_4$—(NC)$_2$-(1,4) | 90 | 123, 124, 126, 159, 164 |
| 2-NC—C$_6$H$_4$—NC | 20 | 64 |
| 4-NC—C$_6$H$_4$—NC | 82 | 164 |
| 2-CH$_3$O—4-NO$_2$—5-Cl—C$_6$H$_2$—NC | 65 | 164 |
| 2,6-Cl$_2$—C$_6$H$_3$—CH$_2$—NC | 49 | 164 |
| 3,4-Cl$_2$—C$_6$H$_3$—CH$_2$—NC | 46 | 164 |
| 2-CH$_3$O—4,5-Cl$_2$—C$_6$H$_2$—NC | 50 | 164 |
| 4-HOOC—C$_6$H$_4$—NC | 34 | 146 |
| 4-Cl—C$_6$H$_4$—CH$_2$—NC | 54 | 164 |
| 2-CH$_3$—3-Cl—C$_6$H$_3$—NC | 79 | 164 |
| 2-CH$_3$—6-Cl—C$_6$H$_3$—NC | 87 | 166 |
| 3-CH$_3$—4-Cl—C$_6$H$_3$—NC | 48 | 164 |

## 2. Isonitrile Syntheses

TABLE I (*Continued*)

| Compound | Yield (%) | Reference |
|---|---|---|
| 2-CH$_3$O—4-Cl—C$_6$H$_3$—NC | 73 | 164 |
| 2-CH$_3$O—5-Cl—C$_6$H$_3$—NC | 23 | 164 |
| 4-NO$_2$—C$_6$H$_4$—CH$_2$—NC | 84 | 164 |
| 2-CH$_3$—5-NO$_2$—C$_6$H$_3$—NC | 54 | 164 |
| 2-CH$_3$—6-NO$_2$—C$_6$H$_3$—NC | 16 | 164 |
| 3-NO$_2$—4-CH$_3$—C$_6$H$_3$—NC | 59 | 164 |
| 2-CH$_3$O—4-NO$_2$—C$_6$H$_3$—NC | 65 | 164 |
| 2-NO$_2$—4-CH$_3$O—C$_6$H$_3$—NC | 60 | 164 |
| 2-CH$_3$O—5-NO$_2$—C$_6$H$_3$—NC | 87 | 164 |
| C$_6$H$_5$—CH$_2$—NC | 82 | 46, 92, 95, 116, 149, 162, 164–166, 176 |
| 2-CH$_3$—C$_6$H$_4$—NC | 83 | 91, 116, 147, 162, 164–167, 169 |
| 4-CH$_3$—C$_6$H$_4$—NC | 72 | 1, 24, 73, 79, 86, 91, 113, 114, 116, 162, 164, 166 |
| 2-CH$_3$O—C$_6$H$_4$—NC | — | 91 |
| 4-CH$_3$O—C$_6$H$_4$—NC | 63 | 23, 25, 86, 91, 165 |
| 4-CH$_3$—SO$_2$—C$_6$H$_4$—NC | 85 | 164 |
|  (norbornyl)—NC | 84 | 164 |
| c-C$_6$H$_{10}$—(NC)$_2$-(1,3) | 97 | 124, 164 |
| c-C$_6$H$_{10}$—(NC)$_2$-(1,4) | 96 | 123, 124, 159, 164 |
| 1-NC—c-C$_6$H$_{10}$—NC | 73 | 164 |
| $\Delta^3$-c-C$_6$H$_9$—CH$_2$—NC | 87 | 164 |
| CN—(CH$_2$)$_6$—NC | 79 | 29, 124, 159, 164 |
| 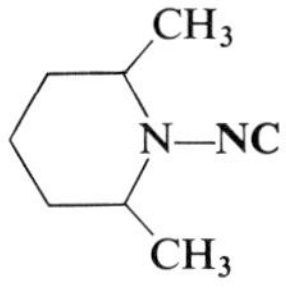 2,6-di-CH$_3$-piperidine-N—NC | 66 | 17 |
| 3-CH$_3$—c-C$_6$H$_{10}$—NC | — | 143 |
| n-C$_7$H$_{15}$—NC | 68 | 92 |
| (C$_2$H$_5$)$_2$N—(CH$_2$)$_3$—NC | 65 | 157 |
| C$_9$ | | |
| 1-CH$_3$—2,5,6-Cl$_3$—C$_6$(NC)$_2$-(2,4) | 67 | 159, 164 |
| C$_6$Cl$_5$—S—(CH$_2$)$_2$—NC | 97 | 164 |

TABLE I (*Continued*)

| Compound | Yield (%) | Reference |
|---|---|---|
| 1-CH$_3$—C$_6$H$_3$—(NC)$_2$-(2,4) | 34 | 159, 164 |
| 1-CH$_3$—C$_6$H$_3$—(NC)$_2$-(2,5) | 72 | 159, 164 |
| 1-CH$_3$—C$_6$H$_3$—(NC)$_2$-(2,6) | 32 | 159, 164 |
| 2-CH$_3$O$_2$C—5-NO$_2$—C$_6$H$_3$—NC | 20 | 164 |
| 3-CH$_3$CO—C$_6$H$_4$—NC | 97 | 164 |
| C$_6$H$_5$—CHCl—CH$_2$—NC | 41 | 50, 52, 54 |
| C$_6$H$_5$—CH=CH—NC | 94 | 50, 52, 54 |
| 3-CH$_3$CO—C$_6$H$_4$—NC | 97 | 164 |
| 4-CH$_3$CO—C$_6$H$_4$—NC | — | 132 |
| 2-CH$_3$O—4-Cl—5-CH$_3$—C$_6$H$_2$—NC | 59 | 164 |
| 2,4-(CH$_3$O)$_2$—5-Cl—C$_6$H$_2$—NC | 40 | 164 |
| 2,5-(CH$_3$O)$_2$—4-Cl—C$_6$H$_2$—NC | 84 | 164 |
| 4-F—C$_6$H$_4$—CH(CH$_3$)—NC | 73 | 164 |
| 2,4-(CH$_3$)$_2$—5-NO$_2$—C$_6$H$_2$—NC | 46 | 164 |
| 2,4-(CH$_3$)$_2$—6-NO$_2$—C$_6$H$_2$—NC | 29 | 164 |
| 2-CH$_3$O—4-NO$_2$—5-CH$_3$—C$_6$H$_3$—NC | 83 | 164 |
| 2,5-(CH$_3$O)$_2$—4-NO$_2$—C$_6$H$_2$—NC | 86 | 164 |
| 2-Cl—4-[(CH$_3$)$_2$N—SO$_2$—]—C$_6$H$_2$—NC | 44 | 164 |
| C$_6$H$_5$—(CH$_2$)$_2$—NC | 48 | 92 |
| C$_6$H$_5$—CH(CH$_3$)—NC | 87 | 164 |
| 2,3-(CH$_3$)$_2$—C$_6$H$_3$—NC | 82 | 164 |
| 2,4-(CH$_3$)$_2$—C$_6$H$_3$—NC | 97 | 164 |
| 2,5-(CH$_3$)$_2$—C$_6$H$_3$—NC | 85 | 62, 164 |
| 2,6-(CH$_3$)$_2$—C$_6$H$_3$—NC | 88 | 164, 166 |
| 3,5-(CH$_3$)$_2$—4-OH—C$_6$H$_2$—NC | 70 | 109 |
| 4-CH$_3$O—C$_6$H$_4$—CH$_2$—NC | 25 | 164 |
| 2-CH$_3$—4-CH$_3$O—C$_6$H$_3$—NC | 27 | 164 |
| 2,4-(CH$_3$O)$_2$—C$_6$H$_3$—NC | 41 | 164 |
| 2,5-(CH$_3$O)$_2$—C$_6$H$_3$—NC | 65 | 164 |
| C$_6$H$_5$—S—(CH$_2$)$_2$—NC | 80 | 164 |
| 4-(CH$_3$)$_2$N—C$_6$H$_4$—NC | 40 | 109 |
| ![norbornene]—CH$_2$—NC | 63 | 164 |
| C$_6$H$_5$(CH$_3$)$_2$Si—NC | 70 | 102 |
| 3-CH$_3$—$\Delta^3$-c-C$_6$H$_8$—NC | 38 | 164 |
| CH$_2$=C(CH$_3$)—CO—O—CH$_2$—C(CH$_3$)$_2$—CH$_2$—NC | 37 | 164 |
| NC—C(CH$_3$)$_2$—CH$_2$—C(CH$_3$)$_2$—NC | — | 128 |
| 2,2,4,4-(CH$_3$)$_4$—c-C$_4$H$_3$—NC | 60 | 88 |
| CH$_2$=C(CH$_3$)—(CH$_2$)$_2$—C(CH$_3$)$_2$—NC | — | 128 |
| CH$_3$—N[(CH$_2$)$_3$—NC]$_2$ | 54 | 164 |
| n-C$_8$H$_{17}$—NC | 89 | 32, 92 |
| (CH$_3$)(C$_2$H$_5$)(n-C$_4$H$_9$)C—NC | — | 128 |

TABLE I (*Continued*)

| Compound | Yield (%) | Reference |
|---|---|---|
| $t$-$C_4H_9$—$CH_2$—$C(CH_3)_2$—NC | 56 | 128, 162 |
| 2-$n$-$C_8H_{17}$—NC | trace | 60 |
| ($n$-$C_4H_9$)$_2$N—NC | 71 | 15, 17 |
| $(C_2H_5)_2P(S)$—$OCH_2$—$C(CH_3)_2$—NC | 22 | 164 |
| $C_{10}$ | | |
| $C_6Cl_4(CH_2$—NC$)_2$-(1,4) | 70 | 164 |
| $C_6H_5$—CH=CH—CH=N—NC | — | 49 |
| 4,6-$(CH_3)_2$—$C_6H_2$—(NC)$_2$-(1,3) | 73 | 164 |
| 4-$CH_3O$—$C_6H_4$—CH=CH—NC | 12.6 | 47, 50, 52, 54 |
| 4-$CH_3O$—$C_6H_4$—CO—$CH_2$—NC | 20 | 47, 50, 52 |
| 4-$C_2H_5$—$O_2C$—$C_6H_4$—NC | 68 | 164 |
| 4-$CH_3O$—$C_6H_4$—CHCl—$CH_2$—NC | — | 47, 52 |
| $CH_3O$—$C_6H_4$—$C(CH_3)$=N—NC | — | 47, 49 |
| 2,4,5-$(CH_3)_3$—6-$NO_2$—$C_6H$—NC | 57 | 164 |
| $C_6H_5$—$C(CH_3)_2$—NC | 45 | 164 |
| 2,3,5-$(CH_3)_3$—$C_6H_2$—NC | 79 | 164 |
| 2,4,5-$(CH_3)_3$—$C_6H_2$—NC | 61 | 164 |
| $C_6H_5$—$C(CH_3)_2$—NC | 45 | 164 |
| 2,4,6-$(CH_3)_3$—$C_6H_2$—NC | 80 | 165, 166 |
| 2-$CH_3O$—5-$C_2H_5SO_2$—$C_6H_3$—NC | 74 | 164 |
| 4-$(CH_3)_2N$—$C_6H_4$—$CH_2$—NC | 81 | 164 |
| $cis$-2,2,4,4-$(CH_3)_4$—$c$-$C_4H_2$—(NC)$_2$-(1,3) | 67 | 88 |
| $trans$-2,2,4,4-$(CH_3)_4$—$c$-$C_4H_2$—(NC)$_2$-(1,3) | 80 | 88 |
| $i$-$C_3H_7$—CH(NC)—CO—NH—$CH_2$—$CO_2C_2H_5$ | 79 | 145, 164 |
| $n$-$C_9H_{19}$—NC | 62 | 92 |
| CN—$(CH_3)_2$—CH[N$(CH_3)_2$]—$(CH_2)_3$—NC | 76 | 164 |
| $(C_2H_5)_2N$—$(CH_2)_3$—$CH(CH_3)$—NC | 50 | 164 |
| $C_{11}$ | | |
| 2,4-$Cl_2$—$C_{10}H_5$—NC-(1) | 79 | 164 |
| 4-Br—$C_{10}H_6$—NC-(1) | 61 | 164 |
| $C_{10}H_7$—NC-(1) | 82 | 164 |
| $C_{10}H_7$—NC-(2) | 70 | 164, 166 |
| 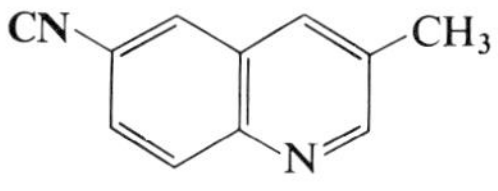 CN-quinoline-CH$_3$ (6-CN, 3-CH$_3$) | 93 | 164 |
| 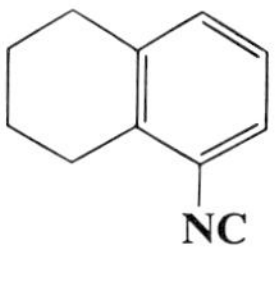 tetrahydronaphthalene-NC | 93 | 164 |
| 2,5-$(C_2H_5O)$—4-$NO_2$—$C_6H_2$—NC | 57 | 164 |
| $C_6H_5$—$C(CH_3)_2$—$CH_2$—NC | 57 | 164 |
| 4-$CH_3$—$C_6H_4$—$CH(C_2H_5)$—NC | 53 | 164 |

TABLE I (*Continued*)

| Compound | Yield (%) | Reference |
|---|---|---|
| 2,6-$(C_2H_5)_2$—$C_6H_3$—NC | 93 | 164 |
| 2,4-$(CH_3)_2$—6-$C_2H_5$—$C_6H_2$—NC | 78 | 164 |
| 4-$(C_2H_5)_2$N—$C_6H_4$—NC | 75 | 109, 166 |
| OC[OCH$_2$—C(CH$_3$)$_2$—NC]$_2$ | 19 | 159, 164 |
| $n$-$C_{10}H_{21}$—NC | 58 | 92 |
| 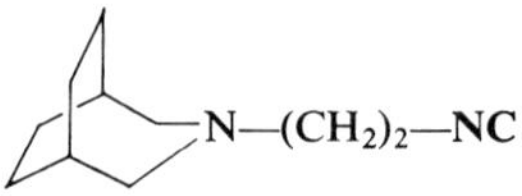 ![bicyclic]—N—(CH$_2$)$_2$—NC | 74 | 164 |

$C_{12}$

| Compound | Yield (%) | Reference |
|---|---|---|
| $C_{10}H_6$—(NC)$_2$-(1,4) | 51 | 159, 164 |
| $C_{10}H_6$—(NC)$_2$-(1,5) | 61 | 159, 164 |
| $C_{10}H_6$—(NC)$_2$-(2,7) | 93 | 164 |
| 5-NC—$C_{10}H_6$—NC-(1) | 69 | 164 |
| 3,4-Cl$_2$—$C_6H_3$—NH—CO$_2$—CH$_2$—C(CH$_3$)$_2$—NC | 87 | 164 |
| 4,6-(CH$_3$)$_2$—$C_6H_2$—(CH$_2$—NC)$_2$-(1,3) | 15 | 164 |
| 2-CH$_3$—5-$i$-C$_3$H$_7$—$C_6H_2$—(NC)$_2$-(1,3) | 24 | 164 |
| 2,3,5,6-(CH$_3$)$_4$—$C_6$—(NC)$_2$ | 82 | 164 |
| 1-CH$_3$—2,6-$(C_2H_5)_2$—$C_6H_2$—NC-(4) | 58 | 164 |
| N[(CH$_2$)$_3$—NC]$_3$ | 70 | 164 |
| $n$-$C_{11}H_{23}$—NC | 56 | 92 |

$C_{13}$

| Compound | Yield (%) | Reference |
|---|---|---|
| 2,4,5-Cl$_3$—$C_6H_2$—O—(4,5-Cl$_2$)—$C_6H_2$—NC-(2) | 84 | 164 |
| 2-$C_6H_5$O—3,5-Cl$_2$—$C_6H_2$—NC | 68 | 164 |
| dibenzofuran-NC | 29 | 164 |
| dibenzofuran-NC | 50 | 164 |
| 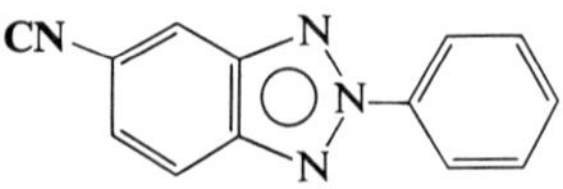 CN-benzotriazole-phenyl | 77 | 164 |
| 2-$C_6H_5$—$C_6H_4$—NC | 70 | 164 |
| acenaphthylene-NC | 97 | 164 |

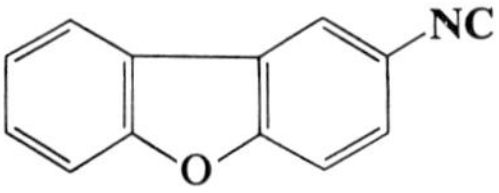

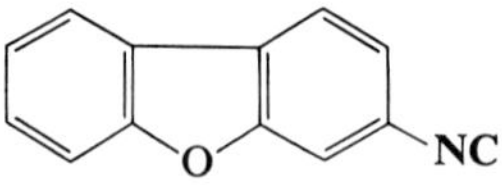

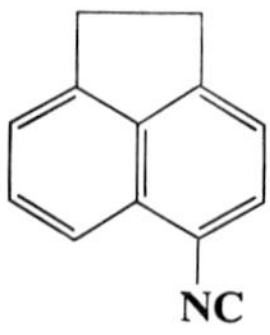

TABLE I (*Continued*)

| Compound | Yield (%) | Reference |
|---|---|---|
| $C_6H_5$—N=N—$C_6H_4$—NC-(4) | 73 | 129, 164 |
| 2-$C_6H_5$—S—$C_6H_4$—NC | 70 | 164 |
| 2-$C_6H_5$—$SO_2$—$C_6H_4$—NC | 68 | 164 |
| 2-$C_2H_5O$—$C_{10}H_6$—NC-(1) | 89 | 164 |
| 2,4-$(C_2H_5)_2$—5-Cl—6-$CH_3$—$C_6$—$(NC)_2$ | 54 | 164 |
| 2,4-$(C_2H_5)_2$—6-$CH_3$—$C_6H$—$(NC)_2$-(1,3) | 51 | 159, 164 |
| 4-$c$-$C_6H_{11}$—$C_6H_4$—NC | 91 | 164 |
| 2,4-$(i$-$C_3H_1)_2$—5-$NO_2$—$C_6H_2$—NC | 72 | 164 |
| 4-$CH_3$—$C_6H_4$—CH($i$-$C_4H_9$)—NC | 68 | 164 |
| 2,4-$(i$-$C_3H_7)_2$—$C_6H_3$—NC | 82 | 164 |
| 2,6-$(i$-$C_3H_7)_2$—$C_6H_3$—NC | 80 | 164 |
| $n$-$C_{12}H_{25}$—NC | 64 | 21, 81, 92, 164 |
| $(n$-$C_4H_9)_3$Sn—NC | 82 | 144 |
| $C_{14}$ | | |
| 4-$C_6Cl_5$—S—$C_6H_3$—$(NC)_2$-(1,3) | 95 | 164 |
| 4-**CN**—2,5-$Cl_2$—$C_6H_2$—N=N(O)$C_6H_2$—$Cl_2$-(2,5)-**NC**-(4) | 65 | 164 |
| 4-**CN**—2,6-$Cl_2$—$C_6H_2$—N=N(O)$C_6H_2$—$Cl_2$-(2,6)-**NC**-(4) | 7 | 164 |
| 4-**CN**—3-Cl—$C_6H_3$—$C_6H_3$—Cl-(3)-**NC**-(4) | 88 | 164 |
| 3-**CN**—6-Cl—$C_6H_3$—N=N—$C_6H_3$—Cl-(6)-**NC**-(3) | 51 | 164 |
| 3-Cl—4-**CN**—$C_6H_3$—N=N(O)$C_6H_3$—Cl-(3)-**NC**-(4) | 67 | 164 |
| 2,4-$Cl_2$—$C_6H_3O$—$C_6H_3$—$(NC)_2$-(1,4) | 79 | 164 |

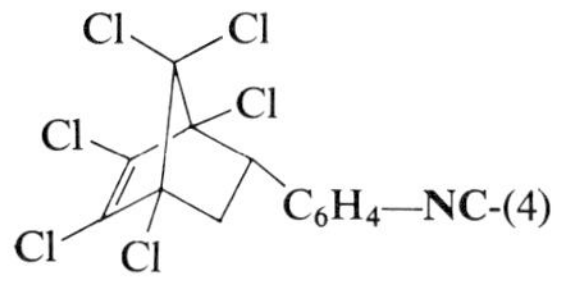

![Structure: tetrachloro bicyclic compound with $C_6H_4$—NC-(4) substituent]  75  164

| 4-Cl—$C_6H_4O$—$C_6H_3$—$(NC)_2$-(2,4) | 67 | 164 |
| 2,5-$Cl_2$—$C_6H_3$—CH(NC)—$C_6H_4$—Cl-(4) | 23 | 164 |
| 2-**CN**—$C_6H_4$—$C_6H_4$—**NC**-(2) | 70 | 164 |
| 2-**CN**—$C_6H_4$—$C_6H_4$—**NC**-(4) | 60 | 164 |
| 4-**CN**—$C_6H_4$—$C_6H_4$—**NC**-(4) | 94 | 164 |
| 4-**CN**—$C_6H_4$—N=N—$C_6H_4$—**NC**-(4) | 35 | 164 |
| 4-**CN**—$C_6H_4O$—$C_6H_4$—**NC**-(4) | 64 | 164 |
| 3-**CN**—$C_6H_4$—N=N(O)—$C_6H_4$—**NC**-(3) | 71 | 164 |
| 4-**CN**—$C_6H_4SO_2$—$C_6H_4$—**NC**-(4) | 93 | 164 |
| 4-Cl—$C_6H_4$—CH(NC)—$C_6H_4$—Cl-(4) | 72 | 164 |

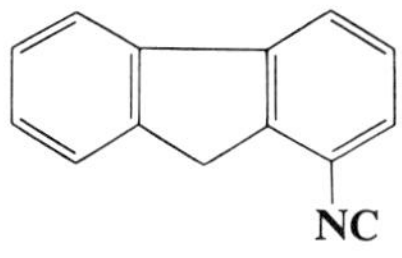

[Fluorene with NC substituent]  97  164

| 4-$C_6H_5$—CO—$C_6H_4$—NC | 86 | 164 |

TABLE I (*Continued*)

| Compound | Yield (%) | Reference |
|---|---|---|
| 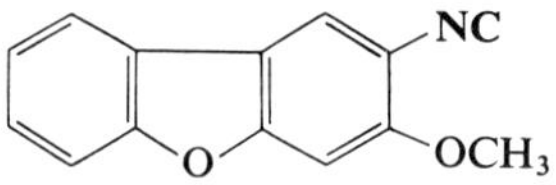 [dibenzofuran with NC and OCH₃ substituents] | 68 | 164 |
| $(C_6H_5)_2Si-(NC)_2$ | 80 | 102 |
| $C_6H_5-CH(C_6H_5)-NC$ | 78 | 164 |
| $2-CH_3-C_6H_4O-C_6H_4-NC-(2)$ | 55 | 164 |
| $4-CH_3-C_6H_4-S-C_6H_4-NC-(4)$ | 83 | 164 |
| $C_6H_5-SO_2-C_6H_3-NC-(3)-CH_3O-(4)$ | 45 | 164 |
| $(C_6H_5)_2P(O)CH_2-NC$ | 80 | 87 |
| $n-C_{11}H_{21}-(NC)_3-(1,6,11)$ | 87 | 164 |
| $C_{15}$ | | |
| $2,4-(CN)_2-C_6H_3-C_6H_4-NC-(4)$ | 83 | 164 |
| $2,4-(CN)_2-C_6H_3O-C_6H_4-NC-(4)$ | 84 | 164 |
| 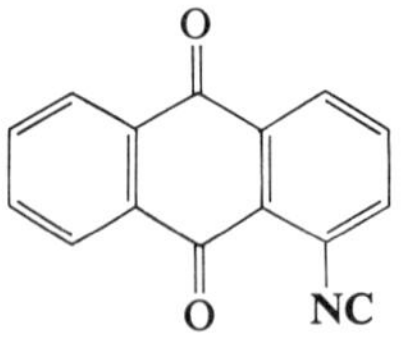 [anthraquinone with NC substituent] | 31 | 164 |
| $3-Cl-4-CN-C_6H_3-Cl-(3)-NC-(4)$ | 20 | 164 |
| $4-CN-C_6H_4-CO-C_6H_4-NC-(4)$ | 72 | 164 |
| $4-CN-C_6H_4O-CO-O-C_6H_4-NC-(4)$ | 76 | 164 |
| 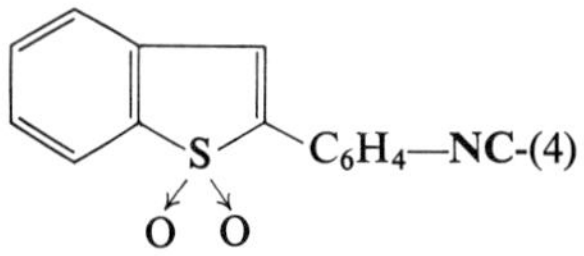 [benzothiophene-SO₂ with $C_6H_4-NC-(4)$] | 60 | 164 |
| $4-CN-C_6H_4-CH_2-C_6H_4-NC-(4)$ | 83 | 159 |
| $4-NO_2-C_6H_4-N=N-C_6H_2-(CH_3)_3-(2,4)-NC-(4)$ | 86 | 164 |
| 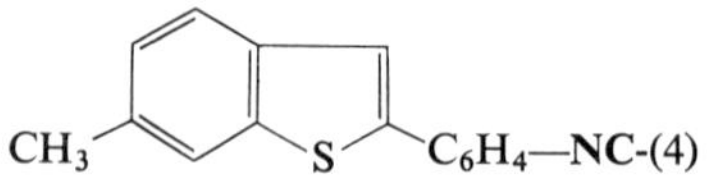 $CH_3$-[benzothiophene]-$C_6H_4-NC-(4)$ | 59 | 164 |
| $2,5-Cl_2-C_6H_4S-CH_2-CH(C_6H_5)-NC$ | 84 | 164 |
| $3,4-Cl_2-C_6H_3S-CH_2-CH(C_6H_5)-NC$ | 98 | 164 |
| $4-Cl-C_6H_4-CH(CH_2-C_6H_5)-NC$ | 35 | 164 |
| $4-Cl-C_6H_4S-CH_2-CH(C_6H_5)-NC$ | 93 | 164 |
| $C_6H_5-CH_2-CH(C_6H_5)-NC$ | 62 | 164 |
| $4-C_6H_5-C_6H_4-CH(CH_3)-NC$ | 47 | 164 |
| $2-CH_3-C_6H_4-N=N-C_6H_3-CH_3-(3)-NC-(4)$ | 80 | 164 |

## 2. Isonitrile Syntheses

**TABLE I** (*Continued*)

| Compound | Yield (%) | Reference |
|---|---|---|
| $C_6H_5S-CH_2-CH(C_6H_5)-NC$ | 83 | 164 |
| $2-CH_3O-5-C_6H_5CH_2-SO_2-C_6H_3-NC$ | 71 | 164 |
| $3,5-(t-C_4H_9)_2-4-OH-C_6H_2-NC$ | 72 | 109 |
| $C_{16}$ | | |
| 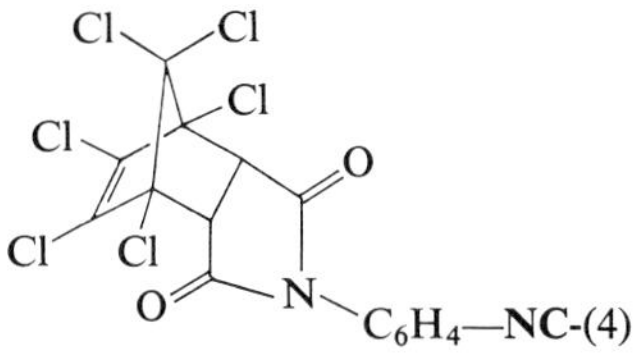 [hexachloro bicyclic dicarboximide with $-C_6H_4-NC-(4)$ on N] | 76 | 164 |
| $2,4-(CN)_2-C_6H_3-C_6H_3-(NC)_2-(2,4)$ | 68 | 164 |
| 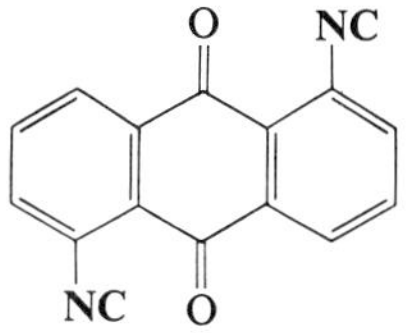 [1,5-diisocyanoanthraquinone] | 46 | 159 |
| 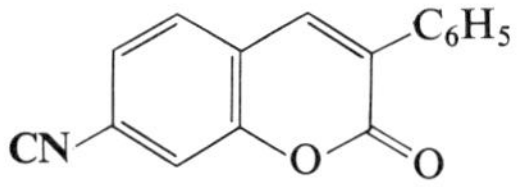 [7-cyano-3-phenylcoumarin] | 41 | 164 |
| $4-CN-3-CH_3-C_6H_3-C_6H_3-CH_3-(3)-NC-(4)$ | 55 | 164 |
| $4-CN-3-CH_3O-C_6H_3-C_6H_3-CH_3O-(3)-NC-(4)$ | 93 | 164 |
| $3-CN-4-CH_3-C_6H_3-N=N(O)-C_6H_3-CH_3-(4)-NC-(3)$ | 51 | 164 |
| $C_6H_5-N(C_2H_5)-CO-C_6H_4-NC-(4)$ | 58 | 164 |
| $4-CH_3O-C_6H_4-CH(NC)-C_6H_4-OCH_3-(4)$ | 88 | 164 |
| $2-CH_3-C_6H_4-S-CH_2-CH(C_6H_5)-NC$ | 88 | 164 |
| $3-CH_3-C_6H_4-S-CH_2-CH(C_6H_5)-NC$ | 96 | 164 |
| $4-CH_3-C_6H_4-S-CH_2-CH(C_6H_5)-NC$ | 91 | 164 |
| $(C_6H_5-CH_2)_2P(O)CH_2-NC$ | 72 | 87 |
| $C_{17}$ | | |
| $4-[2-C_{10}H_7O-]-C_6H_4-NC$ | 89 | 164 |
| $4-[2-C_{10}H_7S-]-C_6H_4-NC$ | 77 | 164 |
| $4-CN-3-Cl-5-CH_3-C_6H_2-CH_2-C_6H_2-Cl-(3)-CH_3-(5)-NC-(4)$ | 37 | 164 |
| $3-CN-4-CH_3O-C_6H_3-O-CO-O-C_6H_3-CH_3O-(4)-NC-(3)$ | 87 | 159 |
| $4-CN-3-CH_3-C_6H_3-CH_2-C_6H_3-CH_3-(3)-NC-(4)$ | 42 | 164 |

## TABLE I (Continued)

| Compound | Yield (%) | Reference |
|---|---|---|

[Structure: 2-(4-isocyano-3-methylphenyl)-5,7-dimethylbenzothiazole] — 64 — 164

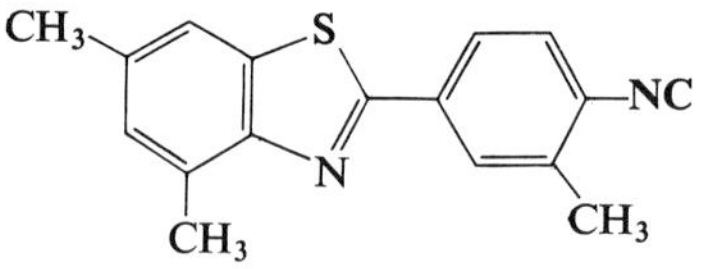

| Compound | Yield (%) | Reference |
|---|---|---|
| 2,4,6-($i$-C$_3$H$_7$)$_3$—C$_6$H—(NC)$_2$-(1,3) | 69 | 159 |
| 3,5-($t$-C$_4$H$_9$)$_2$—4-OH—C$_6$H$_2$—(CH$_2$)$_2$—NC | 70 | 164 |
| **C$_{18}$** | | |
| 4-OH—C$_6$H$_4$—CH=C(NC)—C(NC)=CH—C$_6$H$_4$—OH-(4) | — | 56, 142 |
| 4-CN—3-C$_2$H$_5$O—C$_6$H$_3$—C$_6$H$_3$—C$_2$H$_5$O-(3)-NC-(4) | 65 | 159, 164 |
| 4-CN—3,5-(CH$_3$)$_2$—C$_6$H$_2$—S—C$_6$H$_2$—(CH$_3$)$_2$-(3,5)-NC-(4) | 68 | 164 |
| 4-CN—3-C$_2$H$_5$—C$_6$H$_3$—CH$_2$—C$_6$H$_3$—C$_2$H$_5$-(3)-NC-(4) | 80 | 164 |
| **C$_{19}$** | | |
| (C$_6$H$_5$)$_3$Si—NC | 81 | 11, 102 |
| 4-CN—3,5-(CH$_3$)$_2$—C$_6$H$_2$—CH$_2$—C$_6$H$_2$—(CH$_3$)$_2$-(3,5)-NC-(4) | 72 | 159, 164 |
| $n$-C$_{11}$H$_{25}$CO—C$_6$H$_4$—NC-(4) | 85 | 164 |
| ($n$-C$_4$H$_9$)(C$_6$H$_5$—CH$_2$)—N$^{\oplus}$—[—(CH$_2$)$_3$—NC]$_2$Cl$^{\ominus}$ | 84 | 164 |
| $n$-C$_{12}$H$_{25}$—C$_6$H$_4$—NC-(4) | 55 | 164 |
| $n$-C$_{12}$H$_{25}$—O—C$_6$H$_4$—NC-(4) | 72 | 164 |
| $n$-C$_{18}$H$_{37}$—NC | 95 | 21, 29, 32, 149, 164 |
| **C$_{20}$** | | |
| 4-CH$_3$O—C$_6$H$_4$—CH=C(NC)—C(NC)=CH—C$_6$H$_4$—OCH$_3$-(4) | — | 47, 50, 53, 56 |
| 4-CN—3-CH$_3$—5-C$_2$H$_5$—C$_6$H$_2$—S—C$_6$H$_2$—CH$_3$-(3)-C$_2$H$_5$—(5)—NC-(4) | 66 | 164 |
| $n$-C$_{12}$H$_{25}$—NH—CO—C$_6$H$_4$—NC-(3) | 54 | 164 |
| $n$-C$_{12}$H$_{25}$—N[(CH$_2$)$_3$—NC]$_2$ | 71 | 164 |
| **C$_{21}$** | | |
| (4-CN—C$_6$H$_4$—O)$_3$PS | 13 | 164 |
| 4-Cl—C$_6$H$_4$—CH$_2$—CH(NC)—C$_6$H$_4$—[4-Cl—C$_6$H$_4$]-(4) | 89 | 164 |
| 3,4-Cl$_2$—C$_6$H$_3$—CH$_2$—CH(NC)—C$_6$H$_4$—C$_6$H$_5$-(4) | 92 | 164 |
| 4-CN—3-$i$-C$_3$H$_7$—C$_6$H$_3$—CH$_2$—C$_6$H$_3$—$i$-C$_3$H$_7$-(3)-NC-(4) | 45 | 164 |
| 4-CN—3-CH$_3$—5-C$_2$H$_5$—C$_6$H$_2$—CH$_2$—C$_6$H$_2$—CH$_3$-(3)-C$_2$H$_5$-(5)-NC-(4) | 90 | 159 |
| $n$-C$_{12}$H$_{25}$—S—CH$_2$—CH(C$_6$H$_5$)—NC | 84 | 164 |
| **C$_{22}$** | | |
| 1,1-[4-CN—3-CH$_3$—C$_6$H$_3$]$_2$—$c$-C$_6$H$_{10}$ | 86 | 164 |
| 4-CN—3,5-(C$_2$H$_5$)$_2$—C$_6$H$_2$—S—C$_6$H$_2$—(C$_2$H$_5$)$_2$-(3,5)-NC-(4) | 72 | 164 |

TABLE I (*Continued*)

| Compound | Yield (%) | Reference |
|---|---|---|
| $C_{23}$ | | |
| 4-**CN**—3,5-$(C_2H_5)_2$—$C_6H_2$—$CH_2$—$C_6H_2$—$(C_2H_5)_2$-(3,5)-**NC**-(4) | 71 | 159 |
| $C_{24}$ | | |
| 3-**CN**—4-$CH_3$—$C_6H_3$—O—CO—NH—$(CH_2)_6$—NH—CO—O—$C_6H_3$—$CH_3$-(4)-**NC**-(3) | 42 | 164 |
| $3\beta$-acetoxy—$20\alpha$-**CN**—pregnene-5 | 84 | 62 |
| $C_{25}$ | | |
| [4-**CN**—2,5-$(CH_3)_2$—$C_6H_2]_2$—CH—$C_6H_5$ | 55 | 164 |
| $C_{26}$ | | |
| 2-$CH_3O$—5-$[n\text{-}C_{18}H_{37}$—$N(CH_3)$—$SO_2]$—$C_6H_3$—**NC** | 44 | 164 |
| $C_{28}$ | | |
| $3\alpha$-**CN**—$5\alpha$-Cholestane | 93 | 62, 77 |
| $3\beta$-**CN**—$5\alpha$-Cholestane | — | 77 |
| $6\alpha$-**CN**—$5\alpha$-Cholestane | — | 77 |
| $6\beta$-**CN**—$5\alpha$-Cholestane | — | 77 |
| $7\alpha$-**CN**—$5\alpha$-Cholestane | — | 77 |
| $7\beta$-**CN**—$5\alpha$-Cholestane | — | 77 |
| $C_{33}$ | | |
| $(C_6H_5CH_2)(n\text{-}C_{18}H_{37})\overset{\oplus}{N}[(CH_2)_3\textbf{NC}]_2Cl^{\ominus}$ | 86 | 164 |
| $C_{108}$ | | |
| **CN**—[Gramicidine A]—Ts | — | 145 |

\* The isonitriles are listed according to the number of C, H, Br, Cl, F, N, O, P, and S atoms present.

## REFERENCES

1. Abraham, N. A., and Hajela, N., *C. R. Acad. Sci.* **255**, 3192 (1962).
2. Anderson, G. W., Zimmermann, J. E., and Callahan, F. M., *J. Amer. Chem. Soc.* **88**, 1388 (1966).
3. Arlt, D., Hagemann, H., Hoffmann, P., and Ugi, I., unpublished results (1967).
4. Aronovič, P. M., Bel'skij, N. K., and Michailov, B. M., *Izv. Akad. Nauk SSSR, Ser. Khim.* p. 696 (1956).
5. Beichl, G. J., Colwell, I. E., and Miller, J. G., *Chem. Ind.* (*London*) p. 203 (1960).
6. Bestmann, H. J., Lienert, J., and Mott, L., *Ann. Chem.* **718**, 24 (1968).
7. Betz, W., and Ugi, I., Ger. Publ. Pat. Appl. 1,158,499 (1962–1963) (Farbenfabriken Bayer A. G.).
8. Betz, W., Hoffmann, P., and Ugi, I., unpublished results (1961-1967).
9. Biddle, H. C., *Ann. Chem.* **310**, 1 (1900).
10. Bigorgne, M., and Bouquet, A., *J. Organometal. Chem.* **1**, 101 (1963).
11. Bither, T. A., Knoth, W. H., Lindsey, R. V., and Sharkey, W. A., *J. Amer. Chem. Soc.* **80**, 4151 (1958).

12. Booth, M. R., and Frankiss, S. G., *Chem. Commun.* p. 1347 (1968).
13. Bose, S., *J. Indian Chem. Soc.* **35**, 376 (1958).
14. Brackmann, W., and Smit, P. J., *Rec. Trav. Chim. Pays-Bas* **82**, 757 (1963).
15. Bredereck, H., Föhlisch, B., and Walz, K., *Angew. Chem.* **74**, 388 (1962); *Angew. Chem., Int. Ed. Engl.* **1**, 334 (1962).
16. Bredereck, H., Föhlisch, B., and Walz, K., *Angew. Chem.* **76**, 580 (1964); *Angew. Chem., Int. Ed. Engl.* **3**, 647 (1964).
17. Bredereck, H., Föhlisch, B., and Walz, K., *Ann. Chem.* **686**, 92 (1965).
18. Bredereck, H., Föhlisch, B., and Walz, K., *Ann. Chem.* **688**, 93 (1965).
19. Brügel, W., Daumiller, G., and Rommel, O., *Angew. Chem.* **68**, 440 (1956).
20. Burkhardt, G. N., Lapworth, A., and Robinson, E. B., *J. Chem. Soc.* **127**, 2234 (1925).
21. Bussert, F., U.S. Pat. 3,012,932 (1961) (Standard Oil of Indiana).
22. Casanova, J., Schuster, E. R., and Werner, N. D., *J. Chem. Soc.* p. 4280 (1963).
23. Casanova, J., Werner, N. D., and Schuster, E. R., *J. Org. Chem.* **31**, 3473 (1966).
24. Cotton, F. A., and Zingales, F., *J. Amer. Chem. Soc.* **83**, 351 (1961).
25. Crabtree, E. V., Poziomek, E. J., and Hoy, D. J., *Talanta* **14**, 857 (1967).
26. Craig, A. D., Urenovitch, J. V., and McDiarmid, A. G., *J. Chem. Soc.* p. 548 (1962).
27. Dadieu, A., *Chem. Ber.* **64**, 358 (1931).
28. Davis, T. L., and Yelland, W. E., *J. Amer. Chem. Soc.* **59**, 1998 (1937).
29. Dreyfus, H., U.S. Pat. 2,342,794 (1944); U.S. Pat. 2,347,772 (1944) (Celanese Corp.).
30. Farrar, W. V., *J. Chem. Soc.* p. 3058 (1960).
31. Farrow, M. D., and Ingold, C. K., *J. Chem. Soc.* **125**, 2543 (1924).
32. Feuer, H., Rubinstein, H., and Nielsen, A. T., *J. Org. Chem.* **23**, 1107 (1958).
33. Francesconi, L., and Ciurlo, A., *Gazz. Chim. Ital.* **53**, 327 (1923).
34. Gautier, A., *Ann. Chem.* **142**, 289 (1867).
35. Gautier, A., *Ann. Chem.* **146**, 119 (1868).
36. Gautier, A., *Ann. Chem.* **146**, 124 (1868).
37. Gautier, A., *Ann. Chim. (Paris)* [4] **17**, 103 (1869).
38. Gautier, A., *Ann. Chim. (Paris)* [4] **17**, 203 (1869).
39. Gautier, A., *Ann. Chem.* **149**, 29 (1869).
40. Gautier, A., *Ann. Chem.* **149**, 155 (1869).
41. Gautier, A., *Ann. Chem.* **151**, 239 (1869).
42. Gompper, R., *Angew. Chem.* **76**, 412 (1964); *Angew. Chem., Int. Ed. Engl.* **3**, 560 (1964)
43. Griffith, W. P., *Quart. Rev. (London)* **16**, 188 (1962).
44. Grigat, E., and Pütter, R., *Chem. Ber.* **99**, 2361 (1966).
45. Grisebach, H., and Achenbach, H., *Z. Naturforsch. B* **20**, 137 (1965).
46. Guillemard, M. H., *Ann. Chim. (Paris)* [8] **14**, 311 (1908).
47. Hagedorn, I., *Angew. Chem.* **75**, 305 (1963).
48. Hagedorn, I., and Eholzer, U., *Angew. Chem.* **74**, 215 (1962); *Angew. Chem., Int. Ed. Engl.* **1**, 514 (1962).
49. Hagedorn, I., and Eholzer, U., *Angew. Chem.* **74**, 499 (1962); *Angew. Chem., Int. Ed. Engl.* **1**, 514 (1962).
50. Hagedorn, I., Eholzer, U., and Etling, H., *Chem. Ber.* **98**, 193 and 202 (1965).
51. Hagedorn, I., Eholzer, U., and Lüttringhaus, A., *Chem. Ber.* **93**, 1584 (1960).
52. Hagedorn, I., and Etling, H., *Angew. Chem.* **73**, 26 (1961).
53. Hagedorn, I., and Etling, H., Ger. Publ. Pat. Appl. 1,167,332 (1961–1964) (Farbenfabriken Bayer A. G ).
54. Hagedorn, I., and Etling, H., Ger. Publ. Pat. Appl. 1,168,895 (1961–1964) (Farbenfabriken Bayer A. G.).
55. Hagedorn, I., and Tönjes, H., *Pharmazie* **11**, 409 (1956).

## 2. Isonitrile Syntheses

56. Hagedorn, I., and Tönjes, H., *Pharmazie* **12**, 567 (1957).
57. Hagedorn, I., and Winkelmann, H.-D., *Chem. Ber.* **99**, 850 (1966).
58. Hartley, E. G. J., *J. Chem. Soc.* **109**, 1296 (1916).
59. Hartley, E. G. J., *J. Chem. Soc.* p. 780 (1928).
60. Heldt, W. Z., *J. Org. Chem.* **26**, 3226 (1961).
61. Heldt, W. Z., *Advan. Chem. Ser.* **37**, 99 (1963).
62. Hertler, W. R., and Corey, E. J., *J. Org. Chem.* **23**, 1221 (1958).
63. Herzberg, G., *J. Chem. Phys.* **8**, 847 (1940).
64. Higashino, T., *Chem. Pharm. Bull.* **9**, 635 (1961).
65. Hine, J., and Hine, M., *J. Amer. Chem. Soc.* **74**, 5266 (1952).
66. Hölzl, F., *Z. Electrochem. Angew. Phys. Chem.* **43**, 319 (1937).
67. Hölzl, F., Hauser, W., and Eckmann, M., *Monatsh. Chem.* **48**, 71 (1927).
68. Hölzl, F., and Krakora, J., *Monatsh. Chem.* **64**, 97 (1934).
69. Hölzl, F., Meier-Mohar, T., and Viditz, F., *Monatsh. Chem.* **53/54** (1929).
70. Hoffmann, P., unpublished results (1967).
71. Hoffmann, P., Marguarding, D., and Ugi, I., unpublished results (1967).
72. Hofmann, A. W., *C. R. Acad. Sci.* **65**, 484 (1867).
73. Hofmann, A. W., *Ann. Chem.* **144**, 114 (1867).
74. Hofmann, A. W., *Ann. Chem.* **146**, 107 (1868).
75. Hofmann, A. W., *Ann. Chim. (Paris)* [4] **17**, 210 (1869).
76. Hofmann, A. W., *Chem. Ber.* **3**, 766 (1870).
77. Horobin, R. W., Khan, N. R., McKenna, J., and Hutley, B. G., *Tetrahedron Lett.* p. 5087 (1966).
78. Ingold, C. K., *J. Chem. Soc.* **125**, 87 (1924).
79. Hoy, D. J., and Poziomek, E. J., AD-627229 (1965).
80. Jackson, H. L., and McKusick, B. C., *Org. Syn.* Coll. Vol. IV, p. 438 (1963).
81. Jungermann, E., and Smith, F. W., *J. Amer. Oil Chem. Soc.* **36**, 388 (1959).
82. Kagen, H., and Lillien, I., *J. Org. Chem.* **31**, 3728 (1966).
83. Kaufler, F., and Pomeranz, C., *Monatsh. Chem.* **22**, 492 (1901).
84. Kelso, A. G., and Lacey, A. B., *J. Chromatogr.* **18**, 156 (1965).
85. Klages, F., and Mönkemeyer, K., *Chem. Ber.* **83**, 501 (1950).
86. Krapcho, A. P., *J. Org. Chem.* **27**, 1089 (1962).
87. Kreutzkamp, N., and Lämmerhirt, K., *Angew. Chem.* **80**, 394 (1968).
88. Lautenschläger, F., and Wright, G. F., *Can. J. Chem.* **41**, 863 (1963).
89. Ley, K., and Eholzer, U., unpublished results (1965).
90. Lieke, W., *Ann. Chem.* **112**, 316 (1859).
91. Lindemann, H., and Wiegrebe, L., *Chem. Ber.* **63**, 1650 (1930).
92. Lipp, M., Dallacker, F., and Meier zu Köcker, T., *Monatsh. Chem.* **90**, 41 (1959).
93. Lorenz, D. H., and Becker, E. I., *J. Org. Chem.* **28**, 1707 (1963).
94. Makarov, S. P., Euglin, M. A., Videijko, A. F., and Nikolajeva, T. V., *Zh. Obshch. Khim.* **37**, 2781 (1967).
95. Malatesta, L., *Gazz. Chim. Ital.* **77**, 238 (1947).
96. Malatesta, L., and Bonati, F., "Isocyanide Complexes of Metals," pp. 3 and 4. Wiley (Interscience), New York, 1969.
97. Malz, H., and Kühle, E., Ger. Publ. Pat. Appl. 1,158,501 (1962–1962) (Farbenfabriken Bayer A. G.).
98. Manuel, T. A., *Inorg. Chem.* **3**, 1703 (1964).
99. Martin, D., and Weise, A., *Chem. Ber.* **99**, 976 (1966).
100. Matteson, D. S., and Bailey, R. A., *Chem. Ind. (London)* p. 191 (1967).
101. Matteson, D. S., and Bailey, R. A., *J. Amer. Chem. Soc.* **90**, 3761 (1968).

102. McBride, J. J., *J. Org. Chem.* **24**, 2029 (1959).
103. McBride, J. J., and Beachell, H. C., *J. Amer. Chem. Soc.* **74**, 5247 (1952).
104. McCrosky, C. R., Bergstrom, F. W., and Waitkins, G., *J. Amer. Chem. Soc.* **64**, 722 (1942).
105. Merchx, R., Verhulst, J., and Bruylants, P., *Bull. Soc. Chim. Belg.* **42**, 177 (1933).
106. Meyer, E., *J. Prakt. Chem.* **67**, 147 (1866).
107. Meyer, K. H., and Hopf, H., *Chem. Ber.* **54**, 1709 (1921).
108. Meyer, V., and Jacobson, P., "Lehrbuch der organischen Chemie," Vol. I, Part 1, p. 419. de Gruyter, Leipzig, 1922.
109. Müller, E., and Narr, B., *Z. Naturforsch.* **B 16**, 845 (1961).
110. Müller, E., Kaestner, P., and Rundel, W., *Chem. Ber.* **98**, 711 (1965).
111. Mukaiyama, T., and Hata, T., *Bull. Chem. Soc. Jap.* **33**, 1382 (1960).
112. Mukaiyama, T., and Kodaira, Y., *Bull. Chem. Soc. Jap.* **39**, 1297 (1966).
113. Mukaiyama, T., Nambu, H., and Okamoto, M., *J. Org. Chem.* **27**, 3651 (1962).
114. Mukaiyama, T., Tonooka, K., and Inoue, K., *J. Org. Chem.* **26**, 2202 (1961).
115. Mukaiyama, T., and Yokota, Y., *Bull. Chem. Soc. Jap.* **38**, 858 (1962).
116. Nef, I. U., *Ann. Chem.* **270**, 267 (1892).
117. Nef, I. U., *Ann. Chem.* **280**, 291 (1894).
118. Nef, I. U., *Ann. Chem.* **287**, 265 (1895).
119. Nef, I. U., *Ann. Chem.* **298**, 202 (1897).
120. Nef, I. U., *Ann. Chem.* **298**, 368 (1897).
121. Nef, I. U., *Ann. Chem.* **309**, 126 (1899).
122. Neidlein, R., *Angew. Chem.* **76**, 440 (1964); *Angew. Chem., Int. Ed. Engl.* **3**, 382 (1964).
123. Neidlein, R., *Angew. Chem.* **76**, 500 (1964); *Angew. Chem., Int. Ed. Engl.* **3**, 446 (1964).
124. Neidlein, R., *Arch. Pharm. (Weinheim)* **297**, 589 (1964).
125. Neidlein, R., *Arch. Pharm. (Weinheim)* **299**, 603 (1966).
126. New, R. G. A., and Sutton, L. E., *J. Chem. Soc.* p. 1415 (1932).
127. Oda, R., and Shono, T., *J. Soc. Org. Syn. Chem., Tokyo* **22**, 695 (1964).
128. Otsuka, S., Mori, K., and Yamagami, K., *J. Org. Chem.* **31**, 4170 (1966).
129. Passerini, M., *Gazz. Chim. Ital.* **50**, II, 340 (1920).
130. Passerini, M., *Gazz. Chim. Ital.* **54**, 185 and 633 (1924).
131. Passerini, M., *Gazz. Chim. Ital.* **55**, 555 (1925).
132. Passerini, M., and Banti, G., *Gazz. Chim. Ital.* **58**, 636 (1928).
133. Passerini, M., and Neri, A., *Gazz. Chim. Ital.* **64**, 934 (1934).
134. Peratoner, A., and Palazzo, F. C., *Gazz. Chim. Ital.* **38**, 102 (1908).
135. Petrov, K. A., and Nejmyseva, A. A., *Zh. Obshch. Khim.* **29**, 2165 (1959).
136. Pilgram, K., and Korte, F., *Tetrahedron Lett.* p. 881 (1966).
137. Ploquin, J., *Bull. Soc. Chim. Fr.* [5] **14**, 901 (1947).
138. Powers, J. C., Seidner, R., Parsons, T. G., and Berwin, H. J., *J. Org. Chem.* **31**, 2623 (1966).
139. Poziomek, E. J., and Crabtree, E. V., *Anal. Lett.* **1**, 929 (1968).
140. Reichel, L., and Strasser, O., *Chem. Ber.* **64**, 1997 (1931).
141. Ross, D., Ph.D. Thesis, Universität München, 1957.
142. Rothe, W., *Pharmazie* **5**, 190 (1950).
143. Rupe, H., and Glenz, K., *Ann. Chem.* **436**, 184 (1924).
144. Saegusa, T., Kobayashi, S., Ito, Y., and Yasuda, N., *J. Amer. Chem. Soc.* **90**, 4182 (1968).
145. Sakiyama, F., and Witkop, B., *J. Org. Chem.* **30**, 1905 (1965).
146. Samuel, D., Weinraub, B., and Ginsburg, D., *J. Org. Chem.* **21**, 376 (1956).
147. Sayigh, A. A. R., and Ulrich, H., *J. Chem. Soc.* p. 3146 (1963).

148. Schmidt, U., and Kabitzke, K. H., *Angew. Chem.* **76**, 687 (1964); *Angew. Chem., Int. Ed. Engl.* **3**, 641 (1964).
149. Schneidewind, W., *Chem. Ber.* **21**, 1323 (1888).
150. Seyferth, D., and Damrauer, R., *Tetrahedron Lett.* **2**, 189 (1966).
151. Seyferth, D., and Kahlen, N., *J. Org. Chem.* **25**, 809 (1960).
152. Sheehan, J. C., and Berson, J. H., *J. Amer. Chem. Soc.* **89**, 362, 366 (1967).
153. Sheehan, J. C., and Lengyel, I., *J. Amer. Chem. Soc.* **86**, 746 (1964).
154. Sheehan, J. C., and Lengyel, I., *J. Amer. Chem. Soc.* **86**, 1356 (1964).
155. Shingaki, T., and Takebayashi, M., *Bull. Chem. Soc. Jap.* **36**, 617 (1963).
156. Smith, P. A. S., *in* "Molecular Rearrangements" (P. de Mayo, ed.), p. 457. Wiley (Interscience), New York, 1963.
157. Smith, P. A. S., and Kalenda, N. W., *J. Org. Chem.* **23**, 1599 (1958).
158. Tronov, B. V., and Bardamova, M. I., *Izv. Vyssh. Ucheb. Zaved., Khim. Tekhnol.* **2**, 34 (1959); *Chem. Abstr.* **53**, 21823b (1959).
159. Ugi, I., Ger. Publ. Pat. Appl. 1,158,500 (1962–1963) (Farbenfabriken Bayer A. G.).
160. Ugi, I., unpublished results (1969).
161. Ugi, I., Beck, F., and Fetzer, U., *Chem. Ber.* **95**, 126 (1962).
162. Ugi, I., Betz, W., Fetzer, U., and Offermann, K., *Chem. Ber.* **94**, 2814 (1961).
163. Ugi, I., and Fetzer, U., *Chem. Ber.* **94**, 1116 (1961).
164. Ugi, I., Fetzer, U., Eholzer, U., Knupfer, H., and Offermann, K., *Angew. Chem.* **77**, 492 (1965); *Angew. Chem., Int. Ed. Engl.* **4**, 472 (1965); *in* "Neuere Methoden der präparativen organischen Chemie" (W. Foerst, ed.), Vol. IV, p. 37. Verlag Chemie, Weinheim, 1966.
165. Ugi, I., and Meyr, R., *Angew. Chem.* **70**, 702 (1958).
166. Ugi, I., and Meyr, R., *Chem. Ber.* **93**, 239 (1960).
167. Ugi, I., and Meyr, R., *Org. Syn.* **41**, 101 (1961).
168. Ugi, I., Meyr, R., Fetzer, U., and Steinbrückner, C., *Angew. Chem.* **71**, 386 (1959).
169. Ugi, I., Meyr, R., Lipinski, M., Bodesheim, F., and Rosendahl, F. K., *Org. Syn.* **41**, 13 (1961).
170. Ugi, I., and Rosendahl, F. K., *Ann. Chem.* **666**, 65 (1963).
171. Wade, J., *J. Chem. Soc.* **81**, 1596 (1902).
172. Weith, W., *Chem. Ber.* **6**, 210 (1873).
173. Werner, A., and Detscheff, T., *Chem. Ber.* **38**, 69 (1905).
174. Werner, A., and Piquet, A., *Chem. Ber.* **37**, 4295 (1904).
175. Whitfield, R. C., *Sch. Sci. Rev.* **48**, 103 (1966).
176. Wittmann, R., *Angew. Chem.* **73**, 219 (1961).

# Chapter 3

# Kinetics of the Isonitrile–Nitrile Rearrangement

*Kenneth M. Maloney and B. S. Rabinovitch**

| | |
|---|---|
| I. Introduction | 41 |
| II. Thermal Rearrangement | 42 |
|    A. Mechanism | 42 |
|    B. Nature of the Rearrangement | 42 |
|    C. Gas Phase Studies | 43 |
|    D. Theoretical Rate Calculations | 45 |
|    E. Kinetic Isotope Effects | 54 |
|    F. Effect of Variation of Molecular Parameters on Nonequilibrium Unimolecular Behavior | 56 |
|    G. Rearrangement of Methyl Isocyanide in a Shock Tube | 58 |
|    H. Significance of Gas Phase Results | 59 |
|    I. Solution Studies | 59 |
| III. Internal Excitation | 61 |
|    A. Free Radical Catalyzed Rearrangement of Isonitriles | 61 |
|    B. Rearrangement of Methyl-$t_1$ Isocyanides Resulting from Hot Atom Reaction | 61 |
|    C. Photoisomerization of Methyl Isocyanide | 62 |
|    References | 64 |

## I. INTRODUCTION

This chapter will deal with the quantitative study of the kinetics of isonitrile rearrangement.

The rearrangement of isonitriles to nitriles

$$R—\overset{+}{N}\!\!\equiv\!\!\overset{-}{C} \rightarrow R—C\!\!\equiv\!\!N$$

has been known for half a century,[6] but only in recent years have quantitative studies of this thermal rearrangement reaction appeared.[9,17] The first comprehensive investigation of these was of the thermal gas phase methyl isocyanide system.[22] Recently other thermal studies[19,23] as well as several briefer activation studies have appeared.[24,25,27]

* Work supported by the National Science Foundation.

It is interesting to note that the thermal rearrangement of isonitriles is analogous to the rearrangement of two other isoelectronic species, namely, the rearrangement of the diazonium cation which has been shown to occur to the extent of 1–4% at moderate temperatures in a series of substituted benzene diazonium chlorides[10,11] and the rearrangement of the acetylide anion.[1,28]

## II. THERMAL REARRANGEMENT

### A. Mechanism

The principal resonance structure of the isonitrile molecule is

$$R-\overset{+}{N}\equiv\bar{C}$$

At ordinary temperatures the isonitrile molecular rearrangement is a nucleophilic substitution with isomerization as the only case:

$$R-\overset{+}{N}\equiv\bar{C} \rightarrow R\text{---}\ddot{N}\diagdown_C \rightarrow R-C\equiv N$$

In 1899 Wagner recognized this type of rearrangement to be operative between different members of the dicyclic terpene series. Broader interpretations of the rearrangement by Meerwein showed it to be applicable to other reactions and not exclusive to terpene chemistry. Thus, the isonitrile rearrangement is another example of a Wagner–Meerwein 1–2 shift.[2, 7, 19, 22, 23]

### B. Nature of the Rearrangement

In an effort to delineate the nature of the final intermediates in complex "saturated rearrangement" reactions, Casanova and co-workers[2] made a thorough study of the isonitrile–nitrile rearrangement of a number of aryl and alkyl isocyanides because of the similarity between the stoichiometrically simple isonitrile rearrangement and the final step of a number of saturated rearrangements.

Since no products were found which would have arisen if charge separation had occurred in the transition state of the rearrangement process of cyclobutyl isocyanide (a group of demonstrated propensity for carbon skeleton rearrangement when a cationic carbon is involved, it was concluded by Casanova et al.[2] that the rearrangement proceeds with very little charge separation. This conclusion was further supported by the lack of electronic effect on the rates of isomerization of para-substituted aryl isocyanides (p-chlorophenyl, p-tolyl, and p-methoxyphenyl isocyanide) and the retention of configuration in the rearrangement of optically active sec-butyl isocyanide.

Extended Hückel calculations of Van Dine and Hoffmann (VH)[28] are not in good agreement with the experimental findings of Casanova et al.[2] VH find in every isonitrile studied that there is an increase in positive charge at the migrating carbon atom (R group) in the transition state. However, a cyclic model for the transition state with which the findings of Casanova are consistent had been earlier adopted by Schneider and Rabinovitch (SR)[22,23] for the methyl isocyanide isomerization. The same model for the transition state of ethyl isocyanide was adopted by Maloney and Rabinovitch (MR)[14] and excellent agreement between theory and experiment was obtained (see activated complex below). This model has also been found satisfactory by other workers.[30]

### C. Gas Phase Studies

Quantitative study of the isonitrile–nitrile rearrangement was once mentioned by Ogg.[17] Kohlmaier and Rabinovitch[9] described exploratory kinetic data for tolyl isocyanide. The experimental advantages afforded by isonitrile rearrangement reactions are that they are not attended by production of highly reactive products or chain carriers; nor are they accompanied by pressure change. The stoichiometric simplicity of the rearrangement facilitates theoretical treatment by accurate quantum statistical considerations.

In this section we relate some of the experimental and theoretical findings on the fall-off behavior of the unimolecular rate constants of various isonitriles with pressure.

Until the study of methyl isocyanide fall-off as a function of pressure by Schneider and Rabinovitch (SR),[19,22,23] there was still no example of a unimolecular reaction system whose study approached completeness. This system is the first and only one which extends experimentally from the high to the low pressure region. Since then, in an effort to study the effects of the variation of molecular parameters such as the variation of molecule chain length in a homologous series, a detailed study of ethyl isonitrile has also been made.[14,30]

The methyl isocyanide fall-off was studied at 199.4°, 230.4°, and 259.8°C. Log $k$ vs log $P$ curves are given in Fig. 1; $k_\infty$ was found to be $7.5 \times 10^{-5}$ sec$^{-1}$ at 199.4°C, and $92.5 \times 10^{-5}$ sec$^{-1}$ at 230.4°C; $k_\infty$ at 259.8°C was not determined experimentally but $k$ was extrapolated to $76.7 \times 10^{-4}$ sec$^{-1}$.

The ethyl isocyanide fall-off was studied at 190.0°, 230.9°, and 259.6°C. Log $k$ vs log $P$ curves are given in Fig. 2. In all cases, the experimental data extended into the high pressure region. The values of $k_\infty$ were $0.544 \times 10^{-4}$ sec$^{-1}$ at 190.0°, $15.57 \times 10^{-4}$ sec$^{-1}$ at 230.9°, and $125.7 \times 10^{-4}$ sec$^{-1}$ at 259.6°C.

The classical empirical Kassel fit parameter $s$ and the Slater parameter $n$ for methyl and ethyl isocyanide were determined analytically from the shape

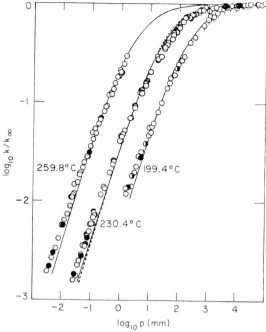

**Fig. 1** Pressure dependence of unimolecular rate constants for $CH_3NC$: log $k/k_\infty$ vs log $p$ at 199.4°, 230.4°, and 259.8°C. For clarity the 260° curve is arbitrarily displaced by one log $k$ unit to the left in the figure, whereas the 199°C curve is displaced the same distance to the right; actually, both of these curves would almost coincide with the 230°C curve. Vertical marks have been placed under the 200°C high pressure points to assist in distinguishing these from the 230° data. The solid curves represent the calculated results for the 300 harmonic model, adjusted on the pressure axis to coincide with the experimental points at log $k/k_\infty = -1$. The dotted curve at 230°C is for the 600 harmonic model, similarly adjusted. Theoretical curves based on the 300 + fig. rot. model coincide almost exactly in shape with the experiments (Table I). (Reproduced from Fig. 1 in Schneider and Rabinovitch.[22])

of the fall-off by a computational procedure.[14] The $s$ and $n$ values were determined from the reduced rate $k/k_\infty$ at various intervals of pressure by comparison with a reference set of entries for the same intervals calculated for a range of $s$ and $n$ values from the theoretical relations of RRK[8]

$$k/k_\infty = \frac{1}{(s-1)!} \int_0^\infty \frac{x^{s-1} e^{-x} dx}{1 + (A/W)x/(x+b)^{s-1}}$$

and Slater[26]

$$I_n(\theta) = \frac{1}{\Gamma(m+1)} \int_0^\infty \frac{x^m e^{-x} dx}{1 + x^m \theta^{-1}}$$

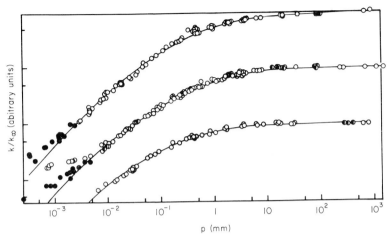

**Fig. 2** Pressure dependence of the unimolecular rate constants for $C_2H_5NC$: $k/k_\infty$ vs pressure at 190.0°, 230.9°, and 259.6°C. The curves are arbitrarily displaced for clarity; the points which deviate markedly from the overall fall-off behavior below $10^{-2}$ mm at 230.9°C are runs performed in a 12-liter vessel and reflect the increasing importance of heterogeneity. The half-filled circles represent runs analyzed by mass spectroscopy. The filled circles are data obtained in a 200-liter vessel and the onset of heterogeneity can be seen in the lower pressure range of these data. The solid curves represent the calculated behavior for the E-300 model, adjusted on the pressure axis to coincide with the experimental points at $k/k_\infty = 0.1$. (Reproduced from Fig. 1 in Maloney and Rabinovitch.[14])

where

$$m = (n-1)/2$$

The mean experimental $s$ and $n$ values averaged over the fall-off curves are given in Table I; theoretical results listed in the table will be discussed later. The Kassel $s$ and the Slater $n$ values increase by 1.2 and 1.5 units, respectively, on going from methyl to ethyl.

### D. Theoretical Rate Calculations

Quantitative theoretical calculation of the rate of isonitrile rearrangement necessitates description of an activated complex in more detail and the statement of the RRKM rate expressions.

1. ACTIVATED COMPLEX

The methyl isocyanide rearrangement involves no formal reaction path degeneracy; the symmetry number $\sigma$ of the molecule is 3 ($C_{3v}$ symmetry), whereas $\sigma^+$ for the activated complex is 1 ($C_1$ or $C_s$ symmetry), so that $\sigma/\sigma^+ = 3$. By contrast $\sigma = \sigma^+ = 1$ for ethyl isocyanide ($C_1$ or $C_s$ symmetry); but in one

## TABLE I

Equivalent Classical Fall-Off Shape Parameters $s$ and $n$ for $CH_3NC$ and $C_2H_5NC$ at 230.9°C

|  |  | $n$ | $s$ |
|---|---|---|---|
| $CH_3NC$ | Expt. | 4.6 | 3.3 |
|  | 300 model[a] | 3.3 | 2.6 |
|  | 300 + fig. rot. model[b] | 4.4 | 3.2 |
| $C_2H_5NC$ | Expt. | 6.1 | 4.5 |
|  | E-300 model[c] | 6.0 | 4.5 |
|  | E-300 + mol. rot. model[d] | 6.4 | 4.8 |

[a] The single out-of-plane twist frequency (~300 cm$^{-1}$) in the vibrational frequency assignment of the activated complex is taken by SR to be coincident with the degenerate molecule bending modes[22] and called the "300 model." All rotations are taken as adiabatic.

[b] The rotational degree of freedom about the figure axis is taken as active.

[c] E-300 represents the "300 model" applied to the ethyl isocyanide molecule. All rotations taken as adiabatic.

[d] A molecular rotation taken active in the E-300 model.

formulation, two diameric transition states may be recognized, one of which has a reaction path degeneracy of two corresponding to enantiomeric forms,

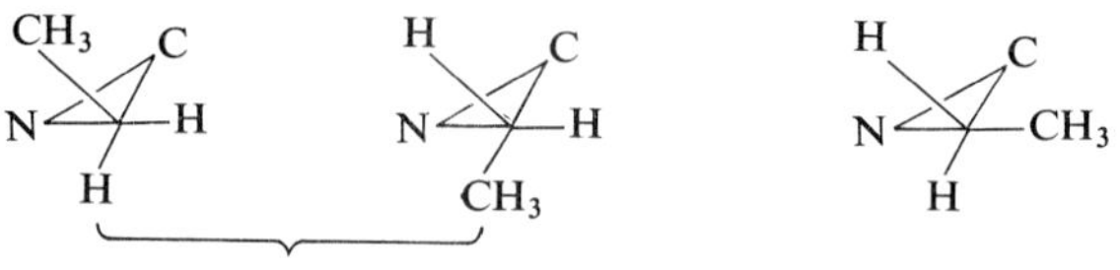

In the calculations, no distinction was made between the diamers and ethyl isocyanide is taken to have a reaction path degeneracy of 3. Alternative transition state configurations may be written in which no eclipse of the "ring" carbon takes place.

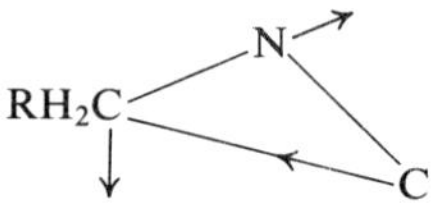

The reaction coordinate adopted by SR is closely related to an asymmetric ring deformation of a triangular species where the bond length parameters for the activated complex correspond to the development of little charge separation in the transition state. The only significantly arbitrary variable is the vibrational frequency assignments for the activated complex is a single out-of-plane twist frequency which in the case of methyl isocyanide, was ascribed by SR in the "300 model" as coincident with the degenerate molecule bending modes; this frequency was varied up to 600 cm$^{-1}$ ("600 model"). The 300 model was found to be superior[14,19,22,23,30] and the same transition state model was also used by Maloney and Rabinovitch in the ethyl isocyanide case; it is then designated as the E-300 model.

## 2. Quantum Statistical Rate Formulation

The rearrangement of methyl and ethyl isocyanide have been treated on the basis of the RRKM quantum statistical formulation.[15,16] An RRKM-type expression for $k_{uni}$ appropriate for internal vibrational and rotational degrees of freedom and without explicit consideration of centrifugal effects is given on simplistic grounds as an absolute rate theory type of expression

$$k_{uni} = \frac{I_r \exp(-E_0/RT)}{Q_v h} \int_{E^+=0}^{\infty} \frac{\Sigma P(E_{vr}^+) \exp(-E^+/RT)}{1 + (I_r/\omega)[\Sigma P(E_{vr})/N(E_{vr})]} \quad (1)$$

with

$$N(E_{vr}) = \frac{(8\pi^2/h^2)^{r/2} \pi \prod_i I_i^{d_i/2} \Gamma(d_i/2)}{\Gamma(r/2)(kT)^{r/2}} \Sigma P(E_v)(E - E_v)^{r/2-1}$$

and

$$P(E_{vr}^+) = \frac{P_r^+}{\Gamma(1 + r/2)(kT)^{r/2}} \Sigma P(E_v^+)(E^+ - E_v^+)^{r/2}$$

where

$$r = \sum_{i=1} d_i$$

$$P_r^+ = (8\pi^2 kT/h^2)^{r/2} \pi \prod_i I_i^{d_i/2} \Gamma(d_i/2)$$

and $\omega$ is the specific collision rate per second.
$I_r$ is the inertial ratio and is given by

$$I_r = \frac{\sigma}{\sigma^+} \left(\frac{I_A^+ I_B^+ I_C^+}{I_A I_B I_C}\right)^{1/2}$$

where the $\sigma$'s are symmetry numbers, $I_A$, $I_B$, and $I_C$ are the principal moments of inertia of the molecule and $I_A^+$, $I_B^+$, and $I_C^+$ for the activated complex; $Q_v$ is the vibration partition function; $\Sigma P(E_{vr}^+)$ is the vibrational-rotational degeneracy at the vibrational-rotational energy $E_{vr}^+$ of the activated complex of total active energy $E^+$; $N(E_{vr})$ is the density of internal vibrational-rotational energy levels of the molecule at the active energy $E_{vr}$; $r$ is the number of rotational degrees of freedom and $I_i$ is the $i$th moment of inertia; $d_i = 1$ for 1° of rotational freedom (internal rotation or figure axis rotation), and $d_i = 2$ for 2° (degenerate) of rotation freedom, e.g., for active overall rotations.

TABLE II

VIBRATION FREQUENCIES FOR THE $CH_3NC$ AND $C_2H_5NC$ SYSTEMS

| Molecular frequencies (cm$^{-1}$) | Complex frequencies (cm$^{-1}$) |
|---|---|
| Methyl isocyanide[22] | 300 Model |
| 2998(3) | 2998(3) |
| 2161 | 1990 |
| 1443(3) | 1443(3) |
| 1041(2) | 1041(2) |
| 945 | 600 |
| 270(2) | 270 |
| Ethyl isocyanide[14] | E-300 model |
| 2953(5) | 2953(5) |
| 2161 | 1900 |
| 1441(4) | 1441(4) |
| 1289(2) | 1289(2) |
| 1080(3) | 1080(3) |
| 945 | 781 |
| 783 | 600 |
| 531 | 400 |
| 270 | 224(2) |
| 224(2) | |

Equation (1) is strictly correct below the high pressure limit only when $I_r$ is unity; since the error caused by departure from this condition is negligible, more sophisticated consideration[18,29] need not be given here.

For the methyl isocyanide system, the vibrational model (300) and 300 + fig. rot. are considered here. For ethyl isocyanide, the E-300 model and another in which a molecular rotation is taken as active have been used.

The vibrational frequency assignments for the methyl and ethyl isocyanide molecules and activated complexes are given in Table II.

## 3. HIGH PRESSURE REGION

At infinite pressure the RRKM expression reduces to

$$k_\infty = \kappa d \frac{kT}{h} I_r \frac{Q_v^+}{Q_v} \exp(-E_0/RT)$$

which is the same as the conventional formulation of Eyring theory, with $\kappa$ set equal to unity; $d$ is the reaction path degeneracy and is unity for $C_{3v}$ symmetry of the parent, and is 3 for lower symmetry. Good agreement between the theoretical and experimental values of $k_\infty$ was obtained in both the methyl isocyanide system (measured $9.2 \times 10^{-4}$ sec$^{-1}$, calculated $8.8 \times 10^{-4}$ sec$^{-1}$ at 230.4°C) and the ethyl isocyanide system (measured $15.6 \times 10^{-4}$ sec$^{-1}$, calculated $12.9 \times 10^{-4}$ at 230.9°C). The $A_\infty$ values also agree which suggests that the frequency assignments in the 300 and E-300 models are reasonable: the measured value for methyl isocyanide is $2.50 \times 10^{13}$ sec$^{-1}$ at 230.4°C, calculated is $2.55 \times 10^{13}$ sec$^{-1}$. The measured value for ethyl isocyanide is $3.16 \times 10^{13}$ sec$^{-1}$ at 230.9°C, calculated is $2.62 \times 10^{13}$ sec$^{-1}$.

## 4. LOW PRESSURE REGION

In the low pressure limit Eq. (1) reduces to

$$k_0 = \omega \int_{E_0}^{\infty} N(E_{vr}) \exp(-E_{vr}/RT) dE_{vr}/Q_{vr}$$

Good agreement between experimental and calculated values of $k_{bi}(k_0/M)$ was obtained for methyl isocyanide (measured $2.0 \times 10^3$ cm$^3$ mole$^{-1}$ sec$^{-1}$, calculated $1.18 \times 10^3$ cm$^3$ mole$^{-1}$ sec$^{-1}$ at 230.4°C). The experimental inaccessibility of the second-order region of ethyl isocyanide prevented comparison of a calculated value of $k_{bi}$ with a measured one.

## 5. THE CRITICAL ENERGY AND ACTIVATION ENERGIES

The relationship between the critical energy $E_0$ and the experimental high pressure activation energy $E_a^\infty$ is well known.[5] The difference $E_a^\infty - E_0$ is ~0.63 kcal mole$^{-1}$ for the E-300 models. $E_0$ for ethyl is 37.6 kcal mole$^{-1}$ which is virtually the same as for methyl isocyanide (37.85 kcal mole$^{-1}$).

In both the methyl[22] and ethyl[14] cases, the dependence of the observed activation energy on pressure has been determined continuously from the calculated fall-off curves on the 300 (Fig. 3) and E-300 models (Fig. 4), respectively. Due to heterogeneous effects in the ethyl fall-off, data below $10^{-2}$ mm were not employed. The theoretical curve on the E-300 model agrees moderately well with the experimental curve derived from the continuous curves; the theoretical spread $E_a^\infty - E_{a0}$ (3.1 kcal mole$^{-1}$) is a little less than

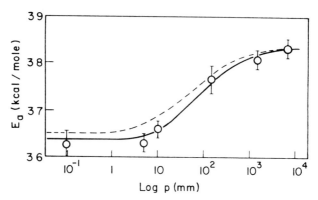

**Fig. 3** "Continuous" Arrhenius activation energies $E_a$ obtained for the isomerization of methyl isocyanide from fall-off curves versus log $p$ (Fig. 1) given by solid curve. Explicit determinations of Arrhenius activation energies given by open circles, with indication of standard deviation; calculated values from the 300 model given by the dashed line. (Reproduced from Fig. 3 in Schneider and Rabinovitch.[22]

the experimental spread of 4.0 kcal (see ref. 14), but this agreement is as good as may be expected. Thus, this model describes the ethyl isocyanide system very well. The spread in $E_a^\infty - E_{a0}$ for ethyl is larger than that for methyl isocyanide. Limiting values of the average energy of reacting molecules for the low and high pressure region are given in Table III. The quasi-constancy of $\langle E^+ \rangle_{p=0}$ with increasing chain length is the sole quantitative vestige of predictions in these systems based on classical statistics according to which $\langle E^+ \rangle_{p=0} \simeq RT$ independent of molecular size, provided $E_0 \gg sRT$.

TABLE III
CALCULATED LIMITING VALUES OF THE AVERAGE EXCESS ENERGY (cm$^{-1}$) OF REACTING MOLECULES IN ISONITRILE SYSTEMS ($T = 230°C$)[a]

|  | $\langle E^+ \rangle_{p=0}$ | $\langle E^+ \rangle_{p=\infty}$ |
|---|---|---|
| $CH_3NC$ | 425 | 870 |
| $C_2H_5NC$ | 480 | 1575 |
| $n\text{-}C_3H_7NC$[b] | 535 | 2300 |
| $i\text{-}C_3H_7NC$[b] | 530 | 2230 |
| $n\text{-}C_4H_9NC$[b] | 590 | 2800 |

[a] Based on the 300 model in all cases.
[b] Assumed critical energy of approximately 38 kcal mole$^{-1}$.

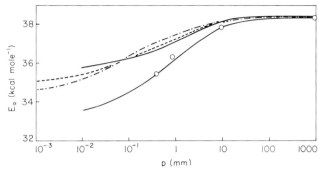

**Fig. 4** Variation of Arrhenius activation energy with pressure as obtained from the experimental fall-off curves for the data of Fig. 2 (solid curves). Separate explicit measurements of $E_a$ are shown by the open circles. Dashed line represents the theoretical behavior for the E-300 model, and (—.) represents the E-300 + mol. rot. model.

### 6. THE PRESSURE FALL-OFF REGION

The experimental fall-off curves of methyl (Fig. 1) and ethyl isocyanide (Fig. 2) are well fitted by the respective 300 models. The pressure correction factors needed to bring the calculated curves into agreement with the observed curves at $k/k_\infty = 0.1$ are listed in Table IV. The value of the diameter, $\sigma = 4.5$ Å,

TABLE IV

PRESSURE CORRECTION FACTORS

| | Methyl isocyanide | | Ethyl isocyanide |
|---|---|---|---|
| $T$ (°C) | 300 Model, $\sigma = 4.5$ Å | $T$ (°C) | E-300 Model, $\sigma = 5.0$ Å |
| 119.4 | 0.91 | 190.0 | 0.88 |
| 230.4 | 1.11 | 230.9 | 1.37 |
| 259.8 | 0.95 | 259.6 | 1.25 |

for methyl isocyanide was obtained from viscosity measurements. A diameter of 5.0 Å was adopted for ethyl isocyanide by MR. Only very modest correction factors between 0.9 and 1.1 for methyl and 0.9 and 1.2 for ethyl were required. Use of larger collision cross sections would result in pressure correction factors of proportionately larger magnitudes.

In the Slater theory,[26] the shape parameter $n$ is independent of temperature. In the classical RRK theory,[8] the shape parameter $s$ is fixed empirically and the theory predicts decreased curvature with temperature rise at not too high temperatures ($E_0 \ll sRT$) for a fixed $s$. Due to the small temperature ranges employed in methyl and ethyl studies, the experimental change [increase in the shape of the fall-off with rise of temperature is slight but shows a consistent increase (Table V)].

TABLE V
EXPERIMENTAL AND THEORETICAL SHAPE PARAMETERS

| Molecule $CH_3NC$ | $T$ (°C) | | | | | |
|---|---|---|---|---|---|---|
| | 199.4 | | 230.4 | | 259.8 | |
| | s | n | s | n | s | n |
| Expt. | 3.2 | 4.4 | 3.3 | 4.6 | — | — |
| 300 Model | 2.5 | 3.1 | 2.6 | 3.3 | 2.7 | 3.5 |
| 300 + fig. rot. | 3.1 | 4.2 | 3.2 | 4.4 | 3.3 | 4.6 |
| Molecule $C_2H_5NC$ | 190.0 | | 230.9 | | 259.6 | |
| Expt. | 4.3 | 5.8 | 4.5 | 6.1 | 4.5 | 6.0 |
| E-300 Model | 4.3 | 5.8 | 4.5 | 6.0 | 4.7 | 6.2 |
| E-300 mol. rot. | 4.6 | 6.3 | 4.9 | 6.5 | 4.9 | 6.6 |

For methyl isocyanide whose figure axis moment increases almost fourfold in the 300 model transition state (Table VI), assumption of an active figure axis rotation was shown to increase the shape parameters $s$ and $n$, and improve the agreement between the calculated and observed values (Tables I and V).[22] By contrast, the assumption of an active overall rotation for ethyl isocyanide moves the fall-off a little to lower pressures together with an increase in $s$ and $n$ (Table I); this model gives poorer fit to the data and, as has been pointed out,[14] there is little reason to treat overall rotation as active for this molecule. The total ratio of products of moments of inertia for the methyl isocyanide complex relative to the molecule is 1.80 (i.e., $I_r/\sigma = 1.34$); the ratio for ethyl isocyanide is smaller ($I_r/\sigma = 1.182$).

For methyl isocyanide at 230.9°C, the pressure at which $k/k_\infty$ equals 0.5 ($p_{1/2}$) is 65 mm. The increase of the chain length to the next higher member in the homologous series, i.e., ethyl isocyanide, reduces $P_{1/2}$ to 0.6 mm.[14]

## 7. ANHARMONICITY EFFECTS

Anharmonicity arises as a result of an enhancement in $N(E)$. In $F_2O$, for which $E_0 \sim 40$ kcal mole$^{-1}$, Rice[21] estimated that anharmonicity causes an increase in rate by a factor of 3–4 at low pressures. SR showed that the energy level density for methyl isonitrile at a similar energy level is probably increased by about ~15%. For the larger ethyl isocyanide molecule with 21

## TABLE VI
### Moments of Inertia (AMU Å$^2$) for Several Isonitrile Molecules[a]

| Molecule | Activated complex model | $I_A$ | $I_B$ | $I_C$ | $I_A^+$ | $I_B^+$ | $I_C^+$ | $I^+/I$ | $(I^+/I)^{3/2}$ |
|---|---|---|---|---|---|---|---|---|---|
| CH$_3$NC | 300 | 3.23 | 50.29 | 50.29 | 11.59 | 31.65 | 40.01 | 1.794 | 1.343 |
| C$_2$H$_5$NC | E-300 | 12.55 | 97.81 | 110.4 | 14.64 | 110.5 | 117.0 | 1.397 | 1.182 |
| n-C$_3$H$_7$NC | nP-300 | 17.90 | 213.8 | 222.0 | 30.88 | 215.7 | 196.1 | 1.536 | 1.240 |
| iso-C$_3$H$_7$NC | iP-300 | 40.33 | 172.5 | 138.7 | 46.03 | 142.6 | 120.7 | 0.821 | 0.906 |
| n-C$_4$H$_9$NC | nB-300 | 26.93 | 382.3 | 367.8 | 32.59 | 356.6 | 374.6 | 1.149 | 1.072 |
| t-C$_4$H$_9$NC | tB-300 | 104.4 | 162.9 | 167.2 | 108.7 | 135.9 | 144.5 | 0.751 | 0.866 |
| CD$_3$NC | 300 | 6.45 | 58.91 | 58.91 | 14.86 | 38.65 | 47.06 | 1.207 | 1.098 |
| C$_2$D$_5$NC | E-300 | 15.36 | 111.0 | 126.3 | 20.33 | 129.3 | 133.3 | 1.627 | 1.276 |
| n-C$_3$D$_7$NC | nP-300 | 25.74 | 245.0 | 251.5 | 40.83 | 247.8 | 229.5 | 1.463 | 1.210 |
| iso-C$_3$D$_7$NC | iP-300 | 55.18 | 200.4 | 158.0 | 60.88 | 169.4 | 139.2 | 0.822 | 0.907 |
| n-C$_4$D$_9$NC | nB-300 | 36.54 | 418.0 | 429.8 | 43.43 | 406.9 | 421.3 | 1.134 | 1.065 |
| t-C$_4$D$_9$NC | tB-300 | 113.1 | 184.1 | 189.5 | 134.09 | 159.1 | 165.5 | 0.760 | 0.872 |

[a] Principal moments of inertia were calculated by means of a revised computer program (4/15/64) of J. H. Schachtschneider.

rather than 12 vibration modes and the same critical energy as methyl isocyanide, the effect is greatly reduced since the average vibrational level populated by each mode is almost halved. For the isonitriles larger than ethyl isocyanide the effect would be insignificant.

### E. Kinetic Isotope Effects

The secondary isotope effect that arises in the isonitrile–nitrile rearrangement reaction is a statistical weight intermolecular isotope effect that occurs in the nonequilibrium region of thermally activated unimolecular reaction systems. The differential quantal effects that occur in unimolecular reaction systems as a consequence of the change in frequency pattern of the molecule upon isotopic substitution, e.g., of H by D, was originally described quantitatively by Rabinovitch, Setser, and Schneider.[20] This is a pure secondary isotope effect of a statistical weight nature, unlike conventional secondary isotope effects studied in equilibrium systems whose origin is basically mechanistic and which involve a change in critical reaction energy $E_0$ upon isotopic substitution; the present effect increases with the increasing degree of isotopic substitution, even at positions remote from the reaction site and is a maximum at the low pressure limit.

A detailed formulation and extensive discussion of the nature of the effect in terms of the RRKM theory has been given elsewhere.[20] It is sufficient to write

$$(k_H/k_D)_0 = \left(\frac{M_D}{M_H}\right)^{1/2} \frac{Q_D \int_{E_{0H}}^{\infty} N(E)_H e^{-E/RT} dE}{Q_H \int_{E_{0D}}^{\infty} N(E)_D e^{-E/RT} dE}$$

where the $M$ terms are the molecular weights, the $Q$ terms are the molecular partition functions for active degrees of freedom and the $N(E)$ are the densities of active energy levels at the energy $E$. In the present cases, $E_{0H} \simeq E_{0D}$, so that mechanistic differences in the rearrangement of the isotopic methyl series and ethyl-$d_5$ isocyanide are absent.

The relative rates of $k_H/k_D$, for the following system rearrangements, $CH_2DNC$–$CH_3NC$,[19] $CD_3NC$–$CH_3NC$,[23] and $C_2D_5NC$–$C_2H_5NC$[13] have been measured over wide pressure ranges. Summary curves of the individual pressure dependence of $k_H/k_D$ for the above-mentioned systems are shown in Fig. 5.

Upon substitution of a D atom for an H atom (in $CH_2DNC$), $P_{1/2}$ is lowered a little from the value of 65 mm for $CH_3NC$[19]; restoration of the symmetry by complete deuteration[23] (i.e., $CD_3NC$) further lowers $P_{1/2}$ to ~30 mm. The pressure correction factors are 1.1 and 1.5, respectively, for the methyl and

3. Isonitrile–Nitrile Rearrangement    55

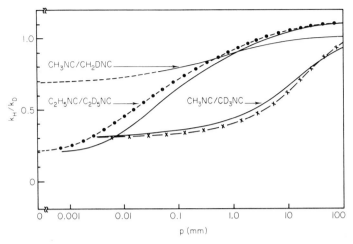

**Fig. 5** Variation of $k_H/k_D$ with pressure. The solid lines are summary curves of the experimental data for the $CH_3/CH_2DNC$, $CH_3NC/CD_3NC$, and $C_2H_5NC/C_2D_5NC$ systems. (X) represents the 300 theoretical curve for the $CD_3NC/CD_3NC$, and (—.) represents the E-300 theoretical curve for the $C_2H_5NC/C_2D_5NC$ system.

methyl-$d_3$ 300 models, with $\sigma = 4.5$ Å at 230°C. The factors for ethyl and ethyl-$d_5$ are 1.375 and 1.5, respectively, for the E-300 models, using $\sigma = 5.0$ Å. The observed limiting high pressure ratios $(k_H/k_D)_\infty$ for $CH_2DNC$, $CD_3NC$, and $C_2D_5NC$ are 1.02, 1.07, and 1.10, respectively, and decline to 0.70, 0.28, and 0.21, respectively, at the lowest pressures.

Measurements of the ratio $(k_H/k_D)$ for $CD_3NC^{23}$ were made over a range of temperatures from 180 to 250°C at 10 atm pressure. A value of $\Delta E_{a\infty} = 0.1$ kcal was taken as the experimental value. However, it was pointed out that a value of $\Delta E_{a\infty}$ of $-0.1$ or $-0.2$ kcal gives better agreement between theory and experiment. Measurements of the ratio for $C_2D_5NC$ were made over a range of temperatures from 190.0 to 259.6°C at 100 mm.[13] $\Delta E_{a\infty}$ was found to be $-0.15$ kcal mole$^{-1}$, i.e., approximately zero.

Thus, one deals with a quantum statistical-weight effect which increases with increasing degree of isotopic substitution, i.e., $N(E)_D$ increases relative to $N(E)_H$. As a result, the order in the isonitrile series is as follows: $C_2D_5NC$ (measured 0.2, calculated 0.19), $CD_3NC$ (measured 0.28, calculated 0.29), $CH_2DNC$ (measured 0.70, calculated 0.67). The agreement between theory and experiment is excellent in all cases (Fig. 5).

Wettaw and Sims (WS)[30] measured the carbon-13 isotope effect in the thermal isomerization of methyl isocyanide to acetonitrile. The carbon-13 isotope effect for $CH_3NC$ at 230°C and 1000 mm has been found[30] to be 1.017. WS obtained very good agreement between RRKM theory and experiment with the use of a ring-complex model (measured 1.017, calculated 1.018 at 1000 mm).

This agreement between the measured and calculated carbon-13 results is instructive in two ways: (1) the carbon-13 kinetic isotope effect is a good mechanistic probe in that vibrations involving CNC are related to the reaction coordinate, and (2) the carbon-13 isotope effect reported by WS is an average of two isotope effects—one with the labeled carbon in the methyl group and the other with it in the isonitrile group.

The fact that the model used by WS is very similar to the "300 model" used by Rabinovitch and co-workers for the interpretation of fall-off behavior and deuterium isotope effects lends additional strong support to the model for the activated complex.

### F. Effect of Variation of Molecular Parameters on Nonequilibrium Unimolecular Behavior

The $n$ and $s$ shape parameters for the methyl–methyl-$d_3$ and ethyl–ethyl-$d_5$ isocyanide systems were determined analytically from the shape of the fall-off by analytical computation[14] (Table VII).

TABLE VII

EQUIVALENT CLASSICAL FALL-OFF SHAPE PARAMETERS $s$ AND $n$ FOR $C_2D_5NC$ AND OTHER MOLECULES (230.9°C)

|  |  | $n$ | $s$ |
|---|---|---|---|
| $CH_3NC$ | Expt. | 4.6 | 3.3 |
|  | 300 | 3.3 | 2.6 |
|  | 300 + fig. rot. | 4.4 | 3.2 |
| $CD_3NC$ | Expt. | 5.2 | 3.7 |
|  | 300 | 3.8 | 2.9 |
|  | 300 + fig. rot. | 5.0 | 3.6 |
| $C_2H_5NC$ | Expt. | 6.1 | 4.5 |
|  | E-300 | 6.0 | 4.5 |
|  | E-300 + mol. rot. | 6.4 | 4.5 |
| $C_2D_5NC$ | Expt. | 6.8 | 5.1 |
|  | E-300 | 7.0 | 5.2 |
|  | E-300 + mol. rot. | 7.5 | 5.6 |

Trideuteration of methyl isocyanide has been shown experimentally to increase the values of $n$ and $s$ by ~0.6 and ~0.4, respectively. Therefore, the replacement of three H atoms by three D atoms has the effect of "adding" one-half an "effective oscillator." An increase in the chain length from methyl to ethyl isocyanide results in an increase of the observed $n$ and $s$ values by

~1.5 and ~1.2, respectively; addition of a methylene group results in approximately three times the effect of substitution by three deuterium atoms. Similarly, an increase in the chain length from methyl-$d_3$ to ethyl-$d_5$ isocyanide causes $n$ and $s$ to increase by ~1.7 and ~1.4, respectively. Pentadeuteration of ethyl results in an increase in $n$ and $s$ of 0.8 and 0.6, respectively. Therefore, substitution of five deuterium atoms adds a little less than one "effective oscillator."

Predicted RRKM curves have also been calculated with use of the 300 model for the propyl and butyl isocyanide thermal isomerizations[14] at a plausible (arbitrary) critical energy of 37 kcal mole$^{-1}$. The dependence of the Slater $n$-value on molecular complexity is shown in Table VIII. The increase in chain length is accompanied by a corresponding increase in the $n$-value.[14]

The observed increase in $n$ from methyl to ethyl is 1.5 $n$-units (Table VIII), which is in excellent agreement with the calculated increase of 1.6 $n$-units when optimum models are compared, i.e., the 300 model for ethyl and the 300 + fig. rot. for methyl isocyanide. No $n$-values have been yet measured for the higher isonitriles; some calculated $n$-values are given in Table VIII; from these, one sees that the increase in $n$ between a straight-chain and a branched-chain isonitrile of the same carbon number is minor and substantially less than the increase observed when the carbon number is increased by one. In the former case, the rearrangement of the skeletal structure of the R group is accompanied by slight changes in the frequency patterns of the molecule; in the latter case, increase in the size of R results in an increase in the number of vibrational modes and also drastically changes the vibrational pattern of the molecule.

The calculations show that $P_{1/2}$ should decrease markedly to ~$10^{-2}$ mm for $n$-propyl and isopropyl isocyanide and to ~$10^{-3}$ mm for $n$-butyl and $t$-butyl

TABLE VIII

CALCULATED AND OBSERVED SLATER $n$-VALUES
AS A FUNCTION OF MOLECULAR COMPLEXITY
at 230.9°C

| Molecule | Observed | Calculated[a] |
|---|---|---|
| $CH_3NC$ | 4.6 | 3.3 (4.4)[b] |
| $C_2H_5NC$ | 6.1 | 6.0 |
| $n$-$C_3H_7NC$ | — | 9.0 |
| $i$-$C_3H_7NC$ | — | 9.1 |
| $n$-$C_4H_9NC$ | — | 15.4 |
| $t$-$C_4H_9NC$ | — | 16.0 |

[a] The calculated value in each case is for the simple 300 vibrational model.
[b] For the 300 + fig. rot. model.

isocyanide. This decrease in curvature together with the shift of the fall-off to lower pressures forecasts the experimental inaccessibility of the fall-off region for still higher isonitrile homologs.

From the calculated values of the average energy of reacting molecules $\langle E^+ \rangle$ for several members of the isonitrile series,[13] it was found that $\langle E^+ \rangle_{p=0}$ is quasi-constant with change in molecular structure, whereas $\langle E^+ \rangle_{p=\infty}$ increases in a manner dependent upon the change in $n$ (Table III).

### G. Rearrangement of Methyl Isocyanide in a Shock Tube

Lifshitz, Carroll, and Bauer (LCB)[12] studied the thermal rearrangement of methyl isocyanide to acetonitrile behind reflected shocks in a single-pulse shock tube over the temperature range 710–860°K. The reaction mixture consisted of 2% methyl isocyanide in argon.

LCB used an Arrhenius extrapolation and Slater's equation for the dependence of $k/k_\infty$ on temperature in order to compare their data with the low temperature data (450–530°K) of SR[22] (Fig. 6). LCB stated that agreement between extrapolated and observed data was well within the experimental error of the temperatures computed for the reflected shocks.

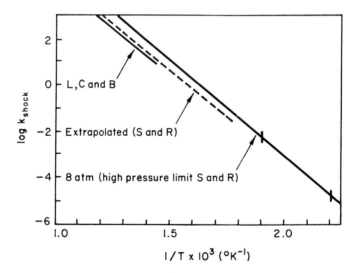

*Fig. 6* Arrhenius plots for the isomerization of methyl isocyanide. The heavy solid line is the line for $k_\infty = 10^{13.4} \exp(-38,400/RT)$ as determined by Schneider and Rabinovitch. The temperature range of their experiments is indicated by the short vertical lines. The line marked L, C, and B is the median line for the shock tube data, while the dashed line is the extrapolated curve (S and R) assuming the argon collision efficiency was 5% of the CH$_3$NC collision efficiency.   (Reproduced from Fig. 1 in A. Lifshitz *et al.*[12])

In view of the fact that the relative efficiency assumed by LCB for argon is about a factor of 4 below the measured value[3,4] at 550°K (an error in the relative efficiency of argon would give rise to an incorrect estimation of the total equivalent pressure of methyl isocyanide, which would in turn give the incorrect value of $k/k_\infty$), the discrepancy between the extrapolated and the median of the observed values of LCB (Fig. 6) is much larger than was known by them.

Since the observed activation energy is a weak function of temperature, we have made calculations to ascertain whether or not there is any appreciable change in the activation energy over the temperature range of the extrapolation. An appreciable change in the activation energy over a given temperature interval would cause the $\log k$ vs $T^{-1}$ plot to be nonlinear over that interval. However, the change in the activation energy over the range of the extrapolation is fortunately very small (0.10 kcal mole$^{-1}$).

It appears that with improved precision, shock tube studies may in future provide important high temperature information on unimolecular reactions.

### H. Significance of Gas Phase Results

Since the formulation of the Lindemann–Christiansen–Hinshelwood mechanism, almost a half century ago, a great deal of experimental work has been done to test the scheme, assumptions, and approximations regarding each step. No convincing comprehensive experimental unimolecular systems existed previously. Now, however, the rearrangement of isonitriles to nitriles provides a very complete example of a gas phase unimolecular reaction system and provides one of the most thorough tests of unimolecular reaction theory. The data for the methyl and ethyl isocyanide systems bear out the general usefulness of the RRKM theory in predicting and describing unimolecular behavior, with regard to fall-off shape, kinetic isotope effects, and pressure region upon lengthening of the chain. The decline of activation energy with pressure also correlates reasonably well with the theory. In short, the isonitrile work bear out the quantum statistical RRKM theory as a good (the best) practical working theory to date.

### I. Solution Studies

Quantitative kinetic studies of the isonitrile–nitrile rearrangement in solution are very sparse. Moreover, kinetic studies of the isonitrile–nitrile rearrangement have the inherent limitation in solution that the totality of quasi-unimolecular behavior is inaccessible; only first-order behavior is observable. Emphasis must therefore be placed on mechanism in such studies.

Kohlmaier and Rabinovitch (KR) examined the thermal rearrangement of $p$-tolyl isocyanide[9] in solution (Nujol solvent). KR reported an activation energy of 36.8 kcal mole$^{-1}$ for $p$-tolyl isonitrile in solution as compared to 33.8 kcal mole$^{-1}$ in the gas phase.

Casanova and co-workers[2] have made some very cogent and important contributions. They measured the rates of rearrangement of three *para*-substituted aryl isocyanides ($p$-chlorophenyl, phenyl, and $p$-methoxyphenyl isocyanide) in diglyme (Table IX). Through the application of the Hammett equation, log $(k/k_0) = \sigma\rho$, which applies to $m$- and $p$-substituted side-chain derivatives of benzene ($\sigma$ is the substituent constant independent of the nature of the reaction and $\rho$ is a proportionality constant, dependent upon the nature of the reaction and the conditions), Casanova *et al.* found that variation in $\sigma$ has virtually no effect on the rate of rearrangement of the *para*-substituted isocyanides. They concluded that this insensitivity of the reaction rate to the polar character of *para*-substituents and to the polarity of the solvent signifies only minor electronic effects on the migrating C atom. Indeed, the study of methyl and ethyl isocyanides demonstrates the absence of any special electronic or energy effects, and the existence of entropy effects that are purely statistical in nature as governed by internal degrees of freedom.

In conclusion, the solution work of Casanova *et al.* (see also Chapter 6) provides important information and confirmation of the nature of the transition state in the isonitrile–nitrile rearrangement reaction: there is no separation of molecular fragments and little charge separation during the course of reaction.

TABLE IX

First-Order Rate Constants for the Thermal Isomerization of *para*-Substituted Isocyanides in Diglyme at 200.0°C[2]

| G, in G—C$_6$H$_4$—N$\equiv$C | $10^5 k_1$, sec$^{-1}$ | $10^5 k_1$, av | $k_1$, rel | $\sigma^a$ |
|---|---|---|---|---|
| H | 28.8 | | | |
| H | 28.0 | | | |
| H | 26.9 | | | |
| H | 26.8 | 27.6 ± 0.8 | 1.13 | 0.000 |
| Cl | 25.1 | | | |
| Cl | 24.3 | | | |
| Cl | 23.9 | | | |
| Cl | 24.3 | 24.4 ± 0.4 | 1.00 | +0.227 |
| CH$_3$O | 32.5 | | | |
| CH$_3$O | 31.0 | | | |
| CH$_3$O | 30.3 | | | |
| CH$_3$O | 31.7 | 31.4 ± 0.9 | 1.29 | −0.268 |

$^a$ Taft.[26a]

## III. EXTERNAL EXCITATION

### A. Free Radical Catalyzed Rearrangement of Isonitriles

SR found that methyl isocyanide rearrangement was catalyzed by the presence of radicals and of oxygen. Shaw and Pritchard (SP)[25] have studied the rearrangement of methyl and ethyl isocyanide catalyzed by methyl radicals produced in the thermal decomposition of di-$t$-butyl peroxide in the gas phase at $\sim 100°C$. A thermal source of methyl radicals was chosen to avoid possible photosensitization of the isomerization.

From exploratory experiments with ethyl isocyanide, SP proposed that the reaction proceeds via a displacement process

$$\overset{\bullet}{C}H_3 + C_2H_5NC \rightarrow CH_3CN + \overset{\bullet}{C}_2H_5$$

wherein the addition of the incoming radical to the divalent carbon of the $-N{\equiv}C$: group is followed by expulsion of the radical originally attached to the N atom; a chain process ensues. On this basis, the rearrangement proceeds via the exothermic unimolecular decomposition of the chemically activated intermediate radical species $CH_3CH_2-N{\equiv}C-CH_3$ whose formation is estimated to be exothermic by 5–10 kcal mole$^{-1}$. SP obtained the reasonable values of the Arrhenius parameters, $E = 7.8 \pm 0.3$ kcal/mole and $A = 10^{12.25}$ mole$^{-1}$ cm$^3$ sec$^{-1}$, for the reaction

$$CH_3NC + \overset{\bullet}{C}H_3 \rightarrow \overset{\bullet}{C}H_3 + CH_3CN$$

Oxygen catalysis may proceed by direct attack but must also entail radical production in the course of oxidation processes. In further efforts to specify the $CH_3 + RNC$ system, SP found that phenyl isocyanide does not react with methyl radicals.

### B. Rearrangement of Methyl-$t_1$ Isocyanide Resulting from Hot Atom Reaction

Ting and Rowland (TR)[27] observed the substitution reaction of hot T atoms,

$$CH_3NC + T^{xx} \rightarrow CH_2TNC^* + H \rightarrow CH_2TNC + H \qquad (1)$$

in the liquid phase but no detectable amount of $CH_2TNC$ was found in the gas phase experiments (Table X); the asterisk represents assumed vibrational excitation. The rearrangement product $CH_2TCN$ was found in both gaseous and liquid phases. TR pointed out that the nature of the product molecules indicated the presence of $CH_2T$ radicals in the gas phase. They assumed that the primary reaction leads to the formation of $CH_2TNC^*$ which rearranges in the gas phase but is collisionally stabilized in the liquid phase. The activation energy for the decomposition of $CH_2TCN^*$ or $CH_2TNC^*$ to $CH_2T + CN$ radicals is $\sim 100$ kcal mole$^{-1}$ and is more readily suppressed by collisional stabilization than is the rearrangement reaction with much lower activation energy (38 kcal mole$^{-1}$).

TR did not take into consideration the fact that $CH_2T$ radicals can catalyze the rearrangement reaction,

$$\dot{C}H_2T + CH_3NC \rightarrow CH_2TCN + \dot{C}H_3 \qquad (2)$$

This reaction requires only 7.8 kcal mole$^{-1}$ to occur and may help to explain why TR did not find detectable amounts of $CH_2TNC$. In addition to (2) above, one may have

$$CH_3NC + T^* \rightarrow \dot{C}H_3 + TCN \qquad (3)$$

$$CH_3NC + \dot{C}H_3 \rightarrow CH_3CN + \dot{C}H_3 \qquad (4)$$

The H atom from (1) above should also initiate the reaction

$$CH_3NC + H \rightarrow \dot{C}H_3 + HCN \qquad (5)$$

These reactions will occur along with the combination reactions of H and T atoms and of methyl radicals, and the decomposition–stabilization reactions of higher formed hydrocarbons. It appears that the rearrangement reaction

$$CH_2TNC^* \rightarrow CH_2TCN \qquad (6)$$

may occur to a minor extent relative to the chemical activation process (2). This is further complicated by the fact that, considering reaction (4), $CH_2TCN$ can also be formed via the substitution reaction

$$CH_3CN + T^* \rightarrow CH_2TCN^* + H \qquad (7)$$

followed by collisional stabilization of $CH_2TCN^*$ to $CH_2TCN$. In general, the system studied by TR is a very complicated radical-catalyzed system.

### C. Photoisomerization of Methyl Isocyanide

SP photolyzed methyl isocyanide vapor with ultraviolet radiation at 2537 Å and found that methyl isocyanide rearranges to acetonitrile without intervention of radicals.

### 3. Isonitrile–Nitrile Rearrangement

The simplest mechanism for methyl isocyanide photoexcitation is

$$CH_3NC + h\nu \rightarrow CH_3NC^\dagger$$

where $CH_3NC^\dagger$ is in some unspecified electronically excited state and is ~112 kcal above the ground state. Excited methyl isocyanide undergoes isomerization to acetonitrile

$$CH_3NC^\dagger \rightarrow CH_3CN^\dagger$$

TABLE X

Relative Yields of Radioactive Products from the Reaction of Recoil Tritium with $CH_3NC$ (Yield of HT = 100)[27]

| Product | $CH_3T$ | $CH_2TOH$ | $C_2H_5T$ | $CH_2TNC$ | $CH_2TCN$ |
|---|---|---|---|---|---|
| Liquid $CH_3NC$ | 7.2 | <0.05 | <0.1 | 19.8 | 14.3 |
| 16 cm $CH_3NC$ | 3.7 | <0.01 | 3.7 | <0.05 | 7.6 |
| 16 cm $CH_3NC$ + 3 cm $O_2$ | 3.6 | 1.3 | <0.06 | <0.05 | 8.4 |
| 16 cm $CH_3NC$ + 5 atm Ar + 3 cm $O_2$ | 4.3 | 1.8 | <0.03 | <0.05 | 15.9 |

where $CH_3CN^\dagger$ now has ~127 kcal of excitation energy. Reaction (9)

$$CH_3NC^\dagger + CH_3NC \rightarrow CH_3NC + CH_3NC \quad (9)$$

represents collisional stabilization of excited parent molecules by unexcited parent molecules, with a similar process for $CH_3CN^\dagger$. Reaction (10) was proposed for deactivation of excited product,

$$CH_3CN^\dagger + CH_3NC \rightarrow CH_3CN + CH_3NC^{\dagger\dagger} \quad (10)$$

where it is possible—although we believe not probable—for $CH_3NC^{\dagger\dagger}$ to have up to ~127 kcal. For the sake of simplicity, SP assumed $CH_3NC^{\dagger\dagger}$ to react in the same manner as $CH_3NC^\dagger$ and thus one has a chain reaction.

Addition of inert gases (Ar and $C_2H_6$) resulted in a decreased rate of isomerization which was further decreased by increasing the pressure of the added gas. However, the addition of oxygen resulted in complete suppression of the isomerization. This suggested that $CH_3NC^\dagger$ is a triplet molecule, although the evidence is inconclusive.

SP also found that addition of various gases ($CO_2$, $N_2O$, and benzene) to $CH_3NC$ photosensitized the isomerization. The mechanism is uncertain; although not much is presently known about the excited state chemistry of isonitriles, this process may have potential as a diagnostic of excited states.

## REFERENCES

1. Casanova, J. Jr., Geisel, M., and Morris, R. N., *J. Amer. Chem. Soc., Preprint.*
2. Casanova, J., Jr., Werner, N. D., and Schuster, R. E., *J. Org. Chem.* **31**, 3473 (1966).
3. Chan, S. C., Spicer, L., and Rabinovitch, B. S., unpublished results (1970) (have found that the values of $\beta$ in Fletcher et al.[4] are systematically low by an average of 20%).
4. Fletcher, F. J., Rabinovitch, B. S., Watkins, K. W., and Locker, D. J., *J. Phys. Chem.* **70**, 2823 (1966).
5. Glasstone, S., Laidler, K. J., and Eyring, H., "The Theory of Rate Processes." McGraw-Hill, New York, 1941.
6. Guillemard, M. H., *C. R. Acad. Sci.* **144**, 141 (1907).
7. Ingold, C. K., "Structure and Mechanisms in Organic Chemistry," Chapter IX. Cornell Univ. Press, Ithaca, New York, 1953.
8. Kassel, L. S., "Kinetics of Homogeneous Gas Reactions." Reinhold, New York, 1932.
9. Kohlmaier, G., and Rabinovitch, B. S., *J. Phys. Chem.* **63**, 1793 (1959).
10. Lewis, E. S., and Holliday, R. E., *J. Amer. Chem. Soc.* **88**, 5043 (1966).
11. Lewis, E. S., and Insole, J. M., *J. Amer. Chem. Soc.* **86**, 32 (1964).
12. Lifshitz, A., Carroll, H. F., and Bauer, S. H., *J. Amer. Chem. Soc.* **86**, 1488 (1964).
13. Maloney, K. M., Pavlou, S. P., and Rabinovitch, B. S., *J. Phys. Chem.* **73**, 2756 (1969).
14. Maloney, K. M., and Rabinovitch, B. S., *J. Phys. Chem.* **73**, 1652 (1969).
15. Marcus, R. A., *J. Chem. Phys.* **20**, 359 (1952).
16. Marcus, R. A., and Rice, O. K., *J. Phys. Colloid Chem.* **55**, 894 (1951).
17. Ogg, R. A., Jr., *J. Chem. Phys.* **7**, 753 (1939).
18. Placzek, D. W., Rabinovitch, B. S., Whitten, G. Z., and Tschuikow-Roux, E., *J. Chem. Phys.* **43**, 4071 (1965).
19. Rabinovitch, B. S., Gilderson, P. W., and Schneider, F. W., *J. Amer. Chem. Soc.* **87**, 158 (1965).
20. Rabinovitch, B. S., Setser, D. W., and Schneider, F. W., *Can. J. Chem.* **39**, 4215 (1962).
21. Rice, O. K., *J. Phys. Chem.* **65**, 1588 (1961).
22. Schneider, F. W., and Rabinovitch, B. S., *J. Amer. Chem. Soc.* **84**, 4215 (1962).
23. Schneider, F. W., and Rabinovitch, B. S., *J. Amer. Chem. Soc.* **85**, 2365 (1963).
24. Shaw, D. H., and Pritchard, H. O., *J. Phys. Chem.* **70**, 1230 (1966).
25. Shaw, D. H., and Pritchard, H. O., *Can. J. Chem.* **45**, 2749 (1967).
26. Slater, N. B., "Theory of Unimolecular Reactions." Cornell Univ. Press, Ithaca, New York, 1959.
26a. Taft, R. W., Jr., *in* "Steric Effects in Organic Chemistry" (M. S. Newman, ed.), p. 571. Wiley, New York, 1956.
27. Ting, C. T., and Rowland, F. S., *J. Phys. Chem.* **72**, 763 (1968).
28. Van Dine, G. W., and Hoffmann, R., *J. Amer. Chem. Soc.* **90**, 3227 (1968).
29. Waage, E. V., and Rabinovitch, B. S., *Chem. Rev.* (1970) (to be published).
30. Wettaw, J. F., and Sims, L. B., *J. Phys. Chem.* **72**, 3440 (1968).

# Chapter 4

# Simple α-Additions

### T. Saegusa and Y. Ito

| | | |
|---|---|---|
| I. | Introduction | 65 |
| II. | α-Addition Reactions of Isonitriles with Reactive Hydrogen Compounds | 67 |
| | A. Reactions with Heteroatom–Hydrogen Bonds Catalyzed by Groups IB and IIB Metals Compounds | 67 |
| | B. Reactions with Reactive Hydrogen Compounds in the Presence of Acid Catalysts | 72 |
| | C. Reactions with Reactive Hydrogen Compounds without Added Catalyst | 73 |
| III. | α-Addition Reactions of Isonitriles with Reactive Halogen Compounds | 76 |
| | A. Acid Chlorides and Related Compounds | 76 |
| | B. Nitrogen–Halogen Compounds | 77 |
| | C. Oxygen–Halogen Compounds | 77 |
| | D. Sulfur–Halogen Compounds | 78 |
| | E. Halogens | 78 |
| IV. | Oxidation of Isonitriles | 79 |
| V. | Reduction of Isonitriles | 80 |
| VI. | Reactions with Carbenes and Nitrenes | 80 |
| VII. | Reaction with Organometallic Compounds | 81 |
| VIII. | Radical Reactions of Isonitriles | 84 |
| | A. Free Radical-Catalyzed Isomerizations | 84 |
| | B. Radical Reactions with Stannanes, Phosphines, and Thiols | 85 |
| | C. Reactions with Peroxide and Nitrosoacetanilide | 86 |
| | D. Electrolysis of Isonitriles | 87 |
| IX. | Reactions Related to α-Additions | 88 |
| | A. Reaction with Benzyne | 88 |
| | B. Reaction with Tropylium Ion | 88 |
| | C. Cationic Isomerization and Oligomerization of Isonitriles | 89 |
| | References | 90 |

## I. INTRODUCTION

This chapter covers the formation of simple α-adducts from isonitriles, according to eq. (1).

$$R-NC + X-Y \longrightarrow R-N=C\!\!<^{X}_{Y} \qquad (1)$$

The two groups, X and Y, of the reactant X—Y are added to the formally divalent carbon atom of an isonitrile. A wide variety of compounds is known to react with an isonitrile in this fashion. Depending on the natures of the reactant and of the catalyst, if one is used, different types of reactions have been observed. The cleavage of the X—Y bond can proceed through either a heterolytic or a homolytic mode. Furthermore, insertion-type reactions are induced by some metal complex catalysts, although the details of the transition states of such reactions have not been elucidated.

There are basically three groups of reactants of the type X—Y, i.e., reactive hydrogen compounds, halides, and organometallic compounds. The first group of reactants are those which contain hydrogen bonded to a heteroatom. These include N—H, P—H, O—H, S—H, Si—H, and hydrogen halides as well as the C—H bonds of certain aromatic compounds. The second group of compounds is characterized by reactive *carbon–halogen* bonds such as those of acyl chlorides and amide chlorides, the *nitrogen–halogen* bonds such as found in *N*-haloamides, the *oxygen–halogen* bonds such as found in hypochlorites, and the *sulfur–halogen* bonds such as found in sulfenyl chlorides. The third group of reactants are organometallic compounds such as Grignard reagents, alkyl-tin amides, and boranes.

In addition to the reactions of isonitriles (1) with the above compounds, the reactions of isonitriles with oxygen, carbenes, and the related species, free radicals and benzyne are included in this chapter; the reactions with carbonyl compounds in the presence of acid catalysts are related to the Passerini reaction and are discussed in Chapter 7.

An isonitrile is considered as a hybrid of the following three resonance structures.

$$R-\overset{\oplus}{N}\equiv\overset{\ominus}{C}: \longleftrightarrow R-\overset{..}{N}=C: \longleftrightarrow \overset{\ominus}{R}=\overset{\oplus}{N}=C:$$
$$\text{(Ia)} \qquad\qquad \text{(Ib)} \qquad\qquad \text{(Ic)}$$

These structures are characterized by the carbon atoms carrying lone-pair electrons. On the basis of the physical properties of isonitriles, structure Ia makes the greatest contribution (see Chapter 1). Isonitriles usually behave as nucleophiles. However, aromatic isocyanides show electrophilic properties in some reactions. The electrophilic reactivity of an aromatic isocyanide (9) may be ascribed to some contribution from structures Ib and Ic.

Some α-additions of isonitriles proceed in the absence of any added catalyst or initiator, whereas others require catalysis. The catalysts and initiators for the α-addition reactions of isonitriles are roughly classified into three groups, viz., the compounds of groups IB and IIB metals (especially copper), acids (both protonic and Lewis), and radical initiators.

## II. α-ADDITION REACTIONS OF ISONITRILES WITH REACTIVE HYDROGEN COMPOUNDS

### A. Reactions with Heteroatom–Hydrogen Bonds Catalyzed by Groups IB and IIB Metals Compounds

Functional groups of heteroatom–hydrogen bonds such as —OH, =PH, and =NH are comparatively inert toward isonitriles. They are reluctant to react with isonitriles except under drastic conditions. For example, ethanol has often been employed as an inert solvent for the reaction of an isonitrile with a third reagent. For the α-addition reactions of the heteroatom–hydrogen bonds to an isonitrile, Saegusa and co-workers have found a group of catalysts among the compounds of groups IB and IIB, particularly copper compounds. The exploration of the catalysts was made on the basis of the fact that isonitriles coordinate with the groups IB and IIB metals and their salts to form the corresponding complexes. The following are the α-addition reactions of heteroatom–hydrogen bonds which are effectively catalyzed by groups IB and IIB metals compounds. The isonitrile carbon is inserted into $=$N—H[39,40] $=$P—H,[36] —O—H,[37,42] —S—H,[45] and $\equiv$Si—H[38] bonds to produce the corresponding derivatives of formimidic acid. All of these catalyzed reactions give products with high selectivities and in high yield, mostly over 80% and sometimes over 95%. These reactions are conveniently utilized for synthetic purposes.

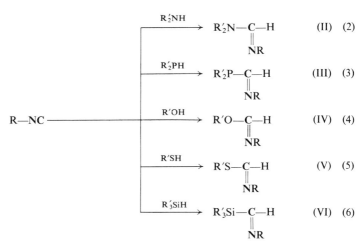

Aliphatic isocyanides do not react with primary and secondary amines in the absence of a catalyst even at temperatures as high as 120°C. Aromatic amine hydrochlorides, however, react with isonitriles to give the corresponding formamidine hydrochlorides.[21] Metallic Cu, Cu(I), and Cu(II) salts (CuCl, $CuCl_2$, $Cu(CN)_2$, Cu(II)-acetylacetonate), AgCl, $ZnCl_2$, and $Cd(OAc)_2$ are quite effective catalysts,[6,39] which cause the amine–isonitrile reaction (2) to produce a formamidine almost quantitatively at 110–120°C. In particular, the AgCl-catalyzed reaction[40] gives the formamidine (II) in quantitative yield in 2.5 hr at 20°C. With regard to the catalytic activity of metallic Cu, it is of interest to note that metallic Cu dissolves in liquid cyclohexyl isocyanide, forming a soluble complex. The IR spectrum of a mixture of metallic Cu and cyclohexyl isonitrile showed a band at 2180 $cm^{-1}$ in addition to the one at 2140 $cm^{-1}$ for the free isonitrile. The band at 2180 $cm^{-1}$ is assigned to the isocyanide which is coordinated with Cu. Chlorides of Ni, Co, and Pd, which are typical transition metal catalysts for several olefin reactions, do not catalyze the amine–isonitrile reaction (2), but induce homopolymerization of isonitriles (see also Millich and Sinclair[20]). $AlCl_3$, a typical Friedel-Crafts catalyst, shows no catalytic activity.

A ternary complex of CuCl, cyclohexyl isocyanide, and piperidine, having the composition [$(CuCl)_2(c$-$C_6H_{11}NC)_2$(piperidine)] was isolated as an unstable white crystalline solid.[40] The IR spectrum of the ternary complex showed absorption bands at 2176 $cm^{-1}$ for the isocyanide component and at 3192 $cm^{-1}$ for the piperidine moiety. The N≡C band of the binary complex of CuCl and cyclohexyl isocyanide appeared at 2192 $cm^{-1}$ and that of liquid isocyanide at 2140 $cm^{-1}$. Further, the =N—H band of free piperidine was observed at 3280 $cm^{-1}$. These observations seem to indicate the formation of a mixed ligand complex in which the isocyanide and piperidine ligands are coordinated with a common Cu(I) ion; they are influenced by one another. The mixed ligand complex decomposed even at room temperature to produce the corresponding formamidine. It is therefore assumed that the reaction has taken place in the ligand sphere of the complex.

The α-addition of a phosphorus–hydrogen bond of a dialkylphosphine to an isonitrile (3) has also been accomplished using groups IB and IIB metal compounds as catalysts.[36] Copper oxide is an excellent catalyst which brings about almost quantitative conversions. Salts of Cu(II), Zn, Cd, and Hg(II) are fairly active. The copper-catalyzed reaction (3) provides a simple synthetic route to $N$-substituted formimidoylphosphines (III), the Schiff bases of formylphosphines. The α-addition of a phosphine to an isonitrile may be compared with the β-addition of a phosphine to an olefin (7) involving a radical mechanism.

Incidentally, Saegusa et al.[45] found that the α-addition of a phosphine to an isonitrile was also induced by a radical initiator (see Section VIII,B,2).

$$R_2PH + {>}C{=}C{<} \rightarrow R_2P-\underset{|}{\overset{|}{C}}-\underset{|}{\overset{|}{C}}-H \qquad (7)$$

Alcohols are rather inert toward isonitriles. An aromatic isocyanide was reported to combine with ethanol in the presence of sodium ethoxide at 120°C to produce an $N,N'$-diaryl formamidine via ethyl $N$-aryl formimidate (VII)[22] (8).

$$C_6H_5-NC + C_2H_5OH \rightarrow [C_6H_5N=CHOC_2H_5] \rightarrow C_6H_5N=CHNHC_6H_5 \quad (8)$$
$$(VII)$$

The alcohol–isonitrile reaction under pressures as high as 8500 atm at 125°C has been reported to give a formamidine (VIII)[4] (9). One possible mechanism involving basic catalysis and α-addition was proposed in this paper and is shown below; however, it is not certainly known. The catalyst is assumed to be a small amount of ethyl amine or ammonia present as a contaminant in the ethyl isocyanide. A radical mechanism for this reaction was ruled out because ethanol and ethyl mercaptan gave different products.

$$ROH \xrightarrow{base} RO^{\ominus} \xrightarrow{C_2H_5-NC} C_2H_5-N=\overset{\ominus}{C}OR \xrightarrow{ROH}$$

$$C_2H_5N=C\overset{OR}{\underset{H}{\diagdown}} \xrightarrow{RO^{\ominus}} C_2H_5-\overset{\ominus}{N}CH(OR)_2 \xrightarrow{C_2H_5-NC}$$

$$C_2H_5-NCH(OR)_2 \xrightarrow{ROH} C_2H_5-N=CH-N-CH(OR)_2 \quad (9)$$
$$\underset{C=NC_2H_5}{|\ominus} \qquad\qquad\qquad\qquad \underset{C_2H_5}{|}$$
$$(VIII)$$

The simple α-addition of an alcohol to an isonitrile (4) is effectively catalyzed by group IB and IIB metal compounds. The catalysts for this reaction are classified into two groups according to their catalytic activities. The first group includes metallic Cu and the oxides of Cu(I), Cu(II), Ag(I), and Hg(II), which accelerate the reactions of isonitriles with various alcohols including saturated, unsaturated, and β-amino alcohols. The second group of catalysts are the chlorides of Cu(I), Ag(I), Zn(II), and Cd(II), which specifically cause the reactions of isonitriles with certain alcohols such as allyl alcohol and $N,N$-dimethylaminoethanol. Of all these catalysts, metallic Cu and Cu oxides are the most active and afford the formimidate product in almost quantitative yields.

The difference in catalytic activity between these two groups of catalysts can be explained by assuming a ternary complex consisting of the catalyst, isonitrile, and alcohol as the site of reaction.[41] For the formation of such a ternary complex, the coordination tendencies of the two ligand components should be comparable. The coordination tendency was examined by IR spectroscopy. The shifts for the isonitrile absorption caused by its coordination with four Cu compounds are the following: metallic Cu, 40 cm$^{-1}$; $Cu_2O$, 41 cm$^{-1}$; CuCl, 52 cm$^{-1}$; CuCN, 61 cm$^{-1}$. It is seen that the coordination of an isonitrile with the first group of catalysts (Cu and $Cu_2O$) is weaker than that with the second group (CuCl and CuCN). On the other hand, the coordination tendency of alcohols with Cu compounds is appreciably dependent on the structures of these alcohols. Allyl alcohol and $N,N$-dimethylaminoethanol dissolve CuCl, whereas saturated alcohols do not. The chelating coordinations of allyl alcohol and $\beta$-amino-ethanol with CuCl have been supported by the position of the C—O band as well as the position of the C=C band of allyl alcohol.

However, the interaction between saturated alcohols and CuCl was not detected by IR studies. On the basis of all of the above observations, the following explanation is offered: Isonitriles are held rather loosely by the first group catalysts, and hence some of the isonitrile will be replaced by alcohol, either saturated alcohols of weak complexing tendencies or alcohols of fairly strong coordinating tendencies. Therefore, the first group of catalysts form ternary complexes of various combinations of alcohols and isonitriles and hence are catalysts for the alcohol–isonitrile reactions. On the other hand, isonitriles are held tightly by the second group of catalysts, and the isonitrile ligand is partly replaced only by those alcohols having strong coordination tendencies. Thus, the second group of catalysts offers a reaction site only to such special alcohols.

The reaction of an isonitrile with an alkane thiol proceeds by two courses.[4, 45]

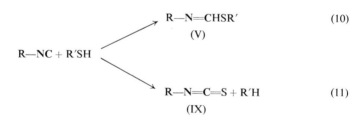

In reaction (10), the isonitrile carbon is inserted into the sulfur–hydrogen bond of the thiol to form the thioformimidate (V), whereas an isothiocyanate (IX) and an alkane are formed in reaction (11). As discussed in Section VIII,B,3, the radical reactions of an isonitrile with a thiol take both courses, depending upon the nature of the thioalkyl group. Metallic copper as well as Cu(I) and Cu(II) salts catalyze only reaction (10) irrespective of the kind of thiol. The copper-catalyzed reaction of a thiol with an isonitrile is not affected by the addition of hydroquinone. Some representative results are listed in Table I.

## TABLE I

COPPER-CATALYZED AND RADICAL-INITIATED REACTIONS OF THIOLS WITH ISONITRILES

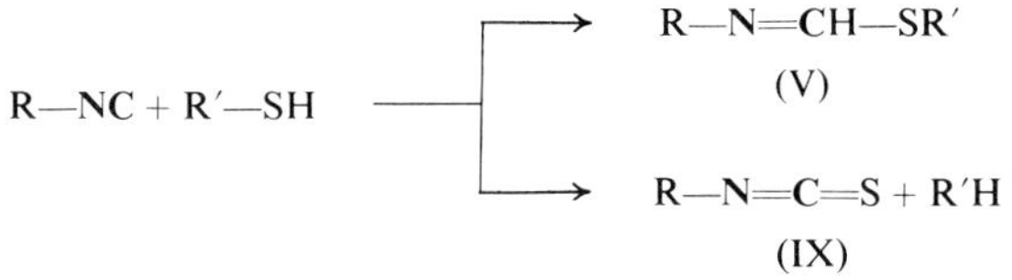

$$R—NC + R'—SH \longrightarrow \begin{cases} R—N{=}CH—SR' \quad (V) \\ R—N{=}C{=}S + R'H \quad (IX) \end{cases}$$

| RNC (mmoles) | R'SH (mmoles) | Additive (mmoles) | Reaction °C | hr | Product (%) V | IX | R'H |
|---|---|---|---|---|---|---|---|
| c-C$_6$H$_{11}$—NC (8.0) | C$_2$H$_5$—SH (10.0) | Cu$_2$O (0.2) | 100 | 3 | 93 | 1 | 0 |
| c-C$_6$H$_{11}$—NC (8.0) | i-C$_3$H$_7$—SH (10.0) | CuO (0.2) | 100 | 1 | 90 | — | — |
| c-C$_6$H$_{11}$—NC (5.0) | t-C$_4$H$_9$—SH (7.0) | CuO and p-BQ$^a$ (0.1) (0.5) | 90 | 3 | 64 | 6 | — |
| t-C$_4$H$_9$—NC (12.0) | C$_6$H$_5$—CH$_2$—SH (10.0) | CuO (0.1) | 100 | 2 | 91 | 0 | 0 |
| c-C$_6$H$_{11}$—NC (8.0) | C$_2$H$_5$—SH (10.0) | AIBN$^b$ (0.01) | 40 | 2 | 85 | — | — |
| c-C$_6$H$_{11}$—NC (8.0) | i-C$_3$H$_7$—SH (10.0) | AIBN$^b$ (0.05) | 70 | 0.2 | 35 | 60 | — |
| C$_6$H$_5$—NC (6.0) | t-C$_4$H$_9$—SH (20.0) | AIBN$^b$ (0.05) | 75 | 0.2 | 0 | 97 | — |
| t-C$_4$H$_9$—NC (12.0) | C$_6$H$_5$—CH$_2$—SH (10.0) | AIBN$^b$ (0.02) | 100 | 0.2 | 0 | 93 | 96 |

$^a$ p-BQ = p-benzoquinone.
$^b$ AIBN = Azobisisobutyronitrile.

The α-addition of a silicon–hydrogen bond to an isonitrile (6) was also accomplished recently with the aid of a catalytically active copper compound, such as Cu-acetylacetonate or CuCl$_2$.[38] This reaction is a novel type of hydrosilation, which offers a convenient method for the preparation of formimidoylsilanes, a new class of organosilicon compounds. The Cu-catalyzed hydrosilation of isonitriles may be compared to the Pt-catalyzed hydrosilation of olefins, which is a β-addition of a silicon–hydrogen bond to an olefin (12).[33]

$$R_3SiH + {>}C{=}C{<} \xrightarrow{\text{Pt complex}} R_3Si{-}\underset{|}{\overset{|}{C}}{-}\underset{|}{\overset{|}{C}}{-}H \qquad (12)$$

## B. Reactions with Reactive Hydrogen Compounds in the Presence of Acid Catalysts

Pyrroles react with phenyl isocyanide at the 2-position in the presence of HCl to give the hydrochlorides of the Schiff base derivatives (XI, XII).[55] Since the formation of formimidoyl chlorides from isonitriles and HCl is known (18),[22] the reactions of equations (13) and (14) are assumed to be electrophilic substitution reactions by a C-protonated isonitrile on pyrrole.

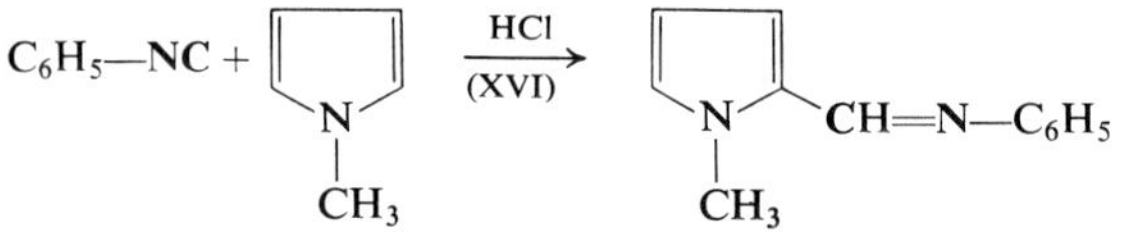

$$C_6H_5-NC + \text{(XVI)} \xrightarrow{HCl} \text{(XI)} \quad (13)$$

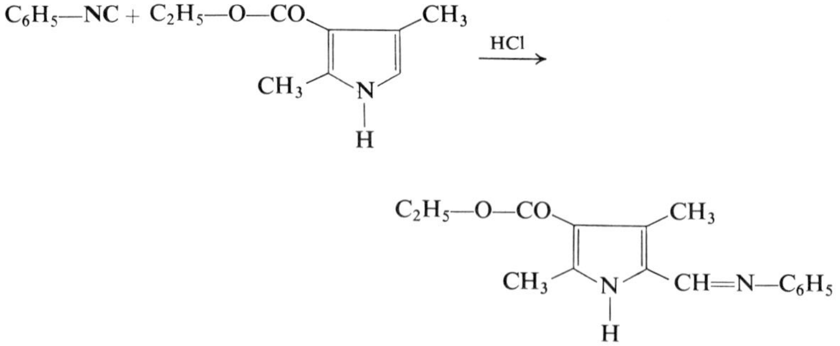

$$C_6H_5-NC + C_2H_5-O-CO-\text{pyrrole} \xrightarrow{HCl} \text{(XII)} \quad (14)$$

Formally, hydroxylamine behaves like an amine in its reaction with phenyl isocyanide,[23] i.e., without any added catalyst, the α-addition produces a formamidine derivative (XIII) (15).

$$C_6H_5-NC + NH_2OH \rightarrow C_6H_5-N=CHNHOH \quad (15)$$
$$\text{(XIII)}$$

On the other hand, the $ZnCl_2$-catalyzed reaction (16) of hydroxylamine with cyclohexyl isocyanide[26] gives cyclohexylurea (XIV).

$$c\text{-}C_6H_{11}-NC + NH_2OH \rightarrow [c\text{-}C_6H_{11}-\overset{\oplus}{N}\!\!=\!\!C-NH_2] \rightarrow c\text{-}C_6H_{11}-NH-CO-NH_2 \quad (16)$$
$$\text{(XIV)}$$

## 4. Simple α-Additions

The hydroxylamine–Lewis acid system is known to form an aminium ion complex $NH_2^\oplus$; reaction (16) may therefore be described as an electrophilic addition of $\overset{\oplus}{NH_2}$ to the isonitrile.

The α-addition of water to an isonitrile gives the corresponding formamide (XV). The hydration of an isonitrile (17) is acid-catalyzed.

$$R-NC + H_2O \rightarrow \left[ R-N=C\underset{OH}{\overset{H}{\diagup}} \right] \rightarrow R-NH-CHO \qquad (17)$$
$$\text{(XV)}$$

The reverse reaction, the dehydration of formamide, is one of the usual routes to isonitriles (see Chapter 2).

### C. Reactions with Reactive Hydrogen Compounds without Added Catalyst

#### 1. Reaction with Hydrogen Halides

A violent reaction occurs between isonitriles and hydrogen halides. In the reaction carried out at −15°C, the salts of formimidoyl halides (XVI) were isolated.[22]

$$R-NC + 2HX \rightarrow \left[ R-N=C\underset{X}{\overset{H}{\diagup}} \right] HX \quad (X = Cl, Br, I) \qquad (18)$$
$$\text{(XVI)}$$

#### 2. Reaction with Carboxylic Acids

The reaction (19) of an isonitrile with a carboxylic acid produces an acid anhydride (XIX) and a formamide[22] (XVIII). The α-addition of a carboxyl group to an isonitrile may be assumed to be the first step. The primary adduct (XVII) then reacts further with a second molecule of the carboxylic acid to form the acid anhydride and formamide:

$$R-NC + R'COOH \rightarrow \left[ R-N=C\underset{OCOR'}{\overset{H}{\diagup}} \right] \xrightarrow{R'COOH} R-NH-CHO + (R'CO)_2O$$
$$\text{(XVII)} \qquad\qquad\qquad \text{(XVIII)} \quad \text{(XIX)}$$
$$(19)$$

The phenyl isocyanide-formic acid reaction is violent even at 0°C, and forms carbon monoxide and formanilide[22] by decomposition of the labile α-adduct (XVII, R = $C_6H_5$—, R' = H—). A stable, isolable α-adduct (XXI) is formed from XX.[10]

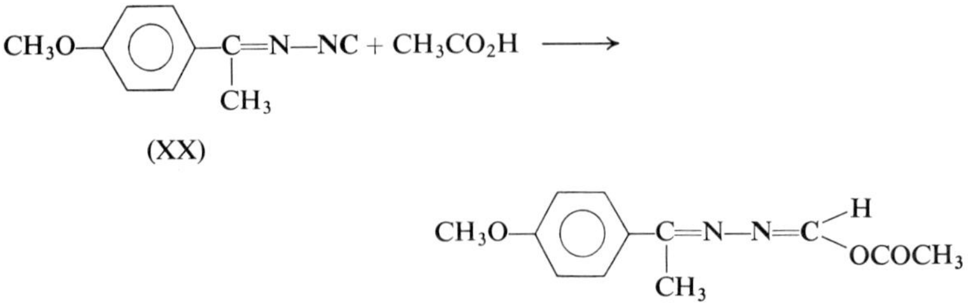

$$CH_3O-\underset{(XX)}{\underset{|}{C_6H_4}}-\underset{CH_3}{\overset{|}{C}}=N-NC + CH_3CO_2H \longrightarrow$$

$$CH_3O-C_6H_4-\underset{\underset{CH_3}{|}}{C}=N-N=C\underset{OCOCH_3}{\overset{H}{\diagup}} \quad (20)$$

(XXI)

The cyclization of the α-adduct (XXII) to the oxazole derivative (XXIII)[10] is a further illustration of the reactivity of carboxylic acid–isonitrile α-adducts.

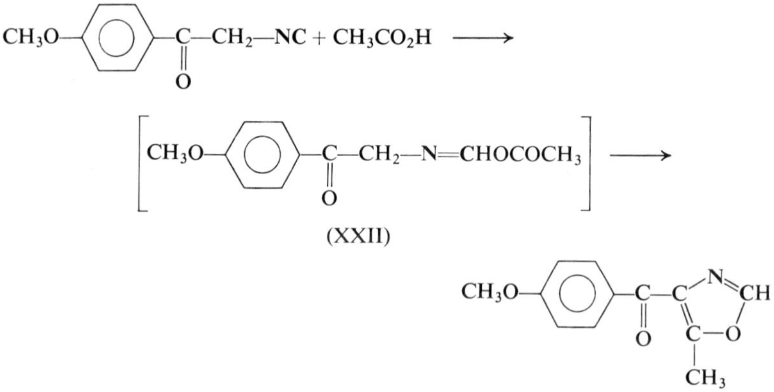

$$CH_3O-C_6H_4-\underset{\underset{O}{\|}}{C}-CH_2-NC + CH_3CO_2H \longrightarrow$$

$$\left[ CH_3O-C_6H_4-\underset{\underset{O}{\|}}{C}-CH_2-N=CHOCOCH_3 \right] \longrightarrow$$

(XXII)

$$CH_3O-C_6H_4-\underset{\underset{O}{\|}}{C}-\underset{\underset{CH_3}{|}}{\overset{N}{\underset{C-O}{C\!\!\!\diagup\!\!\!\diagdown CH}}} \quad (21)$$

(XXIII)

## 3. Reaction with Hydrazoic Acid

The reaction of an isonitrile with hydrazoic acid offers a general synthetic route to 1-substituted tetrazoles (XXV).[3,27]

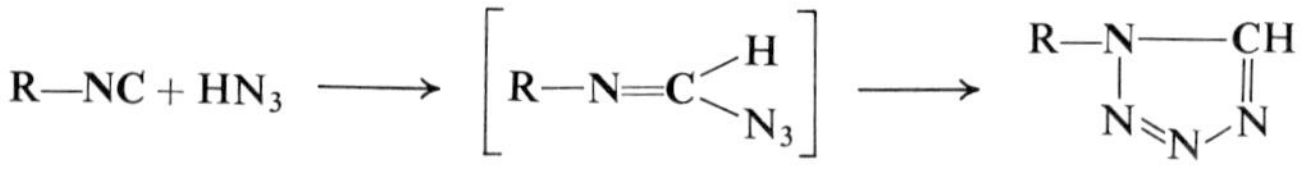

$$R-NC + HN_3 \longrightarrow \left[ R-N=C\underset{N_3}{\overset{H}{\diagup}} \right] \longrightarrow \underset{N\!\!\diagdown\!\!N\diagup\!\!N}{\overset{R-N-CH}{\underset{\|}{|}}} \quad (22)$$

(XXIV) \qquad (XXV)

The α-adduct of hydrazoic acid (XXIV) is an unstable intermediate, which is spontaneously converted into a substituted tetrazole (XXV). Diisonitriles (XXVI) and hydrazoic acid yield bistetrazoles (XXVII).[24]

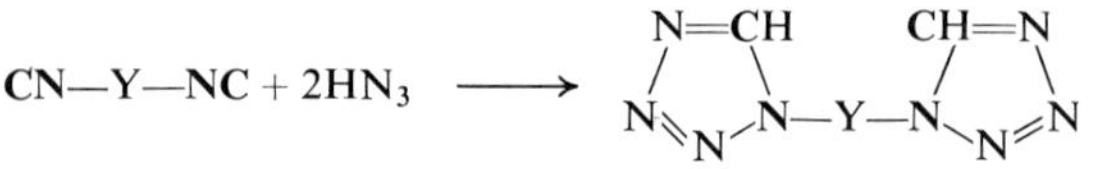

$$CN-Y-NC + 2HN_3 \longrightarrow \underset{N\diagdown N\diagup N-Y-N\diagdown N\diagup N}{\overset{N=CH \quad CH=N}{\diagup\!\!\diagdown\quad\diagup\!\!\diagdown}} \quad (23)$$

(XXVI) \qquad (XXVII)

## 4. Simple α-Additions

### 4. Reaction with Active Methylene Compounds

The active methylene group of pyrazolone derivatives (XXVIII) reacts with phenyl isocyanide to form the α-adduct (XXIX) (24).[18,32]

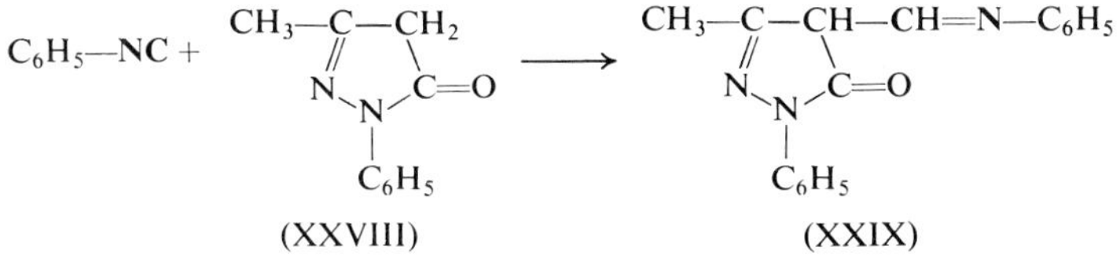

$$C_6H_5\text{---}NC + \underset{(XXVIII)}{\text{pyrazolone}} \longrightarrow \underset{(XXIX)}{\text{adduct}} \quad (24)$$

### 5. Reactions with Phenol and Naphthol

Phenyl isocyanide attacks phenol[29] and naphthols[30,31] without any added catalyst at the position next to the —O—H group, (25)–(27). The reactions, however, are not believed to be simple 1:1 α-additions.

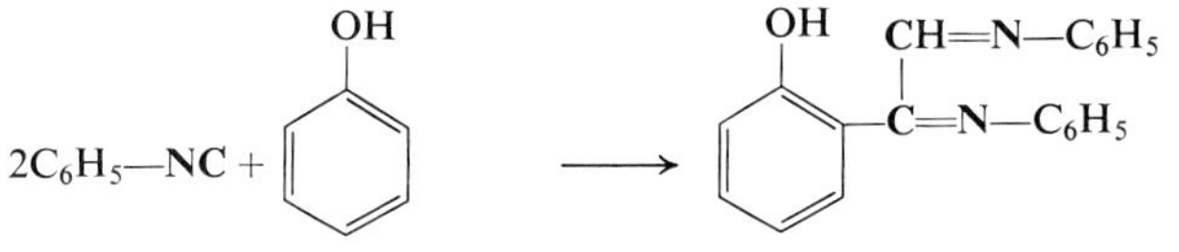

$$2C_6H_5\text{---}NC + \text{phenol} \longrightarrow (XXX) \quad (25)$$

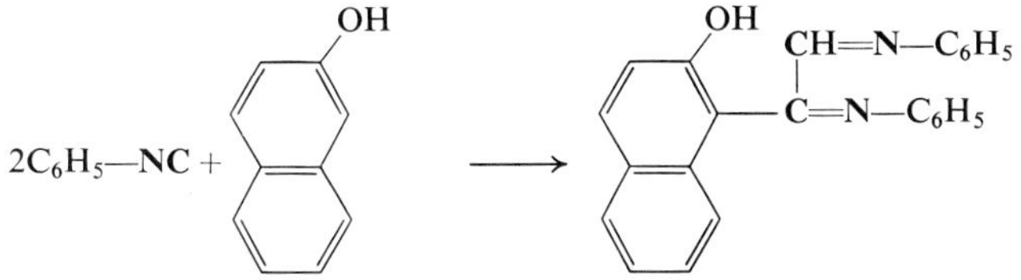

$$2C_6H_5\text{---}NC + \text{2-naphthol} \longrightarrow (XXXI) \quad (26)$$

$$3C_6H_5\text{---}NC + \text{1-naphthol} \longrightarrow (XXXII) \quad (27)$$

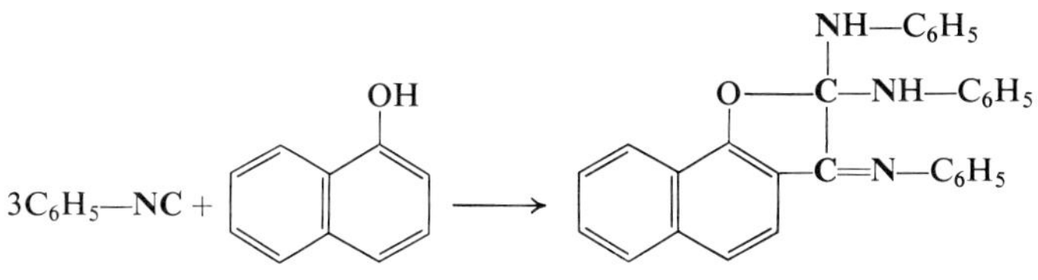

### 6. Reactions with Olefins and Acetylenes

Although olefinic hydrocarbons are inert toward isonitriles, olefins and acetylenes with electron-withdrawing groups such as a carbonyl or nitrile are exceptions. Methyl acrylate reacts with an isonitrile to produce XXXIII.[35]

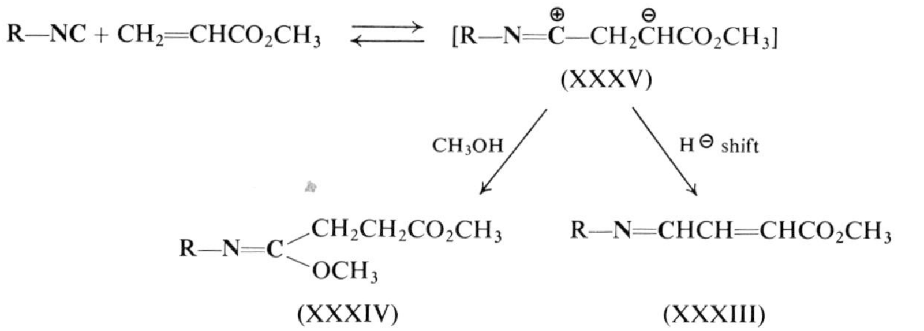

$$R-NC + CH_2=CHCO_2CH_3 \rightleftharpoons [R-N\overset{\oplus}{=}C-CH_2\overset{\ominus}{C}HCO_2CH_3]$$
$$(XXXV)$$

$$CH_3OH \swarrow \qquad \searrow H^{\ominus} \text{ shift} \qquad (28)$$

$$R-N=C\overset{CH_2CH_2CO_2CH_3}{\underset{OCH_3}{\diagdown}} \qquad R-N=CHCH=CHCO_2CH_3$$
$$(XXXIV) \qquad\qquad (XXXIII)$$

The reaction of an isonitrile with methyl acrylate in the presence of methanol affords N-substituted methyl β-carbomethoxypropionimidate (XXXIV) as well as XXXIII. These products may be explained by reaction scheme (28) which involves a 1,3 dipolar intermediate (XXXV). Methyl methacrylate, acrylonitrile, and dicarbomethoxy acetelene react with isonitriles in an analogous fashion.

## III. α-ADDITION REACTIONS OF ISONITRILES WITH REACTIVE HALOGEN COMPOUNDS

### A. Acid Chlorides and Related Compounds

The reaction of an isonitrile with an acid chloride was reported by Nef in 1894.[23] An α-ketoamide (XXXVII) was produced when the reaction mixture was treated with water. The α-addition of acid chlorides to isonitriles was

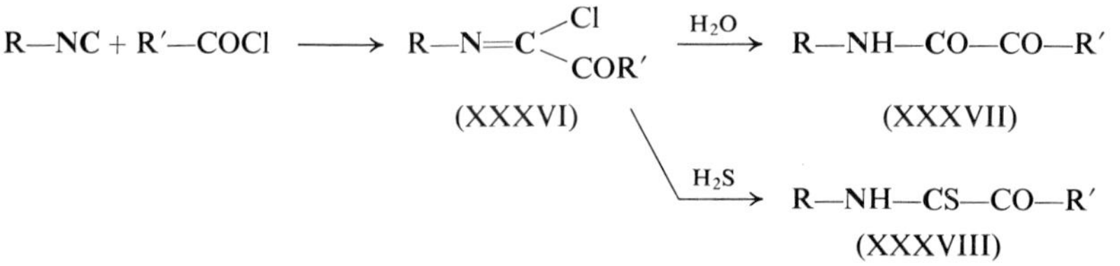

$$R-NC + R'-COCl \longrightarrow R-N=C\overset{Cl}{\underset{COR'}{\diagdown}} \xrightarrow{H_2O} R-NH-CO-CO-R'$$
$$(XXXVI) \qquad\qquad (XXXVII) \qquad (29)$$
$$\xrightarrow{H_2S} R-NH-CS-CO-R'$$
$$(XXXVIII)$$

studied by Ugi[58] who isolated the α-ketoimidoyl chlorides (XXXVI). Treatment of α-ketoimidoyl chlorides with hydrogen sulfide offers a convenient synthetic route to the α-ketothiocarbonamides (XXXVIII).[60] The reaction of 2 moles of an isonitrile with phosgene followed by hydrolysis gives XXXIX.

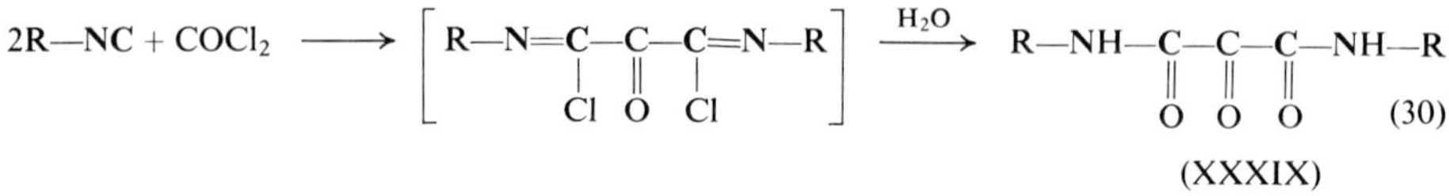

$$2R-NC + COCl_2 \longrightarrow \left[ \begin{array}{c} R-N=C-C-C=N-R \\ | \quad \| \quad | \\ Cl \quad O \quad Cl \end{array} \right] \xrightarrow{H_2O} R-NH-\underset{O}{\overset{\|}{C}}-\underset{O}{\overset{\|}{C}}-\underset{O}{\overset{\|}{C}}-NH-R$$
$$(XXXIX) \qquad (30)$$

N,N-Dialkylamide chlorides (XL), which are prepared from the corresponding amides and phosgene, the so-called "Vilsmeier reagents," are known to be potent electrophiles. In the reaction of an isonitrile with a formamide chloride

(XL) (R' = H), the 1:1 first adduct (XLI) undergoes a second addition step with another isonitrile molecule. On treatment with water, an α-aminomalonamide (XLIII) is produced. In the case of the amide chlorides of acetic acid and higher carboxylic acids, the reaction ceases at the 1:1 addition stage; the product after hydrolysis is an α-ketoamide (XLIV).[13]

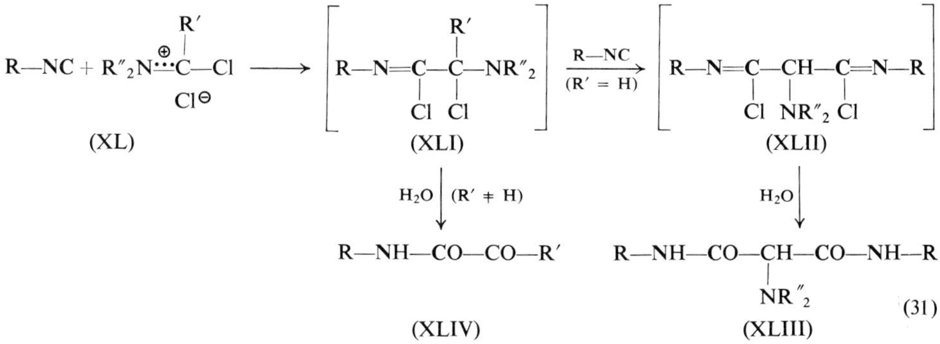

$$R-NC + R''_2\overset{\oplus}{N}\cdots\underset{Cl^{\ominus}}{\overset{R'}{\underset{|}{C}}}-Cl \longrightarrow \left[R-N=\underset{Cl}{\overset{R'}{\underset{|}{C}}}-\underset{Cl}{\overset{|}{C}}-NR''_2\right] \xrightarrow[(R'=H)]{R-NC} \left[R-N=\underset{Cl}{\overset{|}{C}}-CH-\underset{Cl}{\overset{|}{C}}=N-R\right]$$

(XL)　　　　　　　　　　(XLI)　　　　　　　　　　　(XLII)

$$\text{H}_2\text{O} \downarrow (R' \neq H) \qquad\qquad \text{H}_2\text{O} \downarrow$$

R—NH—CO—CO—R'　　　　R—NH—CO—CH—CO—NH—R
　　　　　　　　　　　　　　　　　　　　|
　　　　　　　　　　　　　　　　　　　NR''₂
　　(XLIV)　　　　　　　　　　(XLIII)　　(31)

### B. Nitrogen–Halogen Compounds

$N$-Bromoacetamide and $N$-bromosuccinimide are well known as reagents for "allylic bromination." These $N$-bromoamides react with isonitriles in the presence of a protonic acid via the simple α-addition mechanism.

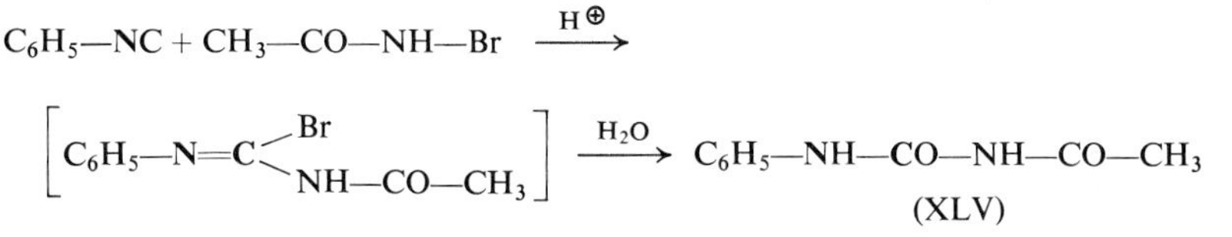

$$C_6H_5-NC + CH_3-CO-NH-Br \xrightarrow{H^{\oplus}}$$

$$\left[C_6H_5-N=C\begin{matrix}{\diagup}Br\\{\diagdown}NH-CO-CH_3\end{matrix}\right] \xrightarrow{H_2O} C_6H_5-NH-CO-NH-CO-CH_3 \quad (32)$$

(XLV)

$$c\text{-}C_6H_{11}-NC + \begin{matrix}CH_2-CO\\|\qquad\quad\diagdown\\CH_2-CO\diagup\end{matrix}N-Br \xrightarrow{H^{\oplus}} \left[c\text{-}C_6H_{11}-N=C\begin{matrix}{\diagup}Br\\{\diagdown}N\begin{matrix}\diagup CO-CH_2\\\diagdown CO-CH_2\end{matrix}\end{matrix}\right] \xrightarrow{H_2O}$$

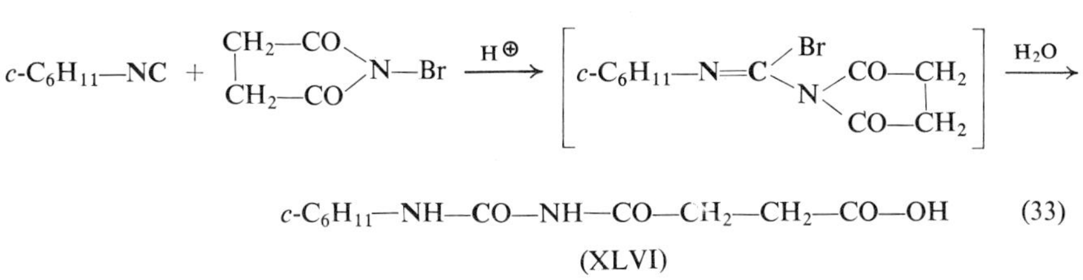

$c\text{-}C_6H_{11}$—NH—CO—NH—CO—CH₂—CH₂—CO—OH　　(33)

(XLVI)

### C. Oxygen–Halogen Compounds

The combination of an alkyl hypochlorite and an acid produces a chloronium ion. Hydrolysis of the product of $t$-butyl hypochlorite and cyclohexyl isocyanide in the presence of zinc chloride catalyst affords $t$-butyl $N$-cyclohexylcarbamate (XLVII).[26] The $t$-butyl hypochlorite–isonitrile reaction in the absence of an acid catalyst results in the oxidation of the isonitrile to the isocyanate.[26]

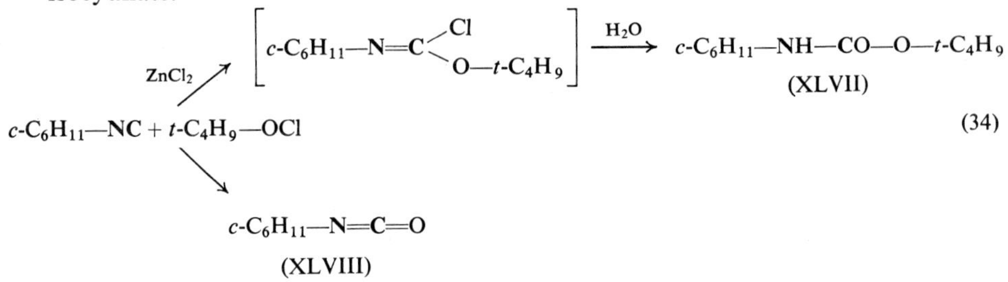

$$c\text{-}C_6H_{11}\text{—}NC + t\text{-}C_4H_9\text{—}OCl \xrightarrow{ZnCl_2} \left[ c\text{-}C_6H_{11}\text{—}N=C\begin{smallmatrix}Cl\\O\text{—}t\text{-}C_4H_9\end{smallmatrix} \right] \xrightarrow{H_2O} c\text{-}C_6H_{11}\text{—}NH\text{—}CO\text{—}O\text{—}t\text{-}C_4H_9$$
(XLVII)  (34)

$$\downarrow$$

$$c\text{-}C_6H_{11}\text{—}N=C=O$$
(XLVIII)

## D. Sulfur–Halogen Compounds

The α-addition of 2,4-dinitrophenylsulfenylchloride to p-tolyl isocyanide has also been reported.[12]

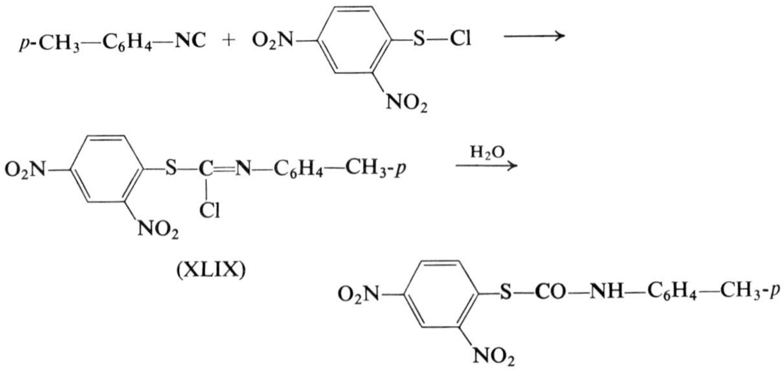

$$p\text{-}CH_3\text{—}C_6H_4\text{—}NC + O_2N\text{—}\underset{NO_2}{\overset{}{C_6H_3}}\text{—}S\text{—}Cl \longrightarrow$$

$$O_2N\text{—}\underset{NO_2}{\overset{}{C_6H_3}}\text{—}S\text{—}\underset{Cl}{\overset{}{C}}=N\text{—}C_6H_4\text{—}CH_3\text{-}p \xrightarrow{H_2O}$$
(XLIX)
$$O_2N\text{—}\underset{NO_2}{\overset{}{C_6H_3}}\text{—}S\text{—}CO\text{—}NH\text{—}C_6H_4\text{—}CH_3\text{-}p$$
(35)

## E. Halogens

Halogens react violently with isonitriles. The low-temperature reactions of isonitriles with the molecular halogens, $Cl_2$, $Br_2$, and $I_2$ yield the corresponding N-alkylimidocarbonyl halides (LI).[22] The α-addition of a halogen to an isonitrile may be regarded as an oxidation of the isonitrile.

$$R\text{—}NC + X_2 \rightarrow R\text{—}N=C\begin{smallmatrix}X\\X\end{smallmatrix} \quad (X = Cl, Br, I)$$
(LI)  (36)

In connection with the halogen–isonitrile reaction, the oxidation–reduction reaction between an isonitrile and $CuCl_2$ should be mentioned, wherein $CuCl_2$ is reduced and the isonitrile is oxidized to an imidocarbonyl chloride.[44]

$$R\text{—}NC + 2CuCl_2 \rightarrow R\text{—}N=C\begin{smallmatrix}Cl\\Cl\end{smallmatrix} + 2\,CuCl$$
(LII)  (37)

## IV. OXIDATION OF ISONITRILES

Various types of oxidizing agents are known to oxidize an isonitrile to the corresponding isocyanate. They are metal oxides (AgO, HgO),[8] ozone,[7] peracids (perbenzoic acid)[52] oxygen,[28, 54] and dimethyl sulfoxide.[14] Soluble homogeneous catalysts of transition metal–oxygen complexes such as $Ni(RNC)_2O_2$[28] and $Pt(P(C_6H_5)_3)_3O_2$[54] catalyze the oxidation of isonitriles with molecular oxygen.

The oxidation of isonitriles by dimethyl sulfoxide and by pyridine N-oxide proceeds only in the presence of halogens or acid catalysts. A mixture of isonitrile, dimethyl sulfoxide, and a small amount of bromine in chloroform gives dimethyl sulfide and the corresponding isocyanate. The reaction cycle (38) including N-substituted imidocarbonyl bromide (LIII) as a key intermediate explains the catalytic activity of bromine.[14]

$$R-NC + Br_2 \rightarrow R-N=CBr_2$$
$$(LIII)$$
$$R-N=CBr_2 + (CH_3)_2SO \rightarrow R-NCO + (CH_3)_2S + Br_2$$
(38)

Chlorine and iodine also catalyze the dimethyl sulfoxide oxidation of isonitriles. The oxidation by pyridine N-oxide likewise requires a halogen catalyst.

Oxidation with dimethyl sulfoxide is also catalyzed by such acids as p-toluenesulfonic acid, hydrochloric and trityl perchlorate.[19] For the proton-catalyzed oxidation, an intermediate (LIV) formed by the protonation of the isonitrile has been assumed.

$$R-NC + (CH_3)_2SO + HX \longrightarrow \left[ R-N=C\begin{matrix}H \\ O-\overset{\oplus}{S}(CH_3)_2\end{matrix} \right] X^\ominus \xrightarrow{-HX}$$

$$(LIV)$$

$$R-N=C=O + (CH_3)_2S \quad (39)$$

The addition of sulfur to isonitriles has been known for a long time. Isothiocyanate (LV) is produced in high yield when a mixture of the isonitrile and flowers of sulfur are heated together.[23, 61]

$$R-NC + S \rightarrow R-N=C=S \quad (40)$$
$$(LV)$$

## V. REDUCTION OF ISONITRILES

Isonitriles are catalytically reduced by hydrogen to the corresponding secondary methyl amines (LVII).[22] The Schiff base (LVI) was assumed to be a possible intermediate, although it has not yet been isolated.

$$R-NC \xrightarrow{H_2} \left(R-N=C{<}^H_H\right) \xrightarrow{H_2} R-NH-CH_3 \quad (41)$$

$$\phantom{R-NC \xrightarrow{H_2}}\quad\quad (LVI) \quad\quad\quad\quad (LVII)$$

Catalytic hydrogenation of the $\alpha,\beta$-diisocyanide (LVIII) gives a derivative of pyrazine (LIX).[10]

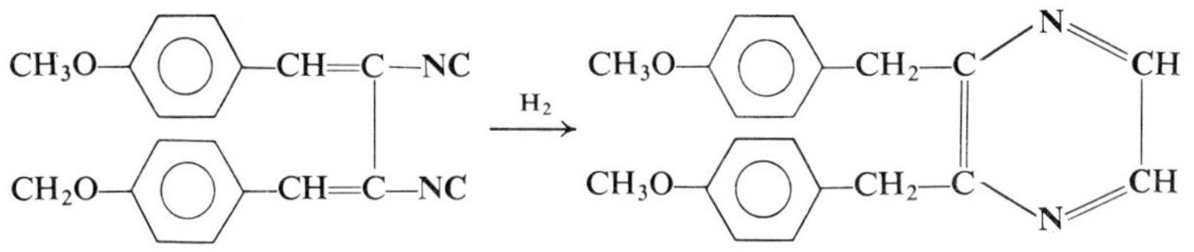

$$\text{(LVIII)} \xrightarrow{H_2} \text{(LIX)} \quad (42)$$

Ugi and Bodesheim[57] reported the reduction of isonitriles with alkali metals or alkaline earth metals in liquid ammonia at $-40$ to $-35°C$, where alkyl isocyanides were quantitatively converted into alkanes and metal cyanides.

$$R-NC + 2\,Me + NH_3 \rightarrow RH + Me(I)CN + Me(I)NH_2 \quad (43)$$

(Me = alkali metal or alkaline earth metal)

## VI. REACTIONS WITH CARBENES AND NITRENES

Reactions of isonitriles with carbenes have been reported.[1,11] Chloroform or ethyl trichloracetate was added dropwise to a mixture of cyclohexyl isocyanide and a potassium alkoxide.[11] The product was $N$-cyclohexyldichloracetoimidate (LXI), which was assumed to have been derived from $N$-cyclohexyldichloroketenimine (LX).

$$c\text{-}C_6H_{11}-NC + {:}CCl_2 \longrightarrow [c\text{-}C_6H_{11}-N=C=CCl_2] \xrightarrow{ROH} c\text{-}C_6H_{11}-N=CCHCl_2$$
$$\phantom{c\text{-}C_6H_{11}-NC + {:}CCl_2 \longrightarrow [c\text{-}C_6H_{11}-N=C=CCl_2] \xrightarrow{ROH}}\quad\quad\;|\;\text{OR}$$
$$\phantom{xxxxxxxxxxxxxxxxxxxxxxx}\text{(LX)}\quad\quad\quad\quad\text{(LXI)} \quad (44)$$

Sodium $p$-toluenesulfochloramide (Chloramine-T) (LXII) is known to be a nitrene precursor. The reaction of cyclohexyl isocyanide with Chloramine-T in methanol produced $N$-($p$-toluenesulfonyl)-$N'$-cyclohexyl-$O$-methyl isourea

(LXIV).[2] As an intermediate, N-(p-toluenesulfonyl-N'-cyclohexylcarbodiimide (LXIII) was assumed to have been formed by the nucleophilic addition of the Chloramine-T anion to the cyclohexyl isocyanide. Another possibility

$$c\text{-}C_6H_{11}\text{—NC} + CH_3\text{—}\langle\bigcirc\rangle\text{—}SO_2\overset{\ominus}{\underset{\overset{\oplus}{Na}}{N}}Cl \longrightarrow \left[CH_3\text{—}\langle\bigcirc\rangle\text{—}SO_2\overset{\curvearrowleft\ominus}{\underset{\underset{Cl}{\curvearrowleft}}{N}}\text{—}C=N\text{—}c\text{-}C_6H_{11}\right] \xrightarrow{-Cl^{\ominus}}$$

$$\left[CH_3\text{—}\langle\bigcirc\rangle\text{—}SO_2N=C=N\text{—}c\text{-}C_6H_{11}\right] \xrightarrow{CH_3OH} CH_3\text{—}\langle\bigcirc\rangle\text{—}SO_2N=\underset{\underset{OCH_3}{|}}{C}\text{—}NH\text{—}c\text{-}C_6H_{11}$$

(LXIII)  (LXIV)  (45)

is that the carbodiimide intermediate (LXIII) is formed directly from the isonitrile and the nitrene.

A three-component reaction of cyclohexyl isocyanide, Chloramine-T, and p-toluenesulfonamide in an acetone–water mixture affords 1,2-bis(p-toluenesulfonyl)-3-cyclohexyl guanidine (LXV).[2] Similarly, the addition of p-toluenesulfonamide to the carbodiimide intermediate (LXIII) was assumed.

$$c\text{-}C_6H_{11}\text{—NC} + CH_3\text{—}\langle\bigcirc\rangle\text{—}SO_2\overset{\ominus}{\underset{\overset{\oplus}{Na}}{N}}Cl + CH_3\text{—}\langle\bigcirc\rangle\text{—}SO_2NH_2 \xrightarrow{-NaCl}$$

$$\begin{array}{c} CH_3\text{—}\langle\bigcirc\rangle\text{—}SO_2NH \\ \phantom{xxxxxxxxxxxxxx} \diagdown \\ \phantom{xxxxxxxxxxxxxxxxx} C\text{—}NH\text{—}c\text{-}C_6H_{11} \\ \phantom{xxxxxxxxxxxxxx} \diagup\!\!\!\!\!\!= \\ CH_3\text{—}\langle\bigcirc\rangle\text{—}SO_2N \end{array}$$ (46)

(LXV)

Iron pentacarbonyl induces the reaction of an isonitrile with an azide, which gives the corresponding unsymmetrical carbodiimide (LXVI) in fairly good yield.[43]

$$R\text{—NC} + R'\text{—}N_3 \xrightarrow{Fe(CO)_5} R\text{—N=C=N—}R' \qquad (47)$$
$$(LXVI)$$

## VII. REACTION WITH ORGANOMETALLIC COMPOUNDS

Two early references[9,34] dealt with the reaction between an isonitrile and a Grignard reagent. It was only recently, however, that this complicated reaction was fully elucidated.[59] Five products (LXVIII–LXXII) were isolated

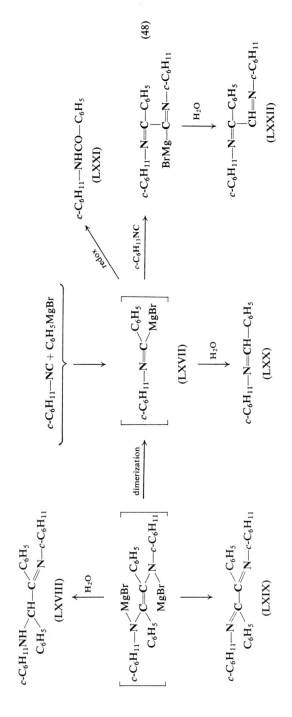

(48)

## 4. Simple α-Additions

in the hydrolysis of the reaction mixture of cyclohexyl isocyanide with phenyl–magnesium bromide kept at 20°C for 40 hr. The main products were LXVIII and LXIX. The intermediate (LXVII) formed by the insertion of the isocyanide into the phenyl–magnesium bond was assumed to be a precursor to these two main products.

The reaction of phenyl isocyanide with triethylborane[5] gives products of successive fissions of one, two, and three boron–ethyl bonds as shown in the following reaction scheme (49) (see Chapter 6).

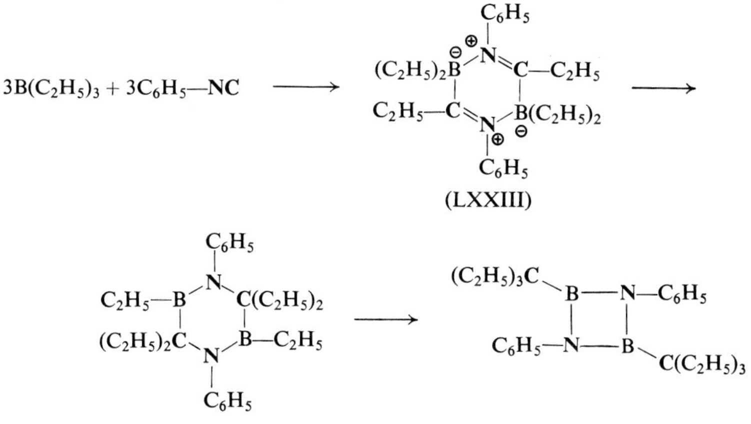

$$3B(C_2H_5)_3 + 3C_6H_5-NC \longrightarrow \text{(LXXIII)} \longrightarrow$$

$$\text{(LXXIV)} \longrightarrow \text{(LXXV)} \quad (49)$$

The intermediate product (LXXIII) was obtained under mild conditions, but rearranged to LXXIV at 200°C and to LXXV at 300°C.

Organotin–amino and organolead–amino compounds are quite reactive and their considerable potential for organic syntheses is being exploited. The simple α-additions of these reagents to isonitriles have been reported.[15,25]

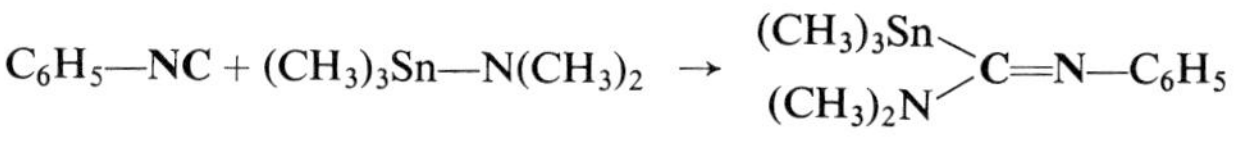

$$C_6H_5-NC + (CH_3)_3Sn-N(CH_3)_2 \rightarrow \underset{(CH_3)_2N}{\overset{(CH_3)_3Sn}{>}}C=N-C_6H_5 \quad (50)$$

(LXXVI)

$$C_6H_5-NC + R_3Pb-N(C_2H_5)_2 \rightarrow \underset{(C_2H_5)_2N}{\overset{R_3Pb}{>}}C=N-C_6H_5 \quad (51)$$

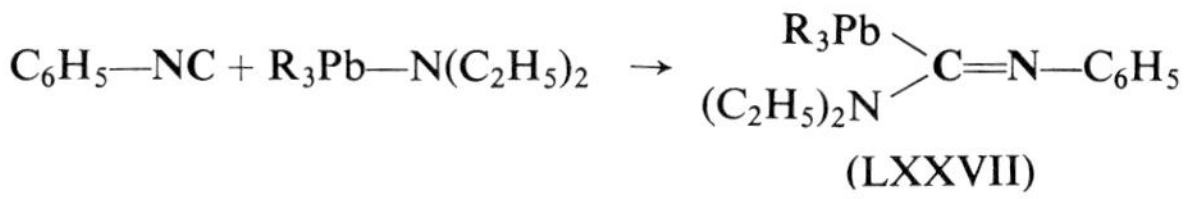

(LXXVII)

(R = —C₂H₅, –n-C₄H₉, —C₆H₅)

These reactions are called the "aminometallation" of isonitriles.

Recently, α-metallation of isonitriles with butyllithium was reported.[50] The reaction of α-lithium isocyanide (LXXVIII) with carbonyl compounds provides a convenient synthetic route to olefins.

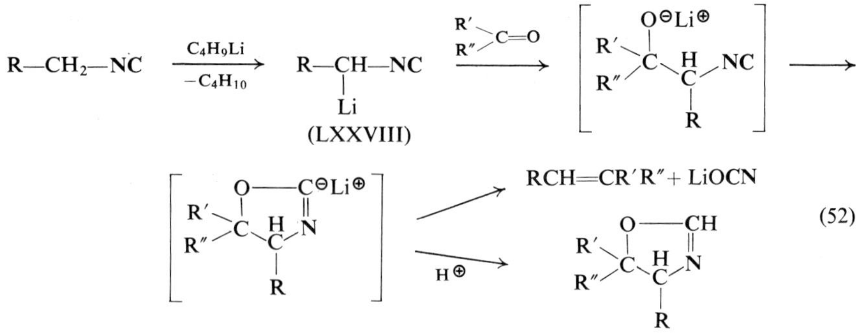

$$R-CH_2-NC \xrightarrow[-C_4H_{10}]{C_4H_9Li} \underset{\text{(LXXVIII)}}{R-\underset{Li}{\underset{|}{CH}}-NC} \xrightarrow{\underset{R''}{R'}>C=O} \left[ \underset{R}{\underset{|}{R''}} \overset{O^{\ominus}Li^{\oplus}}{\underset{|}{\underset{C}{\overset{|}{C}}}} \underset{C}{\overset{H}{\underset{|}{C}}} NC \right] \longrightarrow$$

$$\left[ \underset{R}{\underset{|}{R''}} \overset{O\underline{\quad}C^{\ominus}Li^{\oplus}}{\underset{|}{\underset{C}{\overset{|}{C}}}} \underset{C}{\overset{H}{\underset{|}{C}}} N \right] \begin{matrix} \longrightarrow \\ \\ \xrightarrow{H^{\oplus}} \end{matrix} \begin{matrix} RCH=CR'R'' + LiOCN \\ \\ \underset{R''}{\underset{|}{R'}} \overset{O\underline{\quad}CH}{\underset{|}{\underset{C}{\overset{|}{C}}}} \underset{C}{\overset{H}{\underset{|}{C}}} N \end{matrix} \qquad (52)$$

## VIII. RADICAL REACTIONS OF ISONITRILES

Radical reactions of isonitriles have not been studied extensively. In the following scheme, radical reactions of isonitriles are classified into four types. For all reactions, the imidoyl radical (LXXIX) acts as a common key intermediate. Some of these reactions involve α-additions.

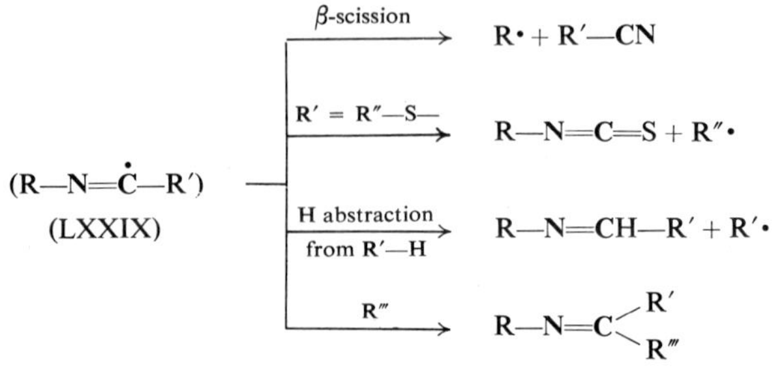

$$R-NC + R'\cdot \longrightarrow (R-N\overset{\cdot}{=}C-R') \qquad (53)$$
$$\text{(LXXIX)}$$

$$(R-N\overset{\cdot}{=}C-R') \begin{cases} \xrightarrow{\beta\text{-scission}} R\cdot + R'-CN & (54) \\ \xrightarrow{R'=R''-S-} R-N=C=S + R''\cdot & (55) \\ \xrightarrow[\text{from } R'-H]{H \text{ abstraction}} R-N=CH-R' + R'\cdot & (56) \\ \xrightarrow{R'''} R-N=C\underset{R'''}{\overset{R'}{<}} & (57) \end{cases}$$

### A. Free Radical-Catalyzed Isomerizations

The radical-initiated vapor phase isomerization of an alkyl isocyanide has been reported.[51] (For further isomerizations of isonitriles, see Chapter 3.) It was initiated by a radical from the decomposition of di-*t*-butyl peroxide. The isomerization is a chain reaction corresponding to (54), where $R = R' = CH_3$ or $C_2H_5$.

$$R-NC + R'\cdot \rightarrow R-N\overset{\cdot}{=}C-R' \rightarrow R\cdot + R'-CN$$
$$R\cdot + R-NC \rightarrow R-N\overset{\cdot}{=}C-R \rightarrow R-CN + R\cdot \qquad (58)$$

## B. Radical Reactions with Stannanes, Phosphines, and Thiols

### 1. Reaction with Trialkyltin Hydrides

The reaction of isonitriles with tri-$n$-butyltin hydride is initiated by azobisisobutyronitrile or di-$t$-butyl peroxide, and produces tri-$n$-butyltin(iso)cyanide (LXXX) and alkane (LXXXI) in high yields.[47] The nature of the reactants and the effect of radical initiators as well as the products are explained by (59).

$$R\cdot + (n\text{-}C_4H_9)_3Sn\text{—}H \rightarrow RH + (n\text{-}C_4H_9)_3Sn\cdot$$

$$(n\text{-}C_4H_9)_3Sn\cdot + R'\text{—}NC \rightarrow (n\text{-}C_4H_9)_3Sn\text{—}\overset{\cdot}{C}=N\text{—}R'$$

$$(n\text{-}C_4H_9)_3Sn\text{—}\overset{\cdot}{C}=N\text{—}R' \rightarrow (n\text{-}C_4H_9)_3Sn\text{—}CN + R'\cdot \quad (59)$$
$$\text{(LXXX)}$$

$$R'\cdot + (n\text{-}C_4H_9)_3Sn\text{—}H \rightarrow (n\text{-}C_4H_9)_3Sn\cdot + R'\text{—}H$$
$$\text{(LXXXI)}$$

$$(R' = c\text{-}C_6H_{11}\text{—}, t\text{-}C_4H_9\text{—}, C_6H_5\text{—}CH_2\text{—})$$

### 2. Reaction with Dialkylphosphines

The copper-catalyzed $\alpha$-addition of the phosphorus–hydrogen bond to an isonitrile has been discussed in Section II,A. The radical-induced reaction of an alkyl isocyanide with diethylphosphine proceeds in two different ways, either according to (54) or (56).[49] The reaction of benzyl isocyanide with diethylphosphine in the presence of azobisisobutyronitrile follows (54) and produces diethylcyanophosphine (LXXXIII) and toluene in almost quantitative yield (60).

$$R\cdot + (C_2H_5)_2P\text{—}H \rightarrow R\text{—}H + (C_2H_5)_2P\cdot$$

$$C_6H_5\text{—}CH_2\text{—}NC + (C_2H_5)_2P\cdot \rightarrow C_6H_5\text{—}CH_2\text{—}N=\overset{\cdot}{C}\text{—}P(C_2H_5)_2$$
$$\text{(LXXXII)}$$
$$(60)$$

$$C_6H_5\text{—}CH_2\text{—}N=\overset{\cdot}{C}\text{—}P(C_2H_5)_2 \rightarrow C_6H_5\text{—}CH_2\cdot + (C_2H_5)_2P\text{—}CN$$
$$\text{(LXXXIII)}$$

$$C_6H_5\text{—}CH_2\cdot + (C_2H_5)_2P\text{—}H \rightarrow C_6H_5\text{—}CH_3 + (C_2H_5)_2P\cdot$$

In the radical reaction of cyclohexyl isocyanide with diethylphosphine, however, diethyl-$N$-cyclohexylformimidoylphosphine (LXXXV) is the main product, presumably via (56), and diethylcyanophosphine is only produced via (62) as a minor by-product.

$$c\text{-}C_6H_{11}\text{—}NC + (C_2H_5)_2P\cdot \longrightarrow c\text{-}C_6H_{11}\text{—}N\!\!=\!\!\overset{\cdot}{C}\text{—}P(C_2H_5)_2$$
$$\text{(LXXXIV)} \qquad (61)$$

$$c\text{-}C_6H_{11}\text{—}N\!\!=\!\!\overset{\cdot}{C}\text{—}P(C_2H_5)_2 + (C_2H_5)_2PH \longrightarrow c\text{-}C_6H_{11}\text{—}N\!\!=\!\!CH\text{—}P(C_2H_5)_2 + (C_2H_5)_2P\cdot$$
$$\text{(LXXXV)}$$

$$c\text{-}C_6H_{11}\text{—}N\!\!=\!\!\overset{\cdot}{C}\text{—}P(C_2H_5)_2 \xrightarrow{\beta\text{-scission}} c\text{-}C_6H_{11}\cdot + (C_2H_5)_2P\text{—}CN$$
$$(62)$$

$$c\text{-}C_6H_{11}\cdot + (C_2H_5)_2P\text{—}H \longrightarrow c\text{-}C_6H_{12} + (C_2H_5)_2P\cdot$$

The difference in the course of reaction may be ascribed to the difference in reactivity between the two imidoyl radicals, LXXXII and LXXXIV. The scission of the β-linkage of the radical LXXXII, producing a benzyl radical of increased stability, is much easier than that of LXXXIV producing the less stable cyclohexyl radical. The latter radical prefers hydrogen abstraction from phosphine to β-scission.

### 3. Reaction with Alkanethiols

Two courses of reaction are possible in the radical reaction of isonitriles with alkanethiols (see Section II,A).[46] The alkyl group of the alkanethiol determines the course of the reaction. The reaction of cyclohexyl isocyanide with 2-methyl-propane-2-thiol (*t*-butyl mercaptan) follows (55) almost exclusively, producing cyclohexyl isothiocyanate (LXXXVII) and isobutane (LXXXVIII) (63) (R′ = *t*-$C_4H_9$—). On the other hand, the radical reaction

$$R'\text{—}SH \xrightarrow{R\cdot} R'S\cdot \xrightarrow{c\text{-}C_6H_{11}\text{—}NC} c\text{-}C_6H_{11}\text{—}N\!\!=\!\!\overset{\cdot}{C}\text{—}S\text{—}R'$$
$$\text{(LXXXVI)}$$

$$\begin{array}{l} \xrightarrow{} R'\cdot + c\text{-}C_6H_{11}\text{—}N\!\!=\!\!C\!\!=\!\!S \xrightarrow{R'\text{—}SH} R'\text{—}S\cdot + R'H \qquad (63)\\ \text{(LXXXVII)} \qquad\qquad\qquad \text{(LXXXVIII)}\\ \xrightarrow{R'\text{—}SH} c\text{-}C_6H_{11}\text{—}N\!\!=\!\!CH\text{—}SR' + R'\text{—}S\cdot \qquad (64)\\ \text{(LXXXIX)} \end{array}$$

with ethanethiol proceeds via (64) and gives *N*-cyclohexyl ethyl thioformimidate (LXXXIX) (R′ = $C_2H_5$—). The products of the reaction with propane-2-thiol corresponds to both of these routes. The dependency of the reaction course upon the thioalkyl group is explained by the reactivity of the intermediate radical (LXXXVI).

### C. Reactions with Peroxide and Nitrosoacetanilide

Reaction between stoichiometric amounts of isonitrile and a radical source is less selective than those reactions mentioned above. However, the products

4. Simple α-Additions    87

are accounted for by assuming the pathways illustrated above. From the reaction of cyclohexyl isocyanide with benzoyl peroxide, for which (65) was postulated, three products, XC, XCI, and XCII, were isolated.[31]

$$BPO \longrightarrow 2C_6H_5-COO\cdot \longrightarrow 2C_6H_5\cdot + 2CO_2$$

$$c\text{-}C_6H_{11}-NC + C_6H_5-COO\cdot \longrightarrow c\text{-}C_6H_{11}-N=\overset{\cdot}{C}OCO-C_6H_5$$

$$\begin{array}{l}\xrightarrow{\text{H abstraction}} c\text{-}C_6H_{11}-N=CHOCO-C_6H_5 \\ \hspace{3cm} (XC) \\ \\ \xrightarrow{C_6H_5\cdot \text{ or BPO}} c\text{-}C_6H_{11}-N=\underset{\underset{C_6H_5}{|}}{C}OCO-C_6H_5 \\ \hspace{3cm} (XCI)\end{array}$$

(65)

$$c\text{-}C_6H_{11}-NC + C_6H_5\cdot \longrightarrow$$
$$c\text{-}C_6H_{11}-N=\overset{\cdot}{C}-C_6H_5 \xrightarrow{\cdot OH} c\text{-}C_6H_{11}-NH-CO-C_6H_5$$
$$(XCII)$$

The presence of the OH radical was presumed to be from water present as an impurity.

In the reaction of cyclohexyl isocyanide with $N$-nitrosoacetanilide, two products, XCIII and XCIV, were isolated.[52]

$$CH_3-CO-N(NO)-C_6H_5 \longrightarrow CH_3-COO\cdot + C_6H_5\cdot + N_2$$
$$\searrow CH_3\cdot + CO_2$$

$$c\text{-}C_6H_{11}-NC + CH_3-COO\cdot \longrightarrow$$
$$c\text{-}C_6H_{11}-N=\overset{\cdot}{C}O-CO-CH_3 \xrightarrow{RH} c\text{-}C_6H_{11}-N=CH-O-CO-CH_3 \quad (66)$$
$$(XCIII)$$

$$c\text{-}C_6H_{11}-NC + CH_3\cdot \longrightarrow$$
$$c\text{-}C_6H_{11}-N=\overset{\cdot}{C}-CH_3 \xrightarrow{\cdot OH} c\text{-}C_6H_{11}-NH-CO-CH_3 \quad (67)$$
$$(XCIV)$$

### D. Electrolysis of Isonitriles

The electrolysis of cyclohexyl isocyanide in methanol containing sodium methoxide gives seven products, XCV–CI, which have been explained by assuming the presence of the methoxy radical as the primary reactive species.[53]

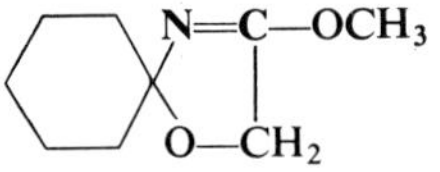

## (Structures)

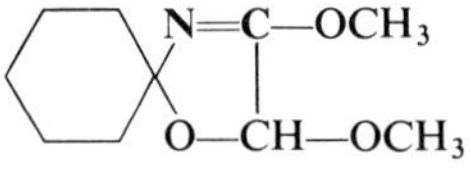

**(XCV)**: cyclohexane spiro with N=C—OCH₃ and O—CH₂

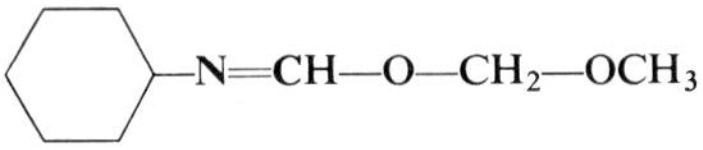

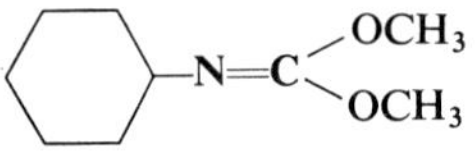

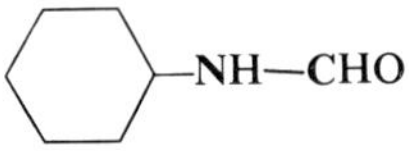

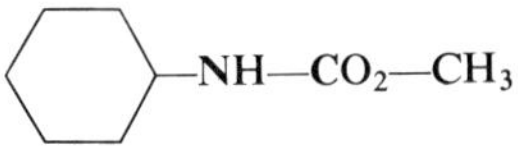

$$\text{(XCV)}$$

**(XCVI)**: cyclohexane spiro with N=C—OCH₃ and O—CH—OCH₃

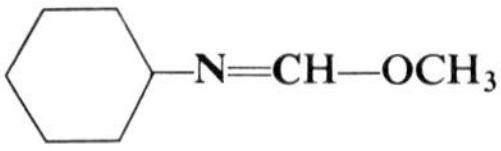

**(XCVII)**: C₆H₁₁—N=CH—OCH₃

**(XCVIII)**: C₆H₁₁—N=CH—O—CH₂—OCH₃

**(XCIX)**: C₆H₁₁—N=C(OCH₃)₂

**(C)**: C₆H₁₁—NH—CHO

**(CI)**: C₆H₁₁—NH—CO₂—CH₃

## IX. REACTIONS RELATED TO α-ADDITIONS

### A. Reaction with Benzyne

The reaction of cyclohexyl isocyanide with benzenediazonium-2-carboxylate (CII) a benzyne precursor, in *t*-butanol produced *N*-cyclohexylbenzamide (CIII) and isobutene along with the evolution of carbon dioxide and nitrogen gasses.[17] The reaction products fit a mechanism involving nucleophilic attack of the isocyanide upon the benzyne.

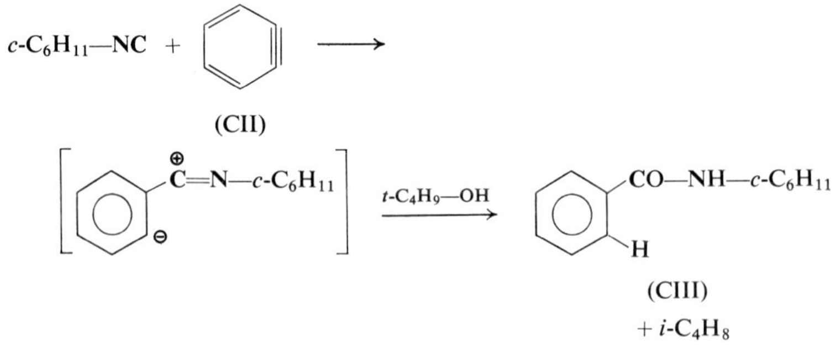

$$c\text{-}C_6H_{11}\text{—NC} + \text{(CII)} \longrightarrow \left[ \begin{array}{c} \text{C}^{\oplus}\text{=N—}c\text{-}C_6H_{11} \\ \text{(aryl anion)} \end{array} \right] \xrightarrow{t\text{-}C_4H_9\text{—OH}} \text{aryl—CO—NH—}c\text{-}C_6H_{11} \quad (68)$$

(CIII) + *i*-C₄H₈

### B. Reaction with Tropylium Ion

Tropylium ion (CIV) reacts with isonitriles in protic solvents to produce the corresponding *N*-substituted amide of cycloheptatriene carboxylic acid (CV).[56] An electrophilic addition of tropylium cation to the isonitrile explains these results.

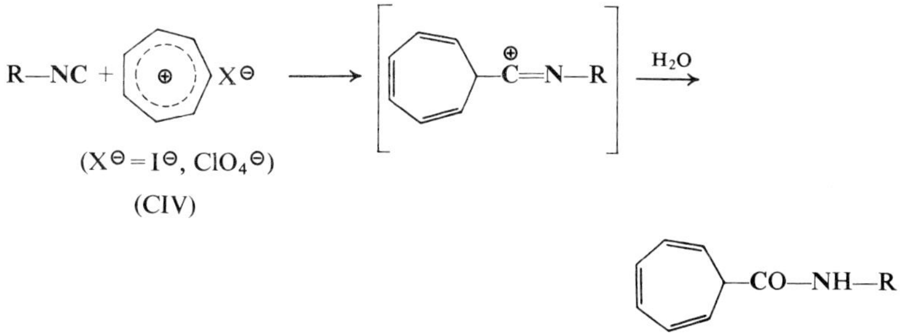

4. Simple α-Additions    89

R—NC + [cycloheptatrienyl cation]⊕ X⊖ ⟶ [⟨cycloheptatrienyl⟩—C⊕=N—R] $\xrightarrow{H_2O}$

$(X^\ominus = I^\ominus, ClO_4^\ominus)$
(CIV)

⟨cycloheptatrienyl⟩—CO—NH—R  (69)

(CV)

## C. Cationic Isomerization and Oligomerization of Isonitriles

A small amount of $BF_3$ catalyzes the dimerization of t-butyl isocyanide.[16]

$$2t\text{-}C_4H_9\text{—NC} \xrightarrow{BF_3} t\text{-}C_4H_9\text{—N=C—}t\text{-}C_4H_9 \quad (70)$$
$$\underset{\text{CN}}{|}$$
(CVI)

Independently, Saegusa et al.[48] investigated the reaction of t-butyl isocyanide in the presence of $BF_3$ as catalyst, in which t-butyl cyanide (CVII), the trimer of t-butyl isocyanide (CVIII), and a mixture of oligomers (CIX) in addition to (CVI) were produced.

$t\text{-}C_4H_9\text{—CN}$         $t\text{-}C_4H_9\text{—C=N—}t\text{-}C_4H_9$
                               $\text{N=C—}t\text{-}C_4H_9$
                                      $|$
                                     $\text{CN}$
(CVII)           (CVIII)

$t\text{-}C_4H_9\text{—(—}C_4H_9NC\text{—)}_n\text{—CN}$

(CIX)

These reaction products are explained by the following scheme of cationic chain reactions which involves N-t-butyl imidoyl cation intermediates. Millich and Sinclair[20] also investigated a polymerization reaction of isonitriles.

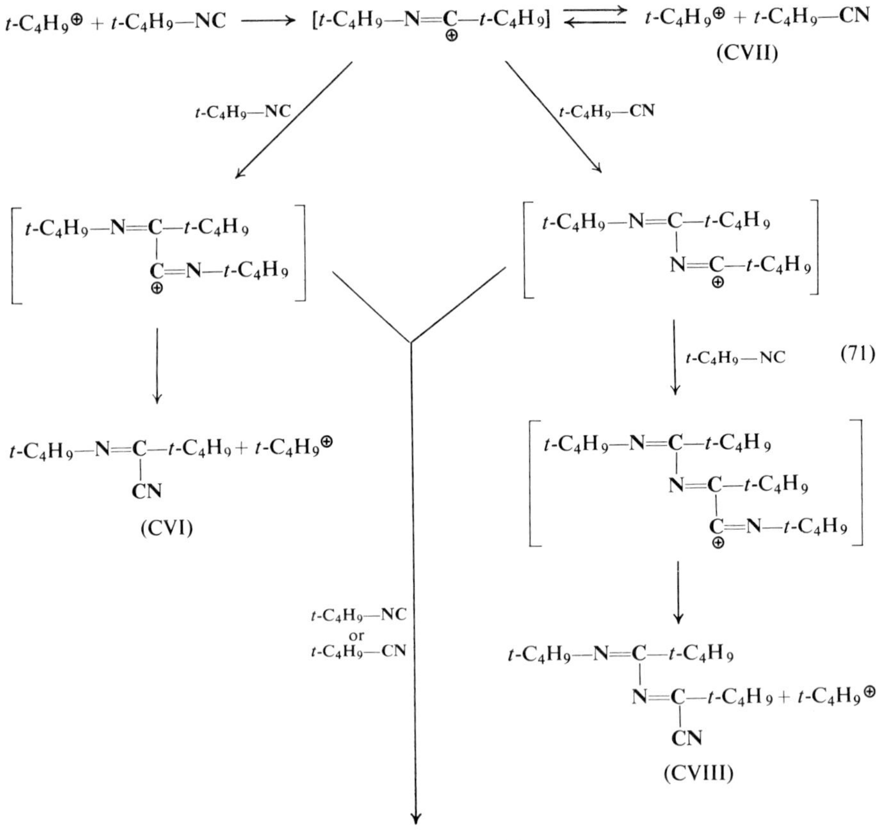

$t\text{-}C_4H_9^{\oplus} + t\text{-}C_4H_9\text{—NC} \longrightarrow [t\text{-}C_4H_9\text{—N}=\overset{\oplus}{C}\text{—}t\text{-}C_4H_9] \rightleftarrows t\text{-}C_4H_9^{\oplus} + t\text{-}C_4H_9\text{—CN}$

(CVII)

$$\begin{bmatrix} t\text{-}C_4H_9\text{—N}=C\text{—}t\text{-}C_4H_9 \\ | \\ \overset{\oplus}{C}=N\text{—}t\text{-}C_4H_9 \end{bmatrix}$$

$$\begin{bmatrix} t\text{-}C_4H_9\text{—N}=C\text{—}t\text{-}C_4H_9 \\ | \\ N=\overset{\oplus}{C}\text{—}t\text{-}C_4H_9 \end{bmatrix}$$

$\downarrow t\text{-}C_4H_9\text{—NC}$ (71)

$t\text{-}C_4H_9\text{—N}=C\text{—}t\text{-}C_4H_9 + t\text{-}C_4H_9^{\oplus}$
$\quad\quad\quad | $
$\quad\quad\quad \text{CN}$
(CVI)

$$\begin{bmatrix} t\text{-}C_4H_9\text{—N}=C\text{—}t\text{-}C_4H_9 \\ | \\ N=C\text{—}t\text{-}C_4H_9 \\ | \\ \overset{\oplus}{C}=N\text{—}t\text{-}C_4H_9 \end{bmatrix}$$

$t\text{-}C_4H_9\text{—NC}$
or
$t\text{-}C_4H_9\text{—CN}$

$t\text{-}C_4H_9\text{—N}=C\text{—}t\text{-}C_4H_9$
$\quad\quad | $
$\quad\quad N=C\text{—}t\text{-}C_4H_9 + t\text{-}C_4H_9^{\oplus}$
$\quad\quad\quad\quad | $
$\quad\quad\quad\quad \text{CN}$
(CVIII)

$t\text{-}C_4H_9\text{−}(C_4H_9NC)_n\text{−CN}$
(CIX)

## REFERENCES

1. Aumann, R., and Fischer, E. O., *Chem. Ber.* **101**, 954 (1968).
2. Aumuller, W., *Angew. Chem.* **75**, 857 (1963).
3. Bredereck, H., Föhlisch, B., and Walz, K., *Justus Liebigs Ann. Chem.* **688**, 93 (1965).
4. Cairns, T. L., Larchar, A. W., and McKurick, B. C., *J. Org. Chem.* **17**, 1497 (1952).
5. Casanova, J., Jr., Kiefer, H. R., Kuwada, D., and Boulton, A. H., *Tetrahedron Lett.* p. 703 (1965).
6. Davis, T. L., and Yelland, W. E., *J. Amer. Chem. Soc.* **59**, 1998 (1937).
7. Feuer, H., Rubinstein, H., and Nielsen, A. T., *J. Org. Chem.* **23**, 1107 (1958).
8. Gautier, A., *Ann. Chim. (Paris)* [4] **17**, 229 (1869).
9. Gilman, H., and Heckert, L. C., *Bull. Soc. Chim. Fr.* [4] **43**, 224 (1928).
10. Hagedorn, I., *Angew. Chem.* **75**, 304 (1963).
11. Halleux, A., *Angew. Chem.* **76**, 889 (1964).

12. Havlik, A., and Wald, M. M., *J. Amer. Chem. Soc.* **77**, 5171 (1955).
13. Ito, Y., Okano, M., and Oda, R., *Tetrahedron* **22**, 447 (1966).
14. Johnson, H. W., Jr., and Daughhetes, P. H., Jr., *J. Org. Chem.* **29**, 246 (1964).
15. Jones, K., and Lappert, M. F., *Organometal. Chem. Rev.* **1**, 67 (1966).
16. Kabbe, H. J., *Angew. Chem.* **80**, 406 (1968).
17. Knorr, R., *Chem. Ber.* **98**, 4038 (1956).
18. Losco, G., *Gazz. Chim. Ital.* **67**, 553 (1937).
19. Martin, D., and Weise, A., *Angew. Chem.* **79**, 145 (1967).
20. Millich, F., and Sinclair, R., *Chem. Eng. News* **45**, 30 (1967); *Chem. Abstr.* **66**, 46632y (1967).
21. Mitin, J. V., Glushenkova, V. R., and Vlasov, G. P., *Zh. Obshch. Khim.* **32**, 3867 (1962).
22. Nef, J. U., *Justus Liebigs Ann. Chem.* **270**, 267 (1892).
23. Nef, J. U., *Justus Liebigs Ann. Chem.* **280**, 291 (1894).
24. Neidleim, R., *Arch. Pharm.* (*Weinheim*) **298**, 491 (1965).
25. Neumann, W. P., and Kühlein, K., *Tetrahedron Lett.* p. 3423 (1966).
26. Okano, M., Ito, Y., Shono, T., and Oda, R., *Bull. Chem. Soc. Jap.* **36**, 1314 (1963).
27. Oliveri-Mandala, E., and Alagna, B., *Gazz. Chim. Ital.* **40**, II, 441 (1910).
28. Otsuka, S., Nakamura, A., and Tatsuno, Y., *Chem. Commun.* p. 836 (1967).
29. Passerini, M., *Gazz. Chim. Ital.* **54**, 633 (1924).
30. Passerini, M., *Gazz. Chim. Ital.* **54**, 639 and 670 (1924).
31. Passerini, M., *Gazz. Chim. Ital.* **55**, 555 (1925).
32. Passerini, M., and Casini, V., *Gazz. Chim. Ital.* **67**, 332 (1937).
33. Ryan, J. W., Menzie, G. K., and Speier, J. L., *J. Amer. Chem. Soc.* **82**, 3601 (1960).
34. Sachs, F., and Loevy, H., *Chem. Ber.* **37**, 874 (1904).
35. Saegusa, T., Ito, Y., Kinoshita, H., and Tomita, S., unpublished results.
36. Saegusa, T., Ito, Y., and Kobayashi, S., *Tetrahedron Lett.* p. 935 (1968).
37. Saegusa, T., Ito, Y., Kobayashi, S., and Hirota, K., *Tetrahedron Lett.* p. 521 (1967).
38. Saegusa, T., Ito, Y., Kobayashi, S., and Hirota, K., *J. Amer. Chem. Soc.* **89**, 2240 (1967).
39. Saegusa, T., Ito, Y., Kobayashi, S., Hirota, K., and Yoshioka, H., *Tetrahedron Lett.* p. 6121 (1966).
40. Saegusa, T., Ito, Y., Kobayashi, S., Hirota, K., and Yoshioka, H., *Bull. Chem. Soc. Jap.* (1970) (in press).
41. Saegusa, T., Ito, Y., Kobayashi, S., Takeda, N., and Hirota, K., *Can. J. Chem.* **45**, 1217 (1967).
42. Saegusa, T., Ito, Y., Kobayashi, S., Takeda, N., and Hirota, K., *Tetrahedron Lett.* p. 1273 (1967).
43. Saegusa, T., Ito, Y., and Shimizu, T., *J. Org. Chem.* (1970).
44. Saegusa, T., Ito, Y., and Tomita, S., unpublished results (1970).
45. Saegusa, T., Kobayashi, S., Hirota, K., Okumura, Y., and Ito, Y., *Bull. Chem. Soc. Jap.* **41**, 1638 (1968).
46. Saegusa, T., Kobayashi, S., and Ito, Y., *J. Org. Chem.* **35**, 2118 (1970).
47. Saegusa, T., Kobayashi, S., Ito, Y., and Yasuda, N., *J. Amer. Chem. Soc.* **90**, 4182 (1968).
48. Saegusa, T., Taka-ishi, N., and Ito, Y., *J. Org. Chem.* **34**, 4040 (1969).
49. Saegusa, T., Yasuda, N., Kobayashi, S., and Ito, Y., unpublished results (1970).
50. Schöllkopf, U., and Gerhart, F., *Angew. Chem.* **80**, 842 (1968).
51. Shaw, D. H., and Pritchard, H. O., *Can. J. Chem.* **45**, 2749 (1967).
52. Shono, T., Kimura, M., Ito, Y., Nishida, K., and Oda, R., *Bull. Chem. Soc. Jap.* **37**, 635 (1964).
53. Shono, T., and Matsumura, Y., *J. Amer. Chem. Soc.* **90**, 5937 (1968).
54. Takahashi, S., Sonogashira, K., and Hagihara, N., *Nippon Kagaku Zasshi* **87**, 610 (1966).

55. Treibs, A., and Dietl, A., *Chem. Ber.* **94**, 298 (1961).
56. Ugi, I., Betz, W., and Offermann, K., *Chem. Ber.* **97**, 3008 (1964).
57. Ugi, I., and Bodesheim, F., *Chem. Ber.* **94**, 1157 (1961).
58. Ugi, I., and Fetzer, U., *Chem. Ber.* **94**, 1116 (1961).
59. Ugi, I., and Fetzer, U., *Chem. Ber.* **94**, 2239 (1961).
60. Walter, W., and Bode, K. D., *Angew. Chem.* **74**, 694 (1962).
61. Weith, W., *Chem. Ber.* **6**, 210 (1873).

# Chapter 5

# Cyclization Reactions

## H. J. Kabbe

|     |                          |     |
| --- | ------------------------ | --- |
| I.  | Introduction             | 93  |
| II. | Four-Membered Rings      | 95  |
| III.| Five-Membered Rings      | 97  |
| IV. | Six-Membered Rings       | 104 |
| V.  | Seven-Membered Rings     | 106 |
|     | References               | 106 |

## I. INTRODUCTION

Isonitriles are ideal partners for cyclization reactions because they can act simultaneously as nucleophiles and electrophiles. Frequently, in these reactions, a dipolar intermediate, $a^{\oplus}$—$b^{\ominus}$, is involved (2). The electrophilic part of the dipole forms a bond with the lone pair of electrons at the isonitrile carbon atom; this carbon atom then becomes electrophilic and closes the ring with the other, electron-rich part of the dipole.

This simple scheme (2,a) is followed, however, only when the reactant $a^{\oplus}$—$b^{\ominus}$ contains more than two atoms. If the dipole consists of only two atoms (e.g., the carbonyl group), a three-membered ring would be formed according to (2,a). Although rings of this size have been postulated as intermediates,[29a] they have never been isolated.* Instead, a further molecule of either one of the two partners a—b (2,b) or isonitrile (2,c) or even a further dipole $d^{\oplus}$—$e^{\ominus}$ (2,d) is inserted between $a^{\oplus}$—$b^{\ominus}$ and the isonitrile, so that ultimately a ring of at least four members is formed.

---

* In this connection the α-lactam (I), prepared in a different way should be mentioned; even at 75°C (I) decomposes into isonitrile and ketone[32] (1). The isomeric iminoester (II) is assumed to be the intermediate which formally could be conceived of as having been formed from cyclohexanone and the isonitrile. As a matter of fact, however, ketones and isonitriles react to give the oxetanes (III); in this case the ring size is enlarged by a further molecule of isonitrile to yield a four-membered ring.[11] (2,c).

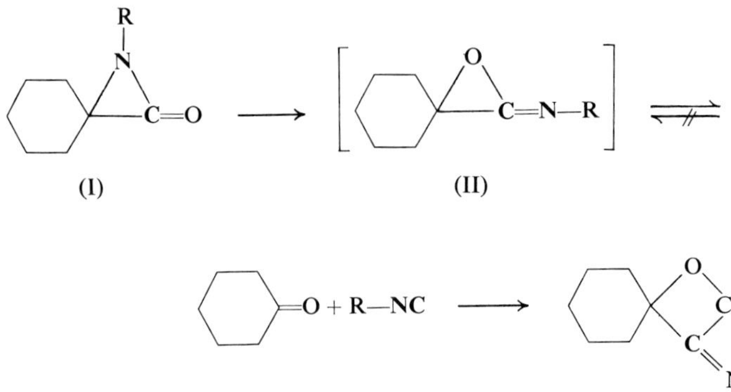

[Scheme (1) showing structures (I), (II), (III) with cyclohexane/aziridine/oxirane rings reacting with R—NC]

(1)

Many ring closures occur without any catalysis; others, only if catalyzed by acids.

Although scheme (2) fairly well represents most of the cycloadditions described below, it should be kept in mind that it is impossible to derive from it any prediction about which of these pathways the reaction might follow. The fact that polyhalocarbonyl compounds (like chloral[11]) react with isonitriles sometimes in a ratio of 1:2 to form oxetanes, whereas others (like

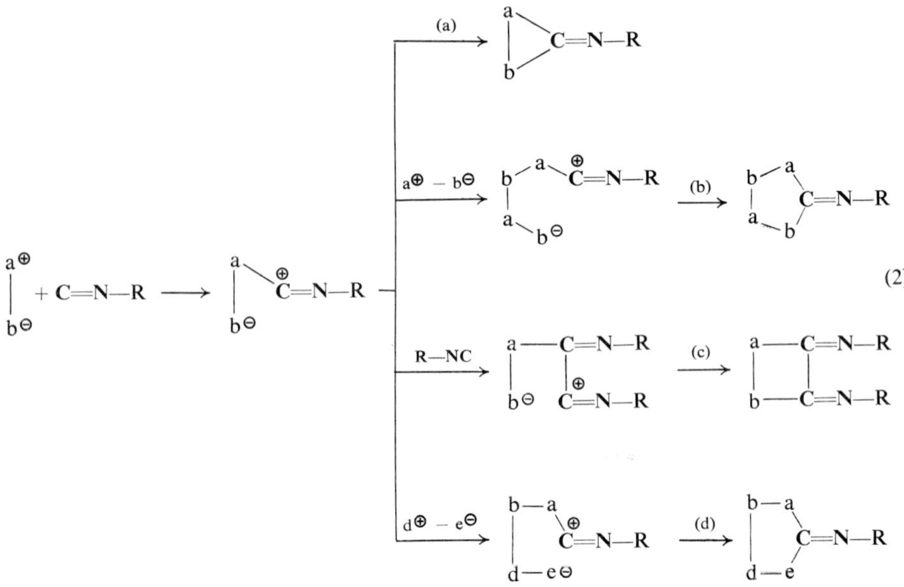

[Scheme (2) showing reaction pathways (a), (b), (c), (d) for cycloaddition of C=N—R with various dipoles]

(2)

5. Cyclization Reactions      95

hexafluoroacetone) in the ratio of 2:1 yield dioxolanes,[3] could hardly have been predicted. As in most of the polar reactants $a^\oplus$—$b^\ominus$, at least one of the reactive sites a or b is a heteroatom, the majority of these cycloadditions lead to heterocyclic compounds. For some of these, the above isonitrile reactions are the easiest and afford the most elegant synthetic access. The reaction mechanism is known for only a few of the cyclizations of isonitriles, so that there are no mechanistic criteria available upon which a systematic presentation might be based. For the following reactions, therefore, the order is based on the ring size and the number of heteroatoms in the products. Heterocyclic compounds that are obtained from isonitriles by the Ugi reaction[34],* are not discussed here, but in Chapter 8.

## II. FOUR-MEMBERED RINGS

1,1-Ditrifluoromethyl-2,2-dicyanoethylene (IV), with its electrophilic double bond, reacts even below 0°C with two equivalents of *t*-butyl isocyanide to form the crystalline cyclobutane derivative (V) which, however, is not stable and decomposes at room temperature within a few hours. Because of this rapid decomposition, its structure could only be proved spectroscopically.[19]

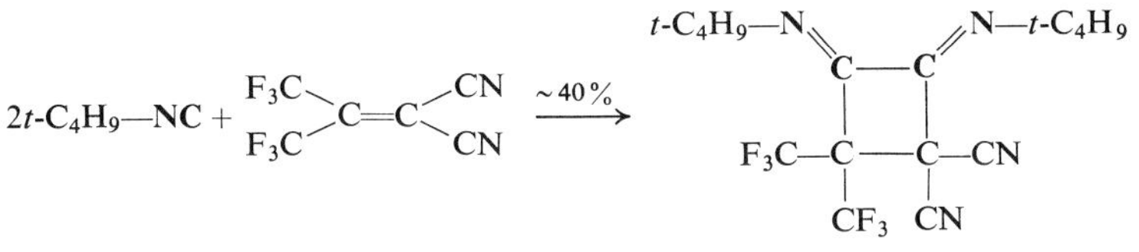

$$2t\text{-}C_4H_9\text{—NC} + \underset{F_3C}{\overset{F_3C}{>}}C=C\underset{CN}{\overset{CN}{<}} \xrightarrow{\sim 40\%} \underset{\underset{CF_3\ \ CN}{|\ \ \ \ \ |}}{\underset{F_3C\text{—}C\text{—}C\text{—}CN}{\overset{t\text{-}C_4H_9\text{—}N\diagdown\ \ \diagup N\text{—}t\text{-}C_4H_9}{\overset{C\text{—}C}{|\ \ \ \ \ \ |}}}} \quad (3)$$

(IV)                          (V)

Kabbe[11] found that a variety of aldehydes and ketones react with isonitriles in a ratio of 1:2 in the presence of a catalytic amount of Lewis acid to form the 2,3-diiminooxetanes (VI) in generally high yield; two cases of this oxetane formation were reported independently by Saegusa *et al.*[29a,b] Some of these oxetanes can be isomerized to the unsaturated iminoamides (VII) which may be hydrolyzed under the influence of acids to form ketoamides, e.g., VIIIb.[12] Ketoamides like VIIIa had already been obtained by Müller and Zeeh[21] from ketones and isonitriles in the presence of approximately equimolar quantities of boron trifluoride.

---

* The term "Ugi reaction" is being used by some authors[1,18,25,33] for "four-component condensations."

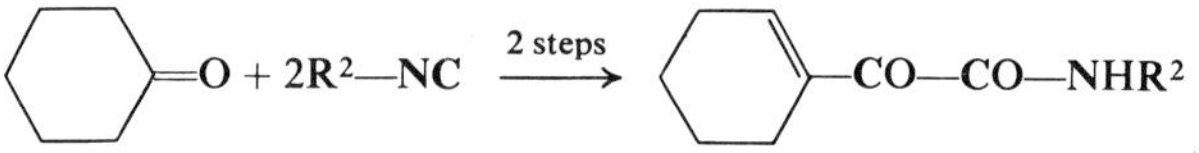

$$\text{cyclohexanone} = O + 2R^2-NC \xrightarrow{\text{2 steps}} \text{cyclohexenyl}-CO-CO-NHR^2 \quad (4)$$

(VIIIa)

$$\underset{\underset{CH_3}{|}}{CH_3-C} \overset{O-C=NR^2}{\underset{C=NR^2}{|}} \longrightarrow \underset{CH_3}{CH_2} \overset{\diagdown}{\underset{\diagup}{C-C}} \overset{CO-NHR^2}{\underset{NR^2}{\diagdown}} \longrightarrow \underset{CH_3}{CH_2} \overset{\diagdown}{\underset{\diagup}{C}}-CO-CO-NHR^2 \quad (5)$$

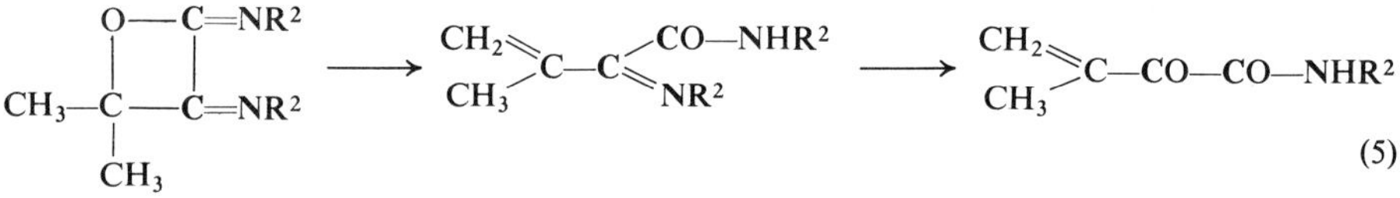

(VIa) (VII) (VIIIb)

Ring opening of VI can also be effected by addition of anhydrous or aqueous acids, e.g., hydrogen halides or carboxylic acids and water give the β-substituted amides (IX).[12]

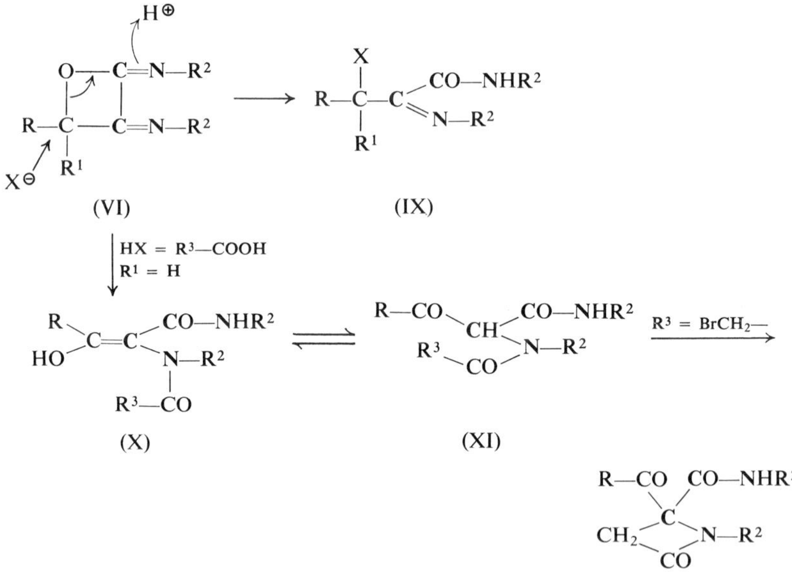

(6)

(VI) (IX)

HX = R³—COOH
R¹ = H

(X) ⇌ (XI) R³ = BrCH₂— →

(XII) (7)

When VI stems from an aldehyde and is reacted with a carboxylic acid, the unsaturated hydroxyacrylamide (X) is formed as a result of an O→N-acyl migration.[12] If one chooses bromoacetic acid as HX, hydrogen bromide is split off from the corresponding ketoform (XI) to give the β-lactam (XII) as the sole product; no intermediates could be isolated even in the absence of base.[13]

5. Cyclization Reactions      97

There is a third reaction (8) of type C [see reaction (2)]. Without further experimental details, structure XIV is assigned to the product obtained by reacting nitrosotrifluoromethane (XIII) and methyl isocyanide at 25°C. This structure is supported by the decomposition of XIV into $N$-trifluoromethyl-$N'$-methylcarbodiimide (XV) and methyl isocyanate (XVI) at 350–400°C.[17]

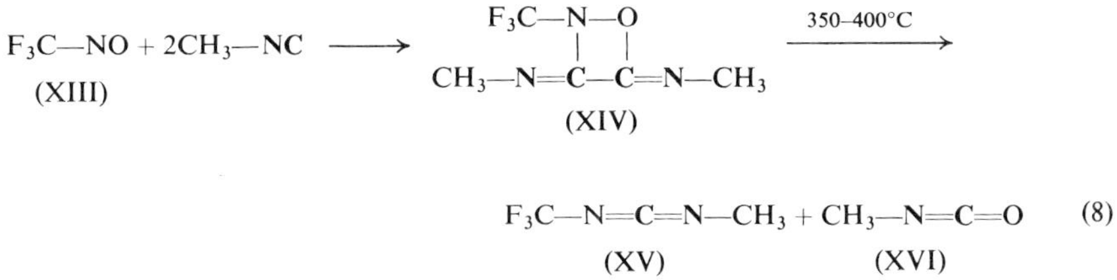

$$F_3C\text{—}NO + 2CH_3\text{—}NC \longrightarrow \underset{(XIV)}{\overset{F_3C\text{—}N\text{—}O}{\underset{CH_3\text{—}N=C\text{—}C=N\text{—}CH_3}{|\quad\quad|}}} \xrightarrow{350\text{–}400°C}$$

(XIII)

$$F_3C\text{—}N=C=N\text{—}CH_3 + CH_3\text{—}N=C=O \quad (8)$$
$$\quad\quad(XV) \quad\quad\quad\quad\quad (XVI)$$

## III. FIVE-MEMBERED RINGS

A variety of tetrazoles are obtained from isonitriles and hydrazoic acid subsequent to an α-addition[24] and by a modified Passerini reaction.[35] With hydrazoic acid as the acid component, the Ugi reaction also yields tetrazoles.

Ugi and Rosendahl found[36] that ketenes (XVII) react with isonitriles in a ratio of 2:1 to form derivatives of cyclopentanetrione (XIX).

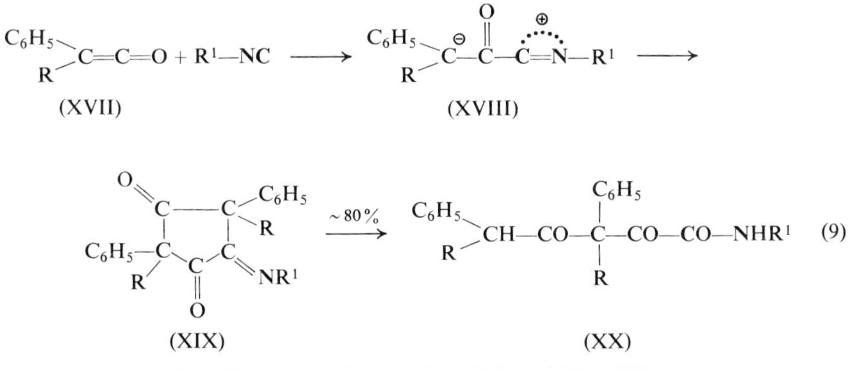

$$\underset{R}{\overset{C_6H_5}{>}}C=C=O + R^1\text{—}NC \longrightarrow \underset{R}{\overset{C_6H_5}{>}}\overset{O}{\underset{}{C\text{—}\overset{\ominus}{C}\text{—}\overset{\oplus}{C}=N\text{—}R^1}} \longrightarrow$$
(XVII) (XVIII)

(XIX) → ~80% → (XX)    (9)

$C_6H_5$–CH–CO–C($C_6H_5$)–CO–CO–NHR$^1$ with R groups

(R=H, $C_6H_5$; R$^1$ = $C_2H_5$, $n$-$C_4H_9$, $c$-$C_6H_{11}$, $C_6H_5$—$CH_2$)

Dipole XVIII is proposed to be the intermediate, which closes the ring with further ketene or with another dipolarophile like chloral. The acidic hydrolysis of XIX leading to the diketoamides XX also confirms the structure. The dipole XVIII can also react with carboxylic acids, in which case diketoamides (XXI) are the products of an acyl migration (see Chapter 7).

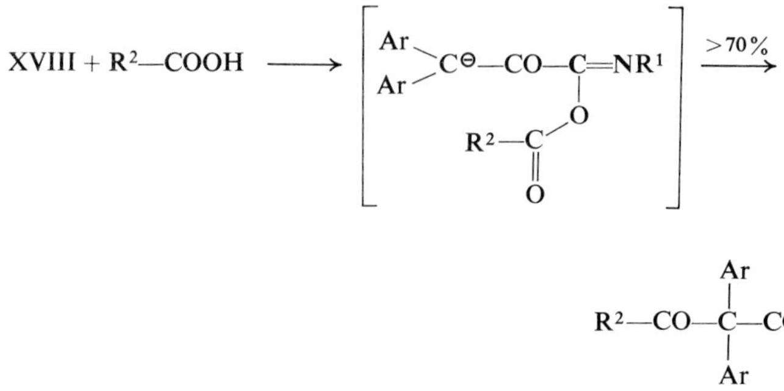

XVIII + R²—COOH ⟶ [Ar₂C⁻—CO—C(=NR¹)—O—C(=O)—R²] —>70%—>

R²—CO—C(Ar)(Ar)—CO—CO—NHR¹ (10)

(XXI)

($R^1 = c\text{-}C_6H_{11}$; $R^2$ = e.g., $CH_3$, $ClCH_2$, $C_6H_5$, $C_6H_5$—$CH_2$)

Two products can be isolated from the reaction of dicarbomethoxyacetylene (XXII) with excess isonitrile[39]; the dipole XXIII, which is formed first, is capable of reacting with further diester XXII both at the triple bond, to form XXIV, as well as at the carbonyl group, to form XXV.

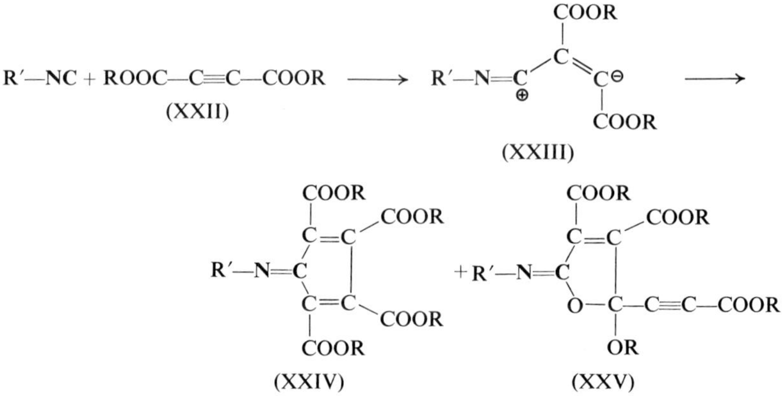

R'—NC + ROOC—C≡C—COOR ⟶ R'—N⁺=C(COOR)—C(COOR)=C⁻—COOR ⟶

(XXII)                              (XXIII)

R'—N=C(COOR)—C(COOR)=C(COOR)—C(COOR)=C(COOR)   + R'—N=C(COOR)—C(COOR)=C(—O—C(OR)—C≡C—COOR)   (11)

(XXIV)                              (XXV)

Passerini investigated the reaction of α-naphthol with phenyl isocyanide[26] which yields the benzocoumarone derivative (XXVI), in addition to other compounds. A confirmation of the structure by modern physicochemical methods appears desirable (see Chapter 4).

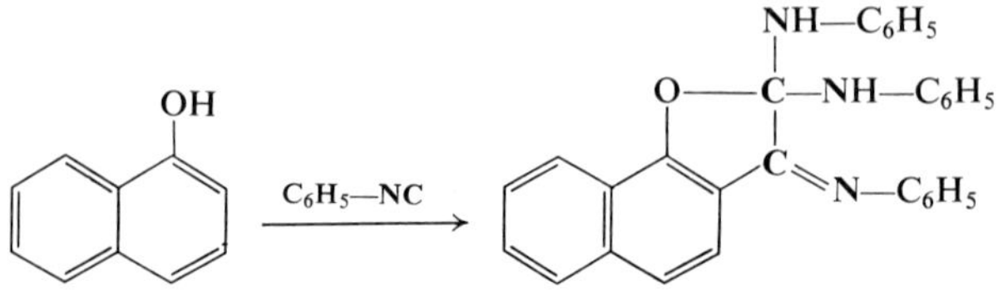

[naphthol-OH] + $C_6H_5$—NC ⟶ [benzocoumarone with C(NH—C₆H₅)(NH—C₆H₅), C=N—C₆H₅] (12)

(XXVI)

Enamines (XXVII) and isonitriles react with isocyanates and isothiocyanates (XXVIII) to form, in a large variety of three-component combinations, the succinimide derivatives (XXX)[16] in nearly quantitative yields. It may be assumed that the 1,4 dipole (XXIX) (see Chapter 8) is an intermediate and the isonitrile is added to the dipole.[15]

(XXVII) (XXVIII) (XXIX) (XXX) (13)

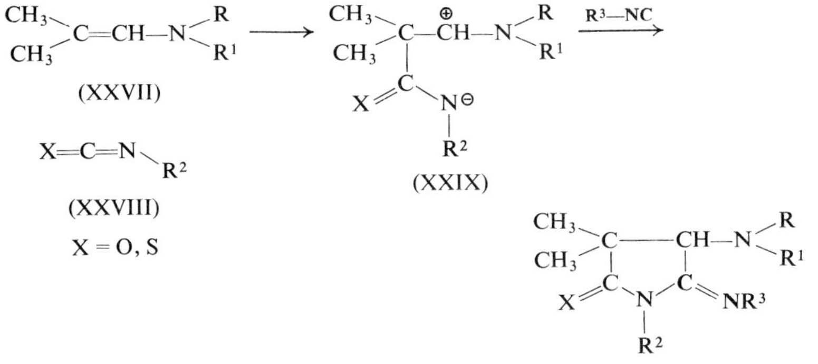

($R + R^1 = (CH_2)_{4,5}$; $R^2 = CH_3$, $c\text{-}C_6H_{11}$, various aryl groups; $R^3 = (CH_3)_3C$, $c\text{-}C_6H_{11}$, aryl)

Yaroslavsky[40] and Knorr[14] independently investigated the reaction of benzene diazonium carboxylate (XXXI) and isonitriles. The reaction probably proceeds through the 1,4 dipolar intermediate (XXII), addition of isonitrile gives the primary product (XXXIII) rearrangement of which accounts for the final product, phthalimide (XXXIV). The yields do not exceed 20%.

(XXXI) (XXXII) (XXXIII) (XXXIV) (14)

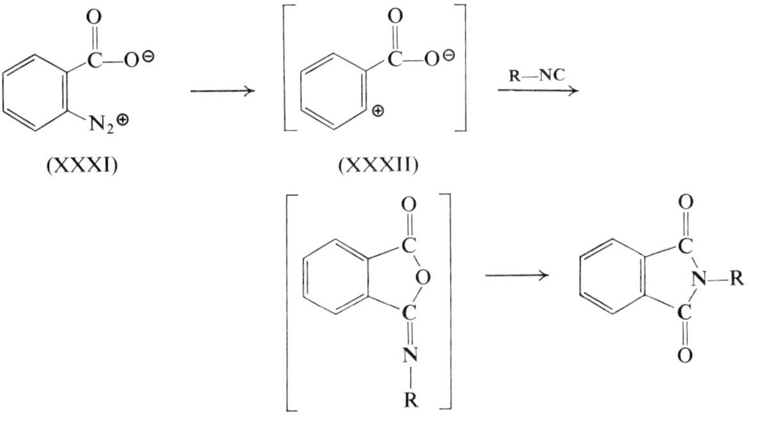

Under conditions which, with aliphatic isocyanides, ordinarily lead to the glyoxylic acid derivatives (VIIIa),[21] Zeeh was able to cyclize aromatic isocyanides with ketones to form the indolenins (XXXVI).[41]

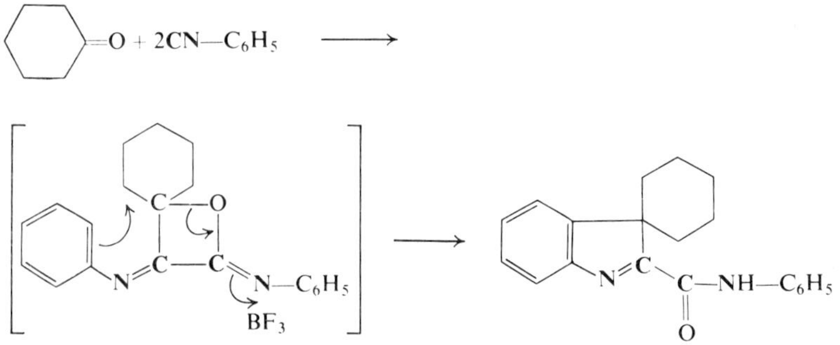

(15)

(XXXV)          (XXXVI)

A further cyclization of carbonyl compounds and isonitriles in the ratio 1:2 was also discovered by Zeeh.[42] Ketones of the acetophenone type (XXXVII) react with $t$-butyl isocyanide to form indole carbonamides (XXXIX); this reaction can also be explained by invoking an oxetane (XXXVIII) as the intermediate.

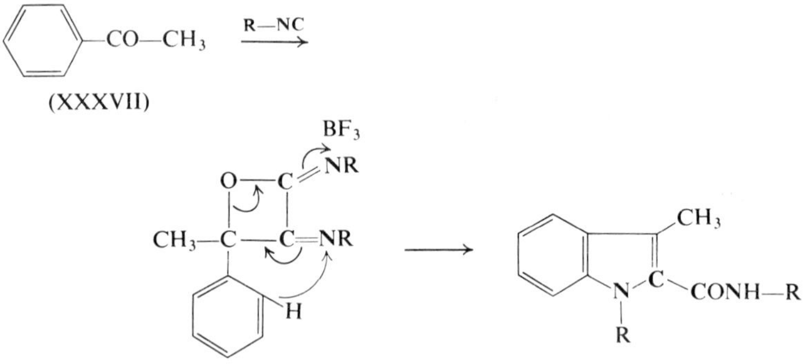

(XXXVII)

(16)

(R = $t$-C$_4$H$_9$)      (XXXVIII)      (XXXIX)

Although isonitriles tend to polymerize, very few oligomers and polymers of isonitriles with a definite, proved structure are known. The tetramer (XL) of phenyl isocyanide which is formed spontaneously from phenyl isocyanide on standing at 0–20°C, was isolated and characterized by Grundmann[7] and shown to be identical with "indigodianil" which is obtained from indigo and aniline[6] (see also the trimer LXIII).

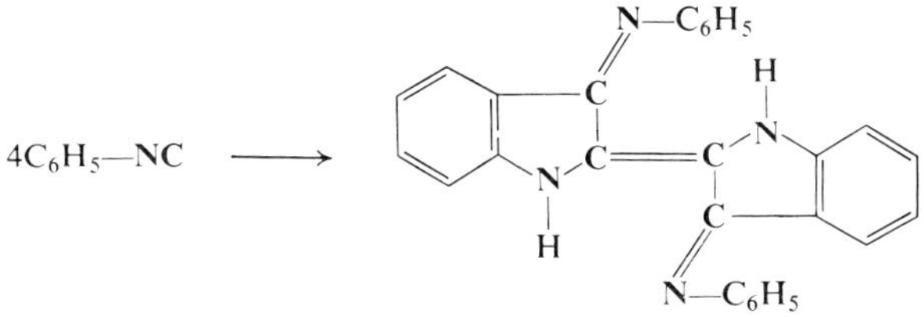

5. Cyclization Reactions 101

$4C_6H_5-NC \longrightarrow$ (XL) (17)

The reaction of hexafluoroacetone with cyclohexyl isocyanide represents another type of isonitrile–ketone cyclization reaction. N-Cyclohexyliminodioxolane (XLI) forms at 0°C in a yield of almost 90%.[2, 3, 20]

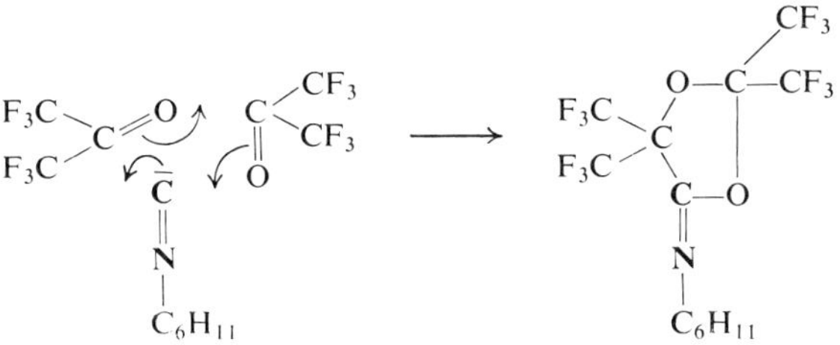

(18)

(XLI)

The acyl imines of hexafluoroacetone are 1,4 dipoles and react as such with isonitriles to form oxazolines, e.g., XLII.[3] The amino acid derivative (XLIII) is obtained from (XLII) after hydrolysis.

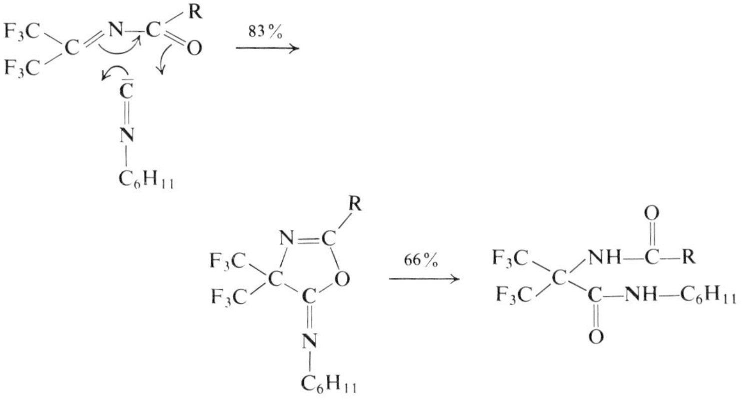

(19)

(XLII) (XLIII)

$(R = C_6H_5-, C_2H_5O-)$

In the acyl imines, incidentally, one single trifluoromethyl group is sufficient to enable this cycloaddition to proceed. Weygand and Steglich[38] found in the reaction of XLIV with the α-isocyano esters (XLV) via the oxazolines (XLVI) a very elegant access to the tripeptides (XLVII) where trifluoroalanine is one of the amino acid components (see Chapters 8 and 9).

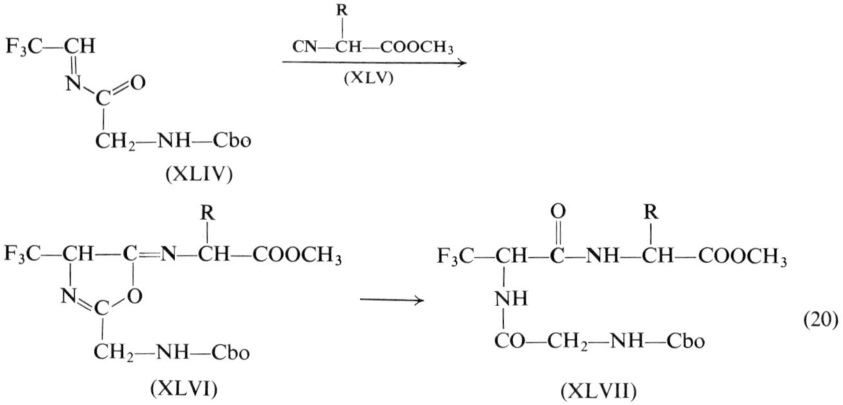

$$\text{(20)}$$

α-Nitroso-β-naphthol (XLVIII) and 4-isocyanoazobenzene (XLIX) yield the oxazoline (L), the structure of which was inferred from the degradation product (LI).[27]

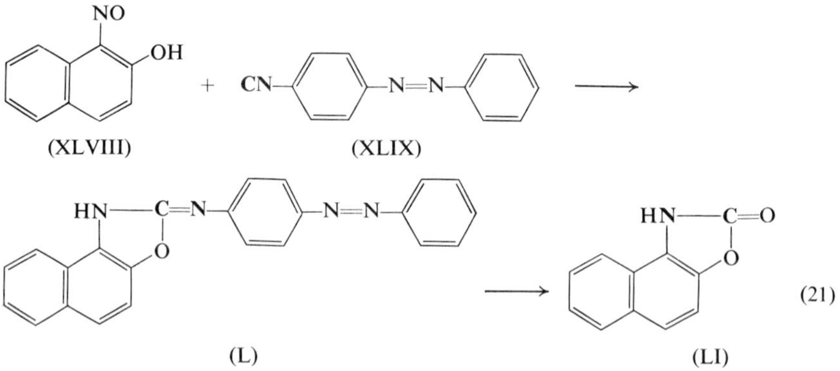

$$\text{(21)}$$

A further oxazolidine synthesis was discovered by Hesse et al.[10] When a mixture of phenyl isocyanide and aromatic aldehyde (Ar—CHO) in ether is added to tri-n-butyl borane at room temperature, the compounds (LIIa–LIIc) are obtained in yields around 70% (see Chapter 6).

$$C_6H_5-NC + 2Ar-CHO + (n-C_4H_9)_3B \longrightarrow \underset{(LII)}{\begin{array}{c} Ar-HC\diagup O \diagdown C \diagup C_4H_9 \\ | \quad\quad\quad \diagdown C_4H_9 \\ Ar-HC-N \diagdown C_6H_5 \end{array}} \quad (22)$$

a: Ar = $C_6H_5$  b: Ar = 4-Cl—$C_6H_4$  c: Ar = 4-$NO_2$—$C_6H_4$

The acyl isocyanates and isothiocyanates (LIII) represent another type of system in which there is a potential 1,4 dipole $a^\oplus$—$b^\ominus$ [method a, (2)]. This type of reaction was reported by Neidlein[22] and later, independently, by Goerdeler and Schenk.[5,30]

$$\underset{(LIII)}{\begin{array}{c} |N-R^1 \\ \| \\ C \\ X \diagup \quad \diagdown \text{electron pair} \\ \| \quad \quad C=Y \\ R-C-N \end{array}} \longrightarrow \underset{(LIV)}{\begin{array}{c} N-R^1 \\ \diagup C \diagdown \\ X \quad \quad C=Y \\ | \quad \quad | \\ R-C=N \end{array}} \quad \begin{array}{l} X = O, S \\ Y = O, S \\ R = C_6H_5-, CHCl_2-, C_6H_5-CH_2-S- \\ R^1 = C_6H_5-, c\text{-}C_6H_{11}-, p\text{-}CH_3O-C_6H_4- \end{array}$$

(23)

Whereas acyl isocyanates (LIII, X=Y=O) are stable compounds, the thioacyl isocyanates (LVI) are produced by decomposition of the thiazole diones (LV) and are capable of reacting with isonitriles to form LIV (X = S, Y = O) in high yields.

$$\underset{(LV)}{\begin{array}{c} N-C=O \\ \| \quad \quad | \\ R-C \diagdown S \diagup C=O \end{array}} \xrightarrow{150°C} \underset{(LVI)}{\left[ \begin{array}{c} R-C-N=C=O \\ \| \\ S \end{array} \right]} \xrightarrow{R^1-NC} \underset{(LIV)}{\begin{array}{c} N-C=Y \\ \| \quad \quad \| \\ R-C \diagdown X \diagup C=N-R^1 \end{array}} \quad (24)$$

Although carbonyl compounds and isonitriles normally react by carbon to carbon addition, p-methoxybenzoylmethyl isocyanide (LVII), when heated, gives an almost quantitative yield of 4-methoxyphenyloxazole (LVIII), i.e., by bond formation between carbon and oxygen.[8]

$$\underset{(LVII)}{CH_3O-\!\!\!\bigcirc\!\!\!-\underset{\underset{O}{\|}}{C}-\underset{\underset{C}{\|||}}{\overset{H_2}{C}}\diagdown N} \longrightarrow \underset{(LVIII)}{CH_3O-\!\!\!\bigcirc\!\!\!-\underset{\underset{O-CH}{|}}{C}\diagdown\overset{H}{\underset{\|}{C}}\diagdown N} \quad (25)$$

When starting with the corresponding hydroxy compound (LIX) Hagedorn et al.[8] obtained the oxazoline (LX).

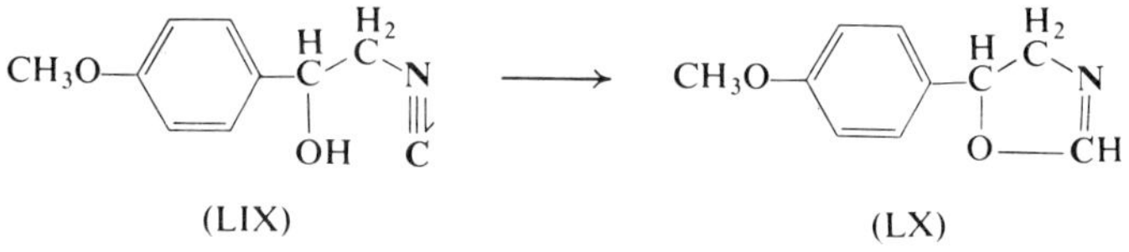

$$\text{(LIX)} \longrightarrow \text{(LX)} \quad (26)$$

In general, carbonyl–isonitrile reactions are catalyzed if necessary by acids or transition metal compounds (see Chapter 3). Schöllkopf and Gerhart,[4] however, were able to add aldehydes and ketones to α-metallated isocyanides. Under extremely mild conditions (−70°C) an oxazoline ring (LXI) is formed in good yield, whereas at higher temperatures, decomposition to olefins (LXII) and metal cyanates predominates.[31]

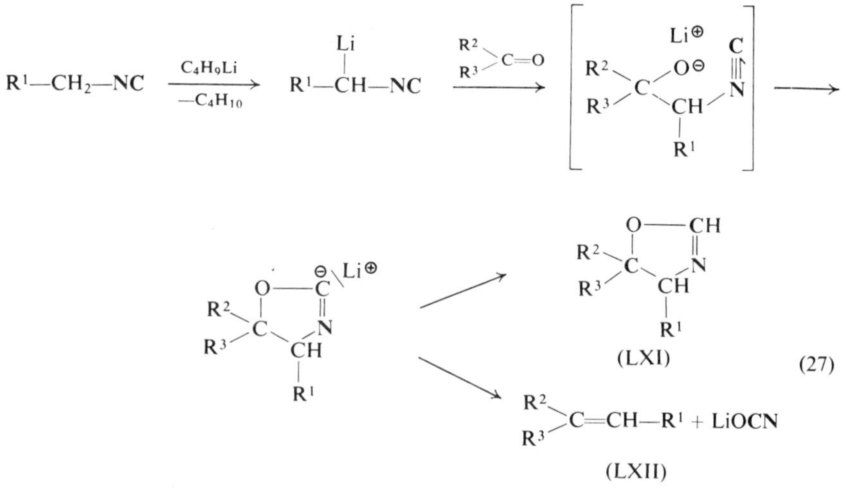

$$\text{R}^1\text{—CH}_2\text{—NC} \xrightarrow[{-\text{C}_4\text{H}_{10}}]{\text{C}_4\text{H}_9\text{Li}} \text{R}^1\text{—CH—NC} \xrightarrow{\text{R}^2\text{R}^3\text{C=O}} \cdots \longrightarrow \text{(LXI) or (LXII) + LiOCN} \quad (27)$$

## IV. SIX-MEMBERED RINGS

Phenyl isocyanide undergoes a different kind of oligomerization [see (17)] in the presence of excess nitrosobenzene.[28] The product formed (in addition to an equivalent amount of diphenyl urea) is the quinoline quinone derivative (LXIII). A further detailed investigation of this complex reaction would probably be rewarding.

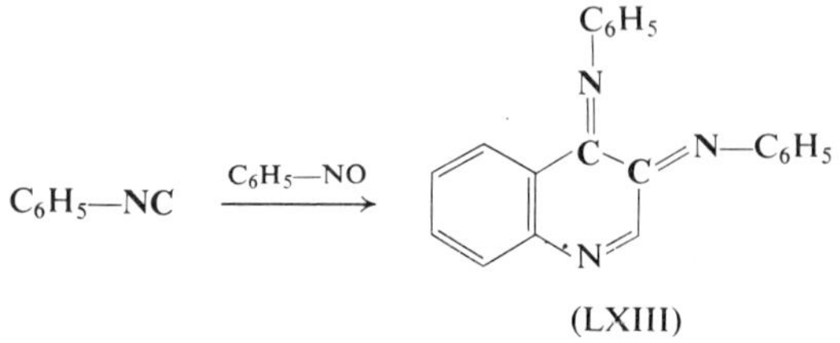

$$\text{C}_6\text{H}_5\text{—NC} \xrightarrow{\text{C}_6\text{H}_5\text{—NO}} \text{(LXIII)} \quad (28)$$

## 5. Cyclization Reactions

Hydrogenation of an α,β-diisocyanide, xanthocyllin dimethyl ether (LXIV), gives the pyrazine (LXV); but experimental details are not given.[9]

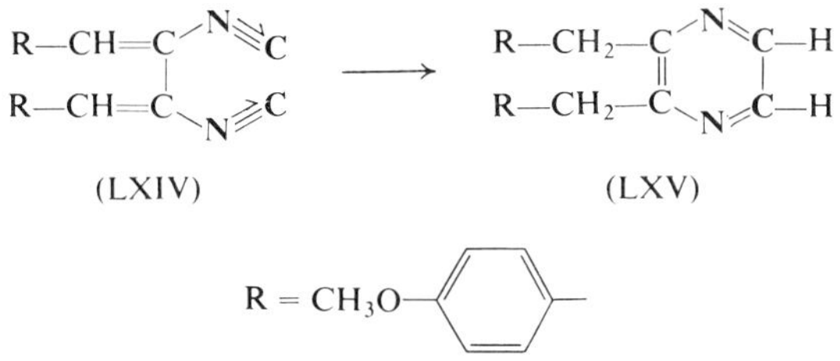

$$\begin{array}{cc}
\text{R—CH=C} \diagup \!\!\!\!{}^{N}\!\!\!\diagdown \text{C} & \text{R—CH}_2\text{—C} \diagup \!\!\!\!{}^{N}\!\!\!\diagdown \text{C—H} \\
\text{R—CH=C} \diagdown \!\!\!\!{}_{N}\!\!\!\diagup \text{C} & \text{R—CH}_2\text{—C} \diagdown \!\!\!\!{}_{N}\!\!\!\diagup \text{C—H} \\
(\text{LXIV}) & (\text{LXV})
\end{array} \quad (29)$$

$R = CH_3O\text{—}C_6H_4\text{—}$

The triazines (LXVI) are obtained in high yield (74–83%) when isonitriles and hydrogen thiocyanate are reacted in a ratio of 1:2 in ether.[23,37] In accordance with investigations by Ugi and Rosendahl,[37] of the two possible pathways (30, a or b), the one which involves intermediate (LXVII) appears to be the more likely. The alkaline ring opening to form the dithiobiurets (LXVIII) is in accord with formula LXVI.

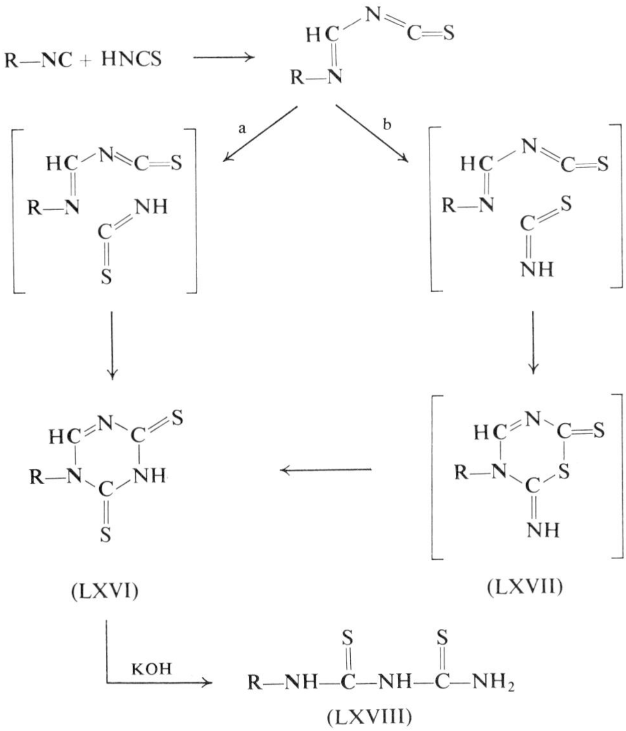

$$R\text{—NC} + \text{HNCS} \longrightarrow \begin{array}{c} HC \diagup\!\!\!\!{}^{N}\!\!\!\diagdown C=S \\ \| \\ R\text{—N} \end{array}$$

(scheme showing pathways a and b leading to LXVI and LXVII, with KOH giving LXVIII)

$$R\text{—NH—}\underset{\|}{\overset{S}{C}}\text{—NH—}\underset{\|}{\overset{S}{C}}\text{—NH}_2$$
(LXVIII)

(30)

## V. SEVEN-MEMBERED RINGS

As a by-product in the acid-catalyzed reaction between cyclohexyl isocyanide and acetaldehyde at $-78°$, a 3:2-addition product is formed to which structure LXIX is assigned.[29b]

(31)

LXIX

## REFERENCES

1. Gambaryan, N. P., *Zh. Vses. Khim. Obshchestva im. D. I. Mendeleeva* **12**, 65 (1967).
2. Gambaryan, N. P., Rokhlin, E. M., Zeifman, Yu. V., Ching-Yun, Ch., and Knunyants, I. L., *Angew. Chem.* **78**, 1008 (1966); *Angew. Chem., Int. Ed. Engl.* **5**, 947 (1966).
3. Gambaryan, N. P., Rokhlin, E. M., Zeifman, Yu. V., Simonyan, L. A., and Knunyants, I. L., *Dokl. Akad. Nauk SSSR* **166**, 864 (1966).
4. Gerhart, F., and Schöllkopf, U., *Tetrahedron Lett.* p. 6231 (1968).
5. Goerdeler, J., and Schenk, H., *Chem. Ber.* **98**, 3831 (1965).
6. Grandmougin, E., and Dessoulavy, E., *Chem. Ber.* **42**, 3636 and 4401 (1909).
7. Grundmann, C., *Chem. Ber.* **91**, 1380 (1958).
8. Hagedorn, I., Eholzer, U., and Etling, H., *Chem. Ber.* **98**, 193 (1965).
9. Hagedorn, I., and Tönjes, H., *Pharmazie* **11**, 409 (1956).
10. Hesse, G., Witte, H., and Gulden, W., *Angew. Chem.* **77**, 591 (1965); *Angew. Chem., Int. Ed. Engl.* **4**, 596 (1965).
11. Kabbe, H. J., *Angew. Chem.* **80**, 406 (1968); *Angew. Chem., Int. Ed. Engl.* **7**, 389 (1968); *Chem. Ber.* **102**, 1404 (1969).
12. Kabbe, H. J., *Chem. Ber.* **102**, 1410 (1969).
13. Kabbe, H. J., and Joop, N., *Justus Liebigs Ann. Chem.* **730**, 151 (1969).
14. Knorr, R., *Chem. Ber.* **98**, 4038 (1965).
15. Ley, K., personal communication.
16. Ley, K., Eholzer, U., and Nast, R., *Angew. Chem.* **77**, 544 (1965); *Angew. Chem., Int. Ed. Engl.* **4**, 519 (1965).
17. Makarov, S. P., Shpansky, V. A., Ginzburg, V. A., Shchekotichin, A. I., Filatov, A. S., Martynova, L. L., Pavlovskaya, I. V., Golovaneva, A. F., and Yakubovich, A. Ya., *Dokl. Akad. Nauk SSSR* **142**, 596 (1962).
18. McFarland, J. W., *J. Org. Chem.* **28**, 2179 (1963).
19. Middleton, W. J., *J. Org. Chem.* **30**, 1402 (1965).
20. Middleton, W. J., England, D. C., and Krespan, C. G., *J. Org. Chem.* **32**, 948 (1967).
21. Müller, E., and Zeeh, B., *Justus Liebigs Ann. Chem.* **696**, 72 (1966).
22. Neidlein, R., *Angew. Chem.* **76**, 500 (1964); *Angew. Chem., Int. Ed. Engl.* **3**, 446 (1964); *Chem. Ber.* **97**, 3476 (1964); *Arch. Pharm. (Weinheim)* **298**, 124 (1965).

## 5. Cyclization Reactions

23. O'Brien, D. E., Baiocchi, F., and Cheng, C. C., *Biochemistry* **2**, 1203 (1963).
24. Oliveri-Mandalá, E., and Alagna, B., *Gazz. Chim. Ital.* **40**, II, 441 (1910).
25. Opitz, G., and Merz, W., *Justus Liebigs Ann. Chem.* **652**, 163 (1962).
26. Passerini, M., *Gazz. Chim. Ital.* **55**, 555 (1925).
27. Passerini, M., *Gazz. Chim. Ital.* **61**, 26 (1931).
28. Passerini, M., and Bonciani, T., *Gazz. Chim. Ital.* **61**, 959 (1931).
29a. Saegusa, T., Taka-ishi, N., and Fujii, H., *Tetrahedron* **24**, 3795 (1968).
29b. Saegusa, T., Taka-ishi, N., and Fujii, H., *Polymer Letters* **5**, 779 (1967).
30. Schenk, H., *Chem. Ber.* **99**, 1258 (1966).
31. Schöllkopf, U., and Gerhart, F., *Angew. Chem.* **80**, 842 (1968).
32. Sheehan, J. C., and Lengyel, I., *J. Amer. Chem. Soc.* **86**, 746 and 1356 (1964).
33. Sjöberg, K., *Sv. Kem. Tidskr.* **75**, 493 (1963).
34. Ugi, I., *Angew. Chem.* **74**, 9 (1962); *Angew. Chem., Int. Ed. Engl.* **1**, 8 (1962).
35. Ugi, I., and Meyr, R., *Chem. Ber.* **94**, 2229 (1961).
36. Ugi, I., and Rosendahl, F. K., *Chem. Ber.* **94**, 2233 (1961); see also Rosendahl, F. K., Ph.D. Thesis, University of München, 1962.
37. Ugi, I., and Rosendahl, F. K., *Justus Liebigs Ann. Chem.* **670**, 80 (1963).
38. Weygand, F., and Steglich, W., *Angew. Chem.* **78**, 640 (1966); *Angew. Chem., Int. Ed. Engl.* **5**, 600 (1966).
39. Winterfeldt, E., *Angew. Chem.* **78**, 757 (1966); *Angew. Chem., Int. Ed. Engl.* **5**, 741 (1966).
40. Yaroslavsky, S., *Chem. Ind. (London)* p. 765 (1965).
41. Zeeh, B., *Tetrahedron Lett.* p. 3881 (1967); *Chem. Ber.* **101**, 1753 (1968).
42. Zeeh, B., *Chem. Ber.* **102**, 678 (1969).

# Chapter 6

# The Reaction of Isonitriles with Boranes

*Joseph Casanova, Jr.*

| | |
|---|---|
| I. Isonitrile–Organoborane Reactions | 109 |
|   A. 1:1 Adducts and Dimeric Isomers | 109 |
|   B. Adducts from Hydrogen Cyanide or Diborane As One Component | 117 |
|   C. Monomer–Dimer Equilibrium | 119 |
|   D. Mechanisms | 120 |
| II. Isonitrile–Organoaluminum and Cyanide–Organoaluminum Reactions | 124 |
| III. Cyanide–Organoborane Reactions | 125 |
| IV. Related Reactions: The Reaction of Carbon Monoxide and Other Lewis Bases with Organoboranes | 126 |
| References | 129 |

The Lewis base character of isonitriles has been known for many years, but has not been explored systematically until only recently. Klages and co-workers[42] examined the reaction of isonitriles with boron halides and some transition metals. Since that time isonitrile complexes of transition metals have been well characterized and the isonitriles in their complexes are shown to be stronger σ-donors than carbon monoxide.[16] Detailed infrared studies by Schleyer and Allerhand[1,63] and by Ferstandig[18,19] revealed clearly that isonitriles formed strong hydrogen bonds to alcohols and phenols through the isonitrile carbon atom (I). As weak a proton source as phenylacetylene was found[19] to give a carbon–hydrogen–carbon type of hydrogen bond of significant strength (II).

$$\overset{\oplus}{RN}\equiv\overset{\ominus}{C}:\cdots HOR' \qquad C_6H_5-CH_2\overset{\oplus}{N}\equiv\overset{\ominus}{C}:\cdots H-C\equiv C-C_6H_5$$

(I)                                   (II)

## I. ISONITRILE–ORGANOBORANE REACTIONS

### A. 1:1 Adducts and Dimeric Isomers

This behavior of isonitriles toward a center of low electron density is particularly manifested in the reactivity of alkyl and aryl isonitriles toward diborane and alkyl or arylboranes. When these species are mixed, in solution or neat, the first member of a series of isomeric 1:1 adducts is produced. The adducts are formed in high yield in an irreversible reaction sequence. These adducts make conveniently accessible several unusual heterocyclic ring systems containing boron and nitrogen. The new rings are very resistant to chemical degradation in acid or base and are stable toward air oxidation. The initial reaction occurs cleanly and rapidly at or below $-78°C$ in most of the cases which have been examined so far.

The earliest evidence of this reaction sequence appeared in 1963 when it was reported[22] that phenyl isocyanide reacted with ethyl- or $n$-butylborane at room temperature, in ether, to produce good yields of 2,5-dibora-2,5-dihydropyrazines, VIIa and VIIb, respectively (see Table I for data). When either VIIa or VIIb was heated above 200°C without solvent, rearrangement of the skeleton occurred (1), producing 2,5-dibora-3,6-dihydropyrazines, VIIIa and VIIIb, isomeric with their respective starting materials. The structural assignments were based on infrared (Table I) and $^1$H NMR data, as well as on the fact that VIIIb gave di-$n$-butyl ketone upon oxidation with chromic

$$RN{\equiv}\overset{\oplus}{C} + B\overset{\ominus}{R'_3} \longrightarrow RN{\equiv}\overset{\oplus}{C}B\overset{\ominus}{R'_3} \longrightarrow \left[ \begin{array}{c} R' \\ | \\ RN{\nearrow}C{\searrow}BR'_2 \end{array} \right] \longrightarrow$$

(III)  (IV)     (V)      (VI)

$$\begin{array}{c} R' \\ | \\ RN{\overset{\oplus}{\nearrow}}\overset{C}{\phantom{x}}{\overset{\ominus}{\searrow}}BR'_2 \\ R'_2B{\overset{\ominus}{\searrow}}\underset{C}{\phantom{x}}{\overset{\oplus}{\nearrow}}NR \\ | \\ R' \end{array} \longrightarrow \begin{array}{c} R'{\searrow}C{\swarrow}R' \\ RN{\phantom{x}}{\phantom{x}}BR' \\ | \phantom{xxx} | \\ R'B{\phantom{x}}{\phantom{x}}NR \\ R'{\nearrow}C{\nwarrow}R' \end{array} \longrightarrow \begin{array}{c} RN{-}B{\diagdown}CR'_3 \\ | \phantom{xxx} | \\ R'B{\diagdown}{\phantom{x}}NR \\ C \\ R'{\phantom{x}}R' \end{array} \quad (1)$$

(VII)      (VIII)      (IX)

anhydride, and also gave $n$-butyl acetate as a product of periodic acid oxidation in acetic acid. These authors further noted the unusual stability of both VII and VIII toward boiling aqueous acid or alkali, a quality which was most unexpected for compounds containing a B—N bond.

In the following year evidence was presented[14] for the formation of a stable simple 1:1 adduct (Vc) between $t$-butyl isocyanide and trimethylborane. This adduct was of the structure anticipated for a simple Lewis acid–Lewis base reaction, and presumably represented a precursor of structures VII and VIII. The 1:1 adduct formed rapidly below $-60°C$ upon mixing ether solutions of the two reagents. The $^1$H NMR spectrum showed the effect of isotropic

## 6. Reaction of Isonitriles with Boranes

### TABLE I

| Compound type | Designation | Substituents R = | R' = | mp (°C) | Infrared (cm⁻¹) | | Yield (%) | Reference |
|---|---|---|---|---|---|---|---|---|
| V | c | $t$-C$_4$H$_9$ | CH$_3$ | | 2247 | $(-\overset{+}{C}{\equiv}\overset{-}{N}-)$ | — | 14 |
|  | d | $c$-C$_6$H$_{11}$ | C$_6$H$_5$ | 126–128 | 2255 | $(-\overset{+}{C}{\equiv}\overset{-}{N}-)$ | 81[a] | 26, 27 |
|  | e | $i$-C$_3$H$_7$ | C$_6$H$_5$ | 105–108 | 2265 | $(-\overset{+}{C}{\equiv}\overset{-}{N}-)$ | 54[a] | 26, 27 |
|  | f | $t$-C$_4$H$_9$ | C$_6$H$_5$ | 150–155 | 2275 | $(-\overset{+}{C}{\equiv}\overset{-}{N}-)$ | 73[a] | 26, 27 |
|  | g | $n$-C$_4$H$_9$ | C$_6$H$_5$ | 94–97 | 2255 | $(-\overset{+}{C}{\equiv}\overset{-}{N}-)$ | 65[a] | 26, 27 |
|  | h | C$_6$H$_5$ | C$_6$H$_5$ | 75–85 | 2225 | $(-\overset{+}{C}{\equiv}\overset{-}{N}-)$ | 52[a] | 26, 27 |
|  | i | $p$-(C$_2$H$_5$)$_2$NC$_6$H$_4$ | C$_6$H$_5$ | 109–111 | 2245 | $(-\overset{+}{C}{\equiv}\overset{-}{N}-)$ | 72[a] | 26, 27 |
|  | p | (C$_6$H$_5$)$_2$CH | C$_6$H$_5$ | 78–82 | 2255 | $(-\overset{+}{C}{\equiv}\overset{-}{N}-)$ | 74[a] | 2 |
| VII | a | C$_6$H$_5$ | C$_2$H$_5$ | 144 | 1560, 1533 | $\overset{+}{\underset{-}{C}{=}\overset{-}{N}}$ | 39[a] | 22, 24 |
|  | b | C$_6$H$_5$ | $n$-C$_4$H$_9$ | 125 | 1560, 1533 | $\overset{+}{\underset{-}{C}{=}\overset{-}{N}}$ | 58[a], 71[a] | 22, 24 |
|  | c | $t$-C$_4$H$_9$ | CH$_3$ | 119–120.5 | 1548 | $\overset{+}{\underset{-}{C}{=}\overset{-}{N}}$ | — | 14 |
|  | d | $c$-C$_6$H$_{11}$ | C$_6$H$_5$ | 339–341 | 1608 | $\overset{+}{\underset{-}{C}{=}\overset{-}{N}}$ | 22[b] | 26, 27 |
|  | h | C$_6$H$_5$ | C$_6$H$_5$ | 285–288 | 1548 | $\overset{+}{\underset{-}{C}{=}\overset{-}{N}}$ | 72[b] | 26, 27 |
|  | i | $p$-(C$_2$H$_5$)$_2$NC$_6$H$_4$ | C$_6$H$_5$ | 275–280 | 1552 | $\overset{+}{\underset{-}{C}{=}\overset{-}{N}}$ | 89[b] | 26, 27 |

[*Table continued overleaf*]

TABLE I (Continued)

| Compound type | Designation | Substituents R = | R' = | mp (°C) | Infrared (Cm$^{-1}$) | | Yield (%) | Reference |
|---|---|---|---|---|---|---|---|---|
| VII | j | $C_6H_5$ | $CH_3$ | 199 | 1399 | $\diagup B-N \diagdown$ | 16$^a$ | 24 |
| | k | $c$-$C_6H_{11}$ | $n$-$C_4H_9$ | 129–131 | — | — | 3$^a$ | 24 |
| | m | $C_6H_5$ | H | dec > 135 | 1589, 2360 | — | — | 4 |
| | o | $C_2H_5$ | $C_2H_5$ | 231–232$^e$ | 1597 | $\diagup C=\overset{+}{N} \diagdown$ | — | 3 |
| | q | $p$-$O_2NC_6H_4$ | $C_2H_5$ | — | — | — | 48$^f$ | 12 |
| | r | $p$-$H_2NC_6H_4$ | $C_2H_5$ | — | — | — | — | 12 |
| | s | $p$-$CH_3CONHC_6H_4$ | $C_2H_5$ | — | — | — | — | 12 |
| VIII | a | $C_6H_5$ | $C_2H_5$ | 204 | 1400 | $\diagup B-N \diagdown$ | ~90$^b$ | 22 |
| | b | $C_6H_5$ | $n$-$C_4H_9$ | 186–189 | 1400 | $\diagup B-N \diagdown$ | ~90$^b$ | 22 |
| | j | $C_6H_5$ | $CH_3$ | 171 | 1410 | $\diagup B-N \diagdown$ | ~50$^c$ | 24 |
| | k | $c$-$C_6H_{11}$ | $n$-$C_4H_9$ | 144 | 1471 | $\diagup B-N \diagdown$ | ~90$^c$ | 24 |

6. Reaction of Isonitriles with Boranes   113

|   |   |   |   |   |   |   |   |
|---|---|---|---|---|---|---|---|
|   | l | p-ClC$_6$H$_4$ | C$_2$H$_5$ | 201–202 | 1397 | $\diagup$B—N$\diagdown$ | 85[d] | 12 |
|   | m | C$_6$H$_5$ | H | dec 80–120 | — | — | — | 64 |
|   | n | C$_2$H$_5$,H | C$_2$H$_5$ | 65 | — | — | 5[a] | 25 |
|   | q | p-O$_2$NC$_6$H$_4$ | C$_2$H$_5$ | — | 1397 | $\diagup$B—N$\diagdown$ | — | 12 |
|   | r | p-H$_2$NC$_6$H$_4$ | C$_2$H$_5$ | 226–228 | 1397 | $\diagup$B—N$\diagdown$ | — | 12 |
|   | s | p-CH$_3$CONH—C$_6$H$_4$ | C$_2$H$_5$ | > 350 | 1397 | $\diagup$B—N$\diagdown$ | — | 12 |
| IX | a | C$_6$H$_5$ | C$_2$H$_5$ | 141–142 | 1385 | $\diagup$B—N$\diagdown$ | 96[d] | 13 |
|   | b | C$_6$H$_5$ | n-C$_4$H$_9$ | 132–133 | 1403 | $\diagup$B—N$\diagdown$ | 50[d] | 69 |
|   | l | p-ClC$_6$H$_4$ | C$_2$H$_5$ | 158–160 | 1389 | $\diagup$B—N$\diagdown$ | 97[d] | 12 |
|   | q | p-O$_2$NC$_6$H$_4$ | C$_2$H$_5$ | 225–228 | — | — | — | 12 |
|   | r | p-H$_2$NC$_6$H$_4$ | C$_2$H$_5$ | — | — | — | — | 12 |

[a] From III and IV.
[b] From V.
[c] From VII.
[d] From VIII.
[e] Reported by Casanova et al.[13]
[f] From nitration of VIIa.
[g] From nitration of VIIIa.

electron density at N and the infrared spectrum (Table I) suggested —C≡N⁺— bonding (2247 cm⁻¹). Compound Vc was unstable in air or water, but in the absence of air or moisture Vc was slowly converted to isomer VIIc in high yield. In contrast to the simple adduct Vc, VIIc was insensitive to air or moisture. Based upon the absence of fragments of $m/e = 138$ in the mass spectrum of VIIc, the monomeric structure VIc was tentatively (but erroneously) assigned at that time.

Hesse and co-workers[26] found that simple adducts of structure V, when prepared from triphenylborane (IV, R' = $C_6H_5$) displayed a distinctive stability, both thermally and toward adventitious oxygen or moisture. Compounds Vd–Vi were prepared in good yields at low temperature in tetrahydrofuran solution. Molecular weight determination in solution showed these compounds clearly to be monomers. By heating compounds Vd, Vh, or Vi for a short time at 80–150°C, rearrangement and dimerization occurred to give VIId, VIIh, or VIIi, respectively, in high yields. Disappearance of the —C≡N⁺— infrared stretch and appearance of a

$$\left( \begin{array}{c} \diagdown \\ \diagup \end{array} C = \overset{\oplus}{N} \begin{array}{c} \diagup \\ \diagdown \end{array} \right)$$

band at 1550–1600 cm⁻¹, as well as analysis and molecular weight measurements, served to substantiate the assignment of structure.

Bittner, Witte, and Hesse[2] have devised a novel synthetic application based on adduct V (2). When benzyhydryl isocyanide (X) is reacted with triphenylborane, the 1:1 adduct is formed (Vp). This adduct, upon treatment with phenyllithium, is converted to ylid (XI). Ylid (XI) is capable of behaving as a 1,3-dipole in addition reactions and adds to aldehydes and ketones, producing oxazolinium betaines (XII) in excellent yields (3), after hydrolysis of the initial product. More vigorous oxidation (3) leads to β-amino alcohols (XIII). If instead of benzyhydryl isocyanide, isopropyl or cyclohexyl isocyanide were employed, the initial adduct (Ve or Vd, respectively) when treated with phenyllithium, produced imidazole derivative (XIV), orange-red in color. Upon brief heating (4) XIVb is isomerized to XVb, and is also converted to cyclohexylidene diphenylmethane in good yield by long refluxing with sulfuric acid.

$(C_6H_5)_2CHN≡C + B(C_6H_5)_3 \longrightarrow$
(X)

$(C_6H_5)_2CHN\overset{\oplus}{≡}\overset{\ominus}{C}B(C_6H_5)_3 \longrightarrow Li(C_6H_5)_2\overset{\oplus}{C}N\overset{\ominus\oplus}{≡}\overset{\ominus}{C}B(C_6H_5)_3$ (2)
(Vp)                                        (XI)

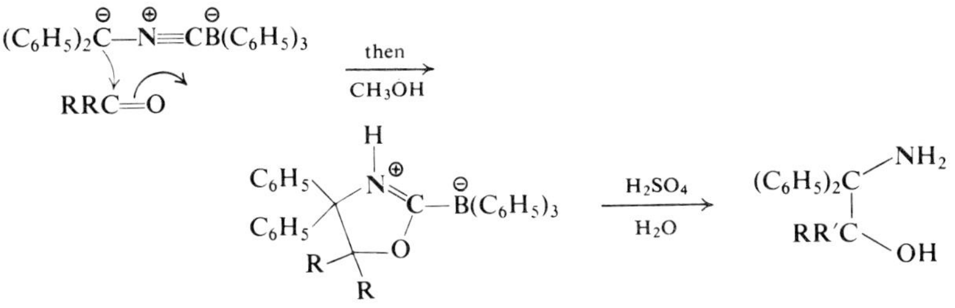

## 6. Reaction of Isonitriles with Boranes  115

$$(C_6H_5)_2\overset{\ominus}{C}-\overset{\oplus}{N}\equiv\overset{\ominus}{C}B(C_6H_5)_3$$
$$RR\overset{\frown}{C}=O$$

$\xrightarrow{\text{then}}$
$\xrightarrow{CH_3OH}$

(XII) → $\xrightarrow[H_2O]{H_2SO_4}$ → (XIII)

(3)

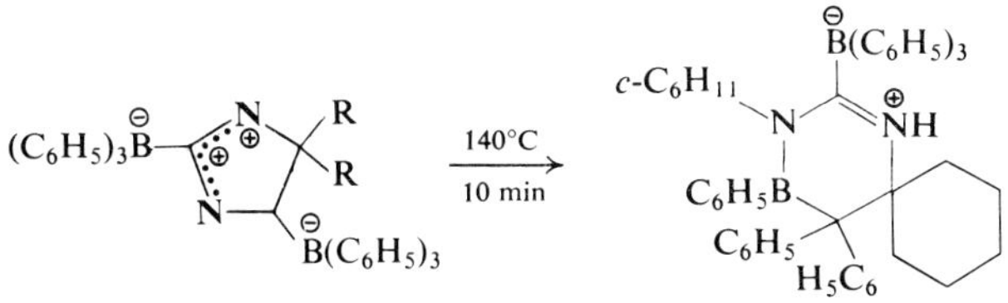

(XIV) $\xrightarrow[10 \text{ min}]{140°C}$ (XVb)

(4)

a: R = CH$_3$—
b: R = —(CH$_2$)$_5$—

A report by Casanova et al.[13] appeared describing the thermal conversion of 2,5-dibora-3,6-dihydropyrazines (VIII) to a new isomeric structure. When VIIIa was heated at 310°C in vacuum for 5 min, a new compound isomeric with VIIIa was formed in nearly quantitative yield. This compound has been shown to possess the 1,3-diaza-2,4-diborolidine structure (IXa), although it was originally thought to be a 1,3-diaza-2,4-diboretidine (XVIa). The N$_2$B$_2$ ring structure (XVI) was assigned originally on the basis of analysis, molecular

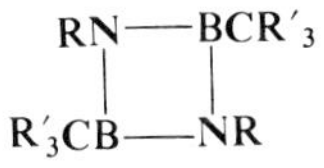

$$\begin{array}{c} RN-BCR'_3 \\ | \quad\quad | \\ R'_3CB-NR \end{array}$$
(XVIa)

weight, mass spectrum, $^1$H and $^{11}$B NMR, IR spectrum, and chemical degradation. Subsequent X-ray crystallographic structure determinations by Tsai and Streib[65] and 32.1 Mc $^{11}$B NMR served to confirm the structure as a five-membered ring (IX) using the closely related p-chlorophenyl series (VIIIl→IXl). One of the interesting results of the work of Tsai and Streib[65] was the observation that the three different B—N bonds were widely disparate in their lengths. Although the N(1)B(2) length (1.430 Å) was similar to the N(3)B(4) length (1.411 Å), both were very much shorter than the B(2)N(3) distance (1.505 Å). This result suggests analogy to the C=C bonds distances of butadiene. Diazadiborolidine (IXb) could also be formed by the action of aluminum chloride on VIIIb in solution at room temperature.[69] Samples prepared by both methods were identical.

Only a few reports have appeared in the literature regarding the synthesis of compounds containing the $N_2B_2$, or 1,3-diaza-2,4-diboretidine ring system. Paetzold[55] obtained a yellow oil from the thermal decomposition of diphenylborazide (XVIII), for which he proposed XIX as a possible structure. Compound XVIII had been prepared from diphenylchloroborane (XVII) and lithium azide (5). Lappert and Majumdar[45,46] found that di($t$-butylamino)-

$$(C_6H_5)_2BCl + LiN_3 \longrightarrow (C_6H_5)_2BN_3 \longrightarrow \begin{bmatrix} C_6H_5B\!\!-\!\!-\!\!NC_6H_5 \\ | \quad\quad | \\ C_6H_5N\!\!-\!\!-\!\!BC_6H_5 \end{bmatrix} \quad (5)$$

(XVII) \quad\quad\quad\quad\quad (XVIII) \quad\quad\quad\quad (XIX)

boron chloride (XX) reacted with $t$-butylamine at 260°C to give 1,3-di-$t$-butyl-2,4-di-$t$-butylamino-1,3-diaza-2,4-diboretidine (XXII) in low yield. The product was characterized by spectroscopic methods, mass spectroscopy being the most important method. These workers also found that the same substance (XXII) could be obtained (6) by heating $t$-butylamine and boron trichloride directly. Bubnov[10] has reported recently that two series of $N_2B_2$ ring compounds (XXIV, XXV) are formed from the reaction of trivinylborane

$$(t\text{-}C_4H_9NH)_2BCl + t\text{-}C_4H_9NH_2 \longrightarrow t\text{-}C_4H_9\!\!-\!\!N[B(NH\!\!-\!\!t\text{-}C_4H_9)_2]_2 \longrightarrow$$
(XX) \quad\quad\quad\quad\quad\quad\quad\quad\quad\quad (XXI)

$$\begin{array}{c} t\text{-}C_4H_9\!\!-\!\!NHB\!\!-\!\!-\!\!N\!\!-\!\!t\text{-}C_4H_9 \\ | \quad\quad\quad | \\ t\text{-}C_4H_9N\!\!-\!\!-\!\!BNH\!\!-\!\!t\text{-}C_4H_9 \end{array} \quad (6)$$

(XXII)

(XXIII) with alkyl nitriles (7). Extended Hückel calculations by Hoffmann[34] pointed out that the two degenerate $\pi$-molecular orbitals of cyclobutadiene, which is the all-carbon analog of the $N_2B_2$ ring, are strongly and symmetrically split in $N_2B_2H_4$. This is due to the fact that the Coulomb integral for nitrogen and boron is substantially different. Hoffmann[34] further noted that the $N_2B_2$ ring system should be square planar, and that the ring system of Lappert and Majumdar[45] was perhaps not a test of his calculation, due to the lone-pair electrons of the nitrogen substituents adjacent to the ring.

6. *Reaction of Isonitriles with Boranes*    117

$$(CH_2{=}CHCH_2)_3B + CH_3CN \longrightarrow \underset{(XXIII)}{} \quad \underset{(XXIV)}{\begin{array}{c} CH_3 \\ | \\ (CH_2{=}CHCH_2)_2B{-\!\!\!-}N{\diagdown}CCH_2CH{=}CH_2 \\ \phantom{xxxxxxxxxxxxx}\ominus \; \oplus \; | \\ \phantom{xxxxxxxxxxxx}N{\underset{\ominus}{\overset{\oplus}{=\!\!=}}}B(CH_2CH{=}CH_2)_2 \\ CH_2{=}CHCH_2C{\diagup} \\ | \\ CH_3 \end{array}} \longrightarrow$$

$$\underset{(XXV)}{\begin{array}{c} CH_3 \\ | \\ CH_2{=}CHCH_2B{-\!\!\!-}NC(CH_2CH{=}CH_2)_2 \\ | \phantom{xxxxxxxxxxx} | \\ (CH_2{=}CHCH_2)_2CN{-\!\!\!-}BCH_2CH{=}CH_2 \\ | \\ CH_3 \end{array}} \quad (7)$$

**B. Adducts from Hydrogen Cyanide or Diborane As One Component**

Two reports have appeared describing the isolation of heterocycles VIIm and VIIIm, which are substituted with hydrogen at boron. Bresadola and co-workers[4] have found by slowly mixing diborane and phenyl isocyanide in dilute solution in petroleum ether at $-68°C$ that an air-sensitive white crystalline solid was formed. The molecular weight, elemental analysis, chemical properties, and spectral behavior were consistent with structure VIIm. The absence of any C—H resonance signal between 1.5 and 4.5 ppm further supported the assignment of structure VIIm. The same workers were unable to duplicate the work of Tanaka and Carter[64] who had claimed the preparation of VIIIm from phenyl isocyanide and diborane in ether or benzene solution. The solid obtained by Tanaka and Carter appeared inhomogeneous from the reported melting range (80–120°C, dec.) and from a "relatively high degree of uncertainty in the molecular weight." Others[11] have also described difficulty using diborane as the Lewis acid. Diborane appears to be very much more reactive toward isonitriles than are the alkylboranes.

Hydrogen cyanide has been used as the CN source in reaction with triethylborane (8). Hesse and co-workers[28] reported that at room temperature this reaction gave VIIn in low yield. Compounds Vn and VIIn, presumably formed first in the reaction, were not isolated. Along with VIIIn the authors isolated 10% of betaine (XXVI). This five-membered ring heterocycle is undoubtedly stabilized by resonance of the type shown in XXVIa, b, c. Compound XXVI may be viewed as the HCN adduct of XXVII. Compound

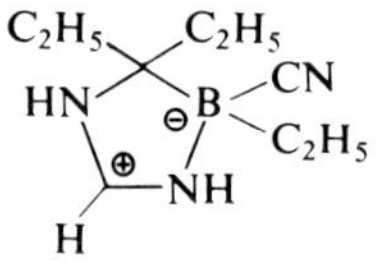

(XXVIa) 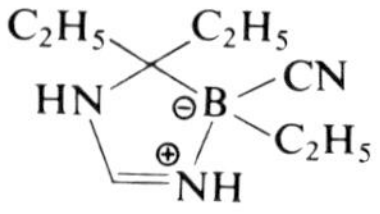    (XXVIb) 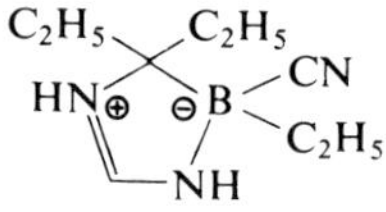    (XXVIc)

XXVII would result from the addition of HCN to VIn.

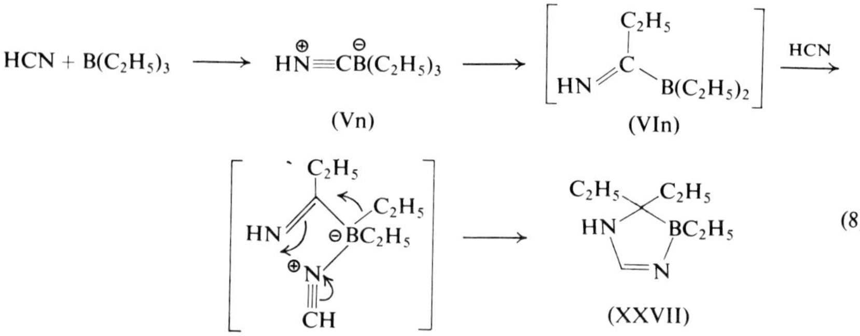

$$HCN + B(C_2H_5)_3 \longrightarrow H\overset{\oplus}{N}{\equiv}\overset{\ominus}{C}B(C_2H_5)_3 \longrightarrow \left[\begin{array}{c}C_2H_5\\ HN{=}C{-}B(C_2H_5)_2\end{array}\right] \xrightarrow{HCN}$$

(Vn)    (VIn)

$$\longrightarrow \left[\text{intermediate}\right] \longrightarrow \text{(XXVII)} \qquad (8)$$

Although direct evidence for the existence of monomer VI has not been obtained, the result with HCN suggests its intermediacy (cf. VIn). Another experiment by Hesse and co-workers[28] offered further evidence for the existence of VI. These authors reacted phenyl isocyanide with tri-$n$-butylborane in the presence of $N$-phenylbenzaldimine (XXVIII) (9). From this reaction a good yield of XXIX was obtained. This product can be viewed as arising

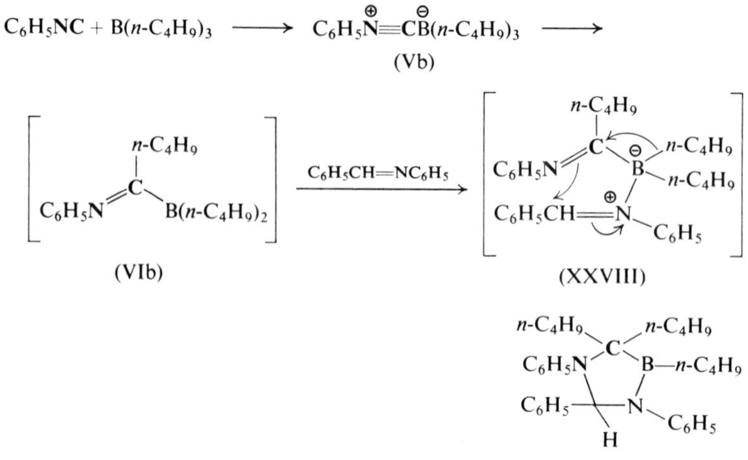

$$C_6H_5NC + B(n\text{-}C_4H_9)_3 \longrightarrow C_6H_5\overset{\oplus}{N}{\equiv}\overset{\ominus}{C}B(n\text{-}C_4H_9)_3 \longrightarrow$$

(Vb)

$$\left[\begin{array}{c}n\text{-}C_4H_9\\C_6H_5N{=}C{-}B(n\text{-}C_4H_9)_2\end{array}\right] \xrightarrow{C_6H_5CH=NC_6H_5} \left[\text{(XXVIII)}\right] \longrightarrow$$

(VIb)

$$\text{(XXIX)} \qquad (9)$$

from 1,3 dipolar addition of VIb and XXVIII. The same authors[27] had earlier reported a similar reaction between phenyl isonitrile and tri-$n$-butylborane in the presence of aryl aldehydes as the trapping agent (10). Again the product XXX, which was obtained in good yield, appeared to arise from 1,3 dipolar addition of the carbonyl group to VI.

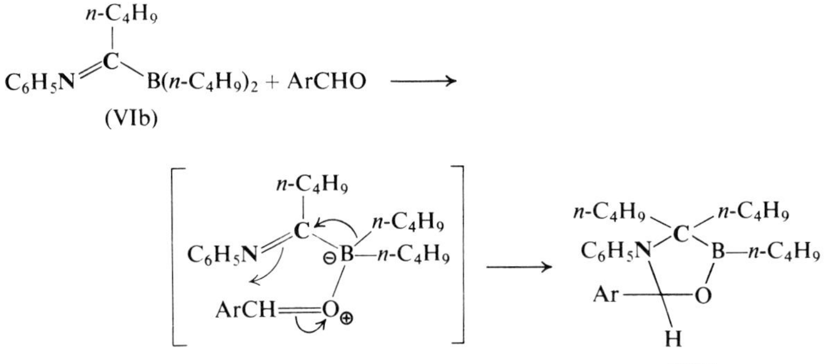

$$C_6H_5N\!\!=\!\!\overset{n\text{-}C_4H_9}{\underset{}{C}}\!\!-\!\!B(n\text{-}C_4H_9)_2 + ArCHO \longrightarrow$$
(VIb)

$$\left[\begin{array}{c} n\text{-}C_4H_9 \\ C_6H_5N\!\!-\!\!\overset{}{\underset{}{C}}\!\!\cdots\!\!\overset{n\text{-}C_4H_9}{\underset{}{B}}\!\!-\!\!n\text{-}C_4H_9 \\ ArCH\!\!=\!\!\overset{\oplus}{O} \end{array}\right] \longrightarrow \begin{array}{c} n\text{-}C_4H_9 \diagdown \overset{}{\underset{}{C}} \diagup n\text{-}C_4H_9 \\ C_6H_5N \diagup \phantom{C} \diagdown B\!\!-\!\!n\text{-}C_4H_9 \\ Ar\!\!-\!\!\underset{H}{\overset{|}{C}}\!\!-\!\!O \end{array} \quad (10)$$
(XX)

## C. Monomer–Dimer Equilibrium

Little direct evidence exists regarding the propensity of a

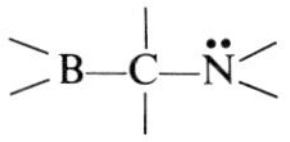

$$>\!\!B\!\!-\!\!\underset{|}{\overset{|}{C}}\!\!-\!\!\ddot{N}\!\!<$$

system for dimerization. The VI→VII transformation suggests that the propensity is high for the

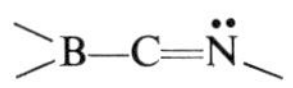

$$>\!\!B\!\!-\!\!C\!\!=\!\!\ddot{N}\!\!\sim$$

arrangement. Schaeffer and Todd[62] have provided good evidence that the equilibrium between monomer and dimer in the saturated case is both facile and temperature-dependent. Aminomethyldimethylborane (XXXI) and its dimer (XXXII) interconverted rapidly, giving an equilibrium value of 24.3 % at 35.25°C and 84.3 % dimer at −25.30°C, by $^{11}$B NMR chemical shift. Miller

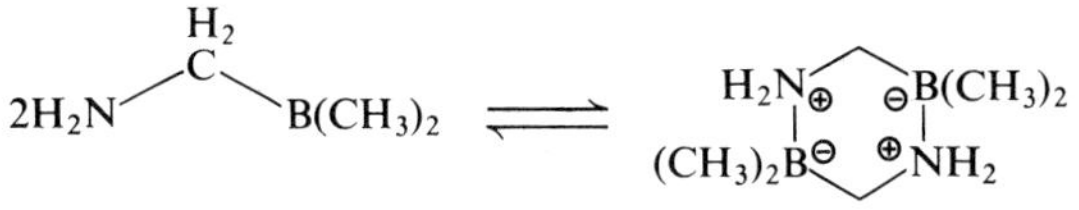

$$2H_2N\!\!-\!\!\overset{H_2}{\underset{}{C}}\!\!-\!\!B(CH_3)_2 \rightleftharpoons \begin{array}{c} H_2\overset{\oplus}{N}\!\!\frown\!\!\overset{\ominus}{B}(CH_3)_2 \\ | \phantom{aaaaaa} | \\ (CH_3)_2\overset{\ominus}{B}\!\!\smile\!\!\overset{\oplus}{N}H_2 \end{array}$$
(XXXI) (XXXII)

and Muetterties[51] prepared XXXIV in low yield by reduction of XXXIII with sodium hydride (11). Compound XXXIV has the same saturated 2,5-diborapiperidine ring as the compound of Schaeffer and Todd (XXXII). However, XXXIV was a stable dimer even up to 100°C. More remotely relevant to the monomer–dimer question are some recent observations of

$$H_2B[N(CH_3)_3]_2Cl^{\ominus}\phantom{}^{\ominus}{}^{\oplus} + NaH \longrightarrow \begin{array}{c}(CH_3)_2N^{\oplus}\phantom{}\phantom{}\phantom{}BH_2 \\ | \phantom{xxx} | \\ H_2B^{\ominus} \phantom{xx} {}^{\oplus}N(CH_3)_2\end{array} \quad (11)$$

(XXXIII)                (XXXIV)

Dorokhov and Lappert[17] and Jennings and co-workers.[41] The former authors reported that di-$n$-butyl($t$-butylaldimino)-borane (XXXV) is as monomeric in the vapor phase at > ~ 70°C. This observation is corroborated by the latter authors who noted that diphenylketiminodiphenyl-borane (XXXVI) was monomeric both in solution and in the mass spectrometer. Pattison and Wade[58] noted further that compound XXXVII, derived from heating the initial product of the reaction of benzophenoneimine with trimethylborane at 160–200°C, showed no fragments larger than monomers in the mass spectrometer.

$$t\text{-}C_4H_9CH{=}N{-}B(n\text{-}C_4H_9)_2 \qquad (C_6H_5)_2C{=}N{-}B(C_6H_5)_2$$
(XXXV)                                        (XXXVI)

$$(C_6H_5)_2C{=}N{-}B(CH_3)_2$$
(XXXVII)

## D. Mechanisms

Noting the structural relationship between isomers V, VII, VIII, and IX, several clear patterns emerge in the thermal rearrangement sequence that connects these compounds. The thermodynamic stability is in the order V < VII < VIII < IX, that is, in the order to be expected in a sequence which involves breaking of B—C bonds and formation of B—N bonds. Hence the sequence is characterized by a series of B to C alkyl (or aryl) migrations. Moreover, each rearrangement occurs from a structure in which the seat of migration, a boron atom, is tetracoordinated and negatively charged, and the terminus of migration, a carbon atom, which may be written in at least one resonance form which bears a positive charge. The pattern persists in mechanisms for the formation of XXXVI, XXIX, and XXX and is also apparent in the reaction between trialkylboranes and other Lewis bases to be discussed later. Such a migration sequence is analogous to the Kuivila[44] mechanism for oxidation of triorganoboranes by alkaline hydrogen peroxide (12). The key step in that mechanism is B to O alkyl (or aryl) migration during the

### 6. Reaction of Isonitriles with Boranes 121

$$HO\overset{\frown}{\underset{(XXXVIII)}{-O-\underset{R}{\overset{\ominus}{B}}\overset{R}{\underset{R}{\diagdown}}}} \quad HO^{\ominus} + ROBR_2 \quad (XXXIX) \quad (12)$$

decomposition of XXXVIII to give dialkylboronate (XXX). The application of this principle to the rearrangement pattern in compounds V through IX leads to resonance which are shown below. A clear pattern for the Wagner–Meerwein type of shift involving B to C alkyl group migration is apparent.

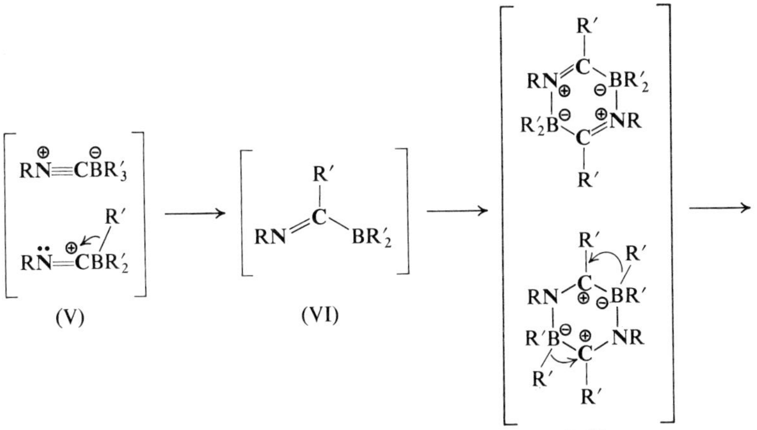

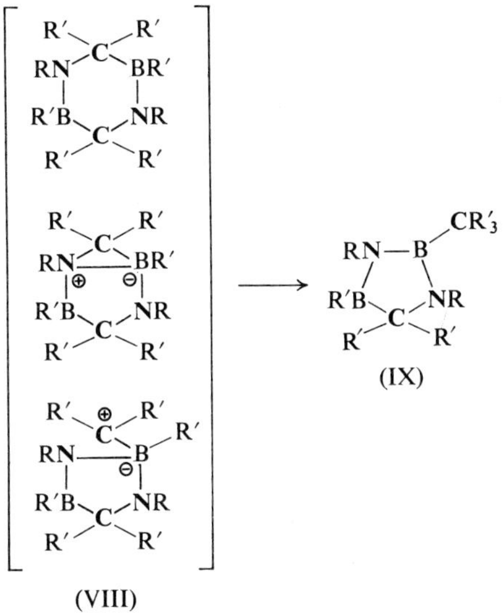

Empirically, other observations emerge from an inspection of the data in Table I. The presence of aromatic substituents on boron (R' = Ar) significantly stabilizes compound V toward further rearrangement. If no substituents are present on boron (R' = H) the reactivity of V is very high. With the exception VIIIk (R = $c$-$C_6H_{11}$) and VIIIn (R = —$C_2H_5$, H), an aryl group is the best substituent at nitrogen for isomerization of VII→VIII. Moreover, no examples of IX are known with R = alkyl. A startling fact emerges when it is noted that VIIIj (R' = $CH_3$) cannot be isomerized to IXj under the same conditions which cause VIIIa (R' = —$C_2H_5$) and VIIIb (R' = —$n$-$C_4H_9$) to undergo isomerization in yields of greater than 90%. An effort to determine the factors which govern the conversion of VIII to IX was carried out.[12] An interesting possibility was presented by the fact that the B—N tricoordinate bond possesses only a 10–15 kcal/mole barrier to internal rotation.[54] Bulky substituents on carbon in compound VIII could lead to a chair rather than a boat conformation for VIII, due to 1,4 diaxial steric repulsion. The chair conformation would be anticipated for best $\pi$-electronic overlap of the B—N bonds. This point is illustrated below. In the chair conformation the $\pi$-interorbital angle between $\sigma$-bonded B—N atoms is >0°, and so $\pi$-bonding is decreased. Transannular 1,3-B—N $\pi$-bonding between nonbonded boron and nitrogen atoms is more favorable under such circumstances, leading to a con-

boat          chair

(VIII)

tribution from structures of VIII such as those thought to be important for rearrangement (*vide infra*). Such an effect on conformation should be apparent through a significant decrease in the B—N IR stretching of VIII–chair compared to VIII–boat. This difference is not observed (cf. VIIIa, VIIIb, and VIIIj, Table I). The only common feature of compounds possessing structure VIII which do undergo isomerization to IX and absent in those cases which fail to isomerize is that the isomerization temperature (305°C) is above the melting point of the former. The physical properties of VII–IX (q,r,s) support this. The several examples of compounds (IX) which have been prepared are highly resistant to chemical attack. Compound IXa is not appreciably affected by either

## 6. Reaction of Isonitriles with Boranes

boiling concentrated hydrochloric acid or boiling alcoholic potassium hydroxide, even after several hours. It is also resistant to oxidation by chromic anhydride. This observation is in marked contrast to the behavior of simple aminoboranes, but it is probably not due to any intrinsic stability of the ring system, but rather to molecular overcrowding, which hinders the approach of an attacking reagent. Such a resistance to chemical attack has been reported for highly hindered borazines by Nagasawa.[53]

A possible clue suggesting the intervention of a different reaction path is afforded by the report of Bresadola and co-workers[3] who examined the products of the reaction of ethyl isocyanide and triethylborane (13). These authors presented evidence that when the reaction was carried out in ether at room temperature two new isomeric compounds resulted. The relative amounts of each compound were not described. One compound was the anticipated 2,5-dibora-2,5-dihydropyrazine (VIIo). The other possessed two ethylidene groups on the basis of its $^1$H NMR, and was assigned structure XL. One plausible explanation of this behavior is suggested by the ketimine–enamine

$$C_2H_5NC + B(C_2H_5)_3 \xrightarrow[\text{ether}]{-20°C} \text{(VIIo)} + \text{(XL)} \quad (13)$$

tautomerism of VIo. Most series which proceed into compound VII possess an $N$-aryl group. The presence of an $N$-aryl substituent on VI (VI, R= Ar) would strongly favor ketimine (LX) over enamine (XLI) in the equilibrium, due to conjugation of the double bond with the aromatic ring. However,

$$\text{(LX)} \rightleftharpoons \text{(XLI)} \longrightarrow \text{(XL)} \quad (14)$$

the work of Schaeffer and Todd[62] suggests that the XLI→XL conversion should be thermally reversible (14). If this were the case, heating XL should produce VIIo. This behavior was not observed by Casanova et al.[13] when XL was heated alone or in the presence of Lewis acids. The question of at what point the normal reaction path is interdicted to give compound XL remains open. Hesse and Witte[24] did not observe any ethylidenyl product during the formation of VIIk. This may be an artifact of the experimental procedure, inasmuch as a yield of only 3% is recorded for its preparation.

Only a few excursions into the reaction of isonitriles with polyhedral boranes have been reported. Graybill and Hawthorne[20] briefly alluded to unpublished work on the reaction of isonitriles with decaborane-14. Hyatt and co-workers[38] later described a detailed study of this reaction, using several different isonitriles. A simple 1:2 adduct (XLIII), akin to structure V, was described as the product of the ligand displacement reaction of ethyl isocyanide on bis(diethyl sulfide)decarborane (XLII) (14). However, direct reaction between the isonitrile and decarborane-14 (XLIV) yielded a rather unusual 1:1 adduct, formulated as XLV by these workers. Hertler[21] reported that most alkyl, cycloalkyl, and aralkyl isocyanides reacted with decaborane-14 to yield intermediates such as XLV, which could be alkylated readily at nitrogen (16).

$$2C_2H_5NC + B_{10}H_{12}[S(C_2H_5)_2]_2 \rightarrow B_{10}H_{12}[CN-C_2H_5]_2 + 2S(C_2H_5)_2 \qquad (15)$$
$$\text{(XLII)} \hspace{4cm} \text{(XLIII)}$$

$$C_2H_5NC + B_{10}H_{14} \rightarrow B_{10}\overset{\ominus}{H}_{12}\overset{\oplus}{C}NH_2-C_2H_5 \qquad (16)$$
$$\text{(XLIV)} \hspace{3cm} \text{(XLV)}$$

## II. ISONITRILE–ORGANOALUMINUM AND CYANIDE–ORGANOALUMINUM REACTIONS

Organoaluminum compounds also have been used as the Lewis acid component, both with isonitriles[23] and with cyanides.[40, 57, 61] Hesse and Witte[23] reported that triphenylaluminum reacted with phenyl isocyanide to form a 1:1 adduct (17) (XLVI), quite analogous to compound V in the isonitrile–borane series.

$$C_6H_5NC + Al(C_6H_5)_3 \rightarrow C_6H_5\overset{\oplus}{N}\equiv\overset{\ominus}{C}Al(C_6H_5)_3 \qquad (17)$$
$$\text{(XLVI)}$$

Pasynkiewicz and co-workers[57] confirmed an earlier report of Reinheckel and Jahnke[61] showing that methyl ketones (XLIX) could be prepared in good yield from hydrolysis of the 1:1 adduct (XLVIII) of nitriles and hexamethyldialuminum (18). Presumably the initial adduct, XLVII, like V, rearranges to XLVIII before hydrolysis. Compound XLVIII is analogous to VI. This synthetic method provides a novel sequence of transformations, the potential of which has not been fully explored. In a similar series of reactions other workers[40] reported similar behavior for t-butyl cyanide and hexamethyl-

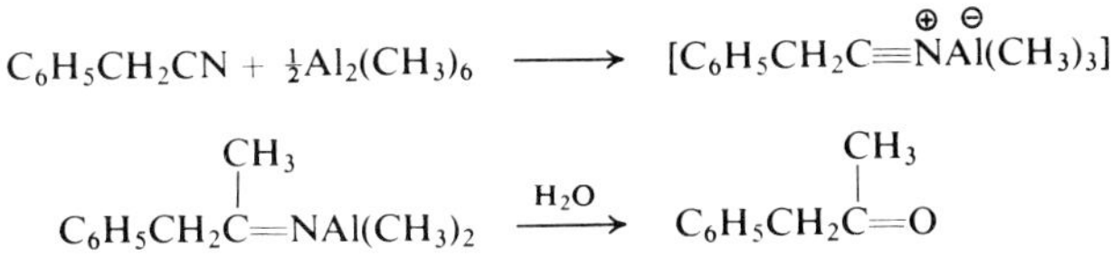

6. *Reaction of Isonitriles with Boranes*    125

$$C_6H_5CH_2CN + \tfrac{1}{2}Al_2(CH_3)_6 \longrightarrow [C_6H_5CH_2C\overset{\oplus}{\equiv}\overset{\ominus}{N}Al(CH_3)_3]$$

$$\underset{(XLVIII)}{C_6H_5CH_2\overset{\overset{CH_3}{|}}{C}=NAl(CH_3)_2} \overset{H_2O}{\longrightarrow} \underset{(XLIX)}{C_6H_5CH_2\overset{\overset{CH_3}{|}}{C}=O} \qquad (18)$$

dialuminum (19). In this reaction the initial product (LI), kept scrupulously anhydrous, went on to dimerize, giving a novel four-membered ring (LII) as the isolable reaction product.

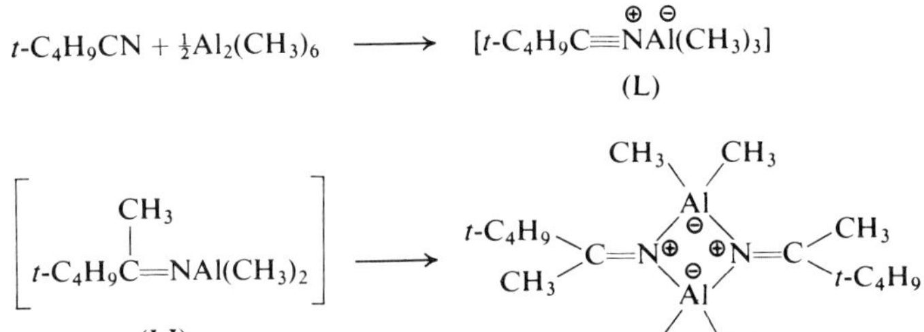

$$t\text{-}C_4H_9CN + \tfrac{1}{2}Al_2(CH_3)_6 \longrightarrow \underset{(L)}{[t\text{-}C_4H_9C\overset{\oplus}{\equiv}\overset{\ominus}{N}Al(CH_3)_3]}$$

$$\underset{(LI)}{\left[ t\text{-}C_4H_9\overset{\overset{CH_3}{|}}{C}=NAl(CH_3)_2 \right]} \longrightarrow \underset{(LII)}{\text{(four-membered Al-N ring structure)}} \qquad (19)$$

### III. CYANIDE–ORGANOBORANE REACTIONS

A variety of other less closely related reactions between nitriles and boranes have been recorded. Lloyd and Wade[47] reported a 1,3-diaza-2,4-diboretidine ring product (LIII), analogous to LII which they obtained with hexamethyldialuminum, from the reaction of acetonitrile and trimethyl- or triethylborane (20). Both the *cis* and *trans* isomers were identified.

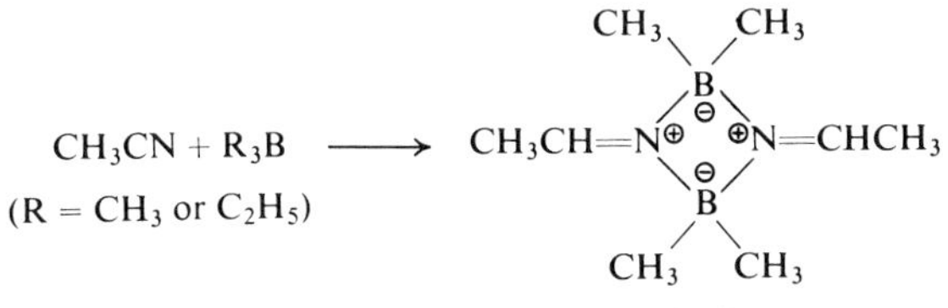

$$\underset{(R\,=\,CH_3\text{ or }C_2H_5)}{CH_3CN + R_3B} \longrightarrow \underset{(LIII)}{CH_3CH=\overset{\oplus}{N}\;\;\overset{\oplus}{N}=CHCH_3} \qquad (20)$$

These workers excluded a six-membered ring as the structure of their product, by infrared and molecular weight measurements. The structure of the *trans* isomer has been confirmed by X-ray crystallographic analysis.[15] Surprisingly, such a six- rather than a four-membered ring structure has been reported for a closely related reaction. Thermal isomerization of the initial adduct of propionitrile and tri-*n*-propylborane is reported[37] to lead to a stable 2,5-diboropyrazine system (LIV) (21).

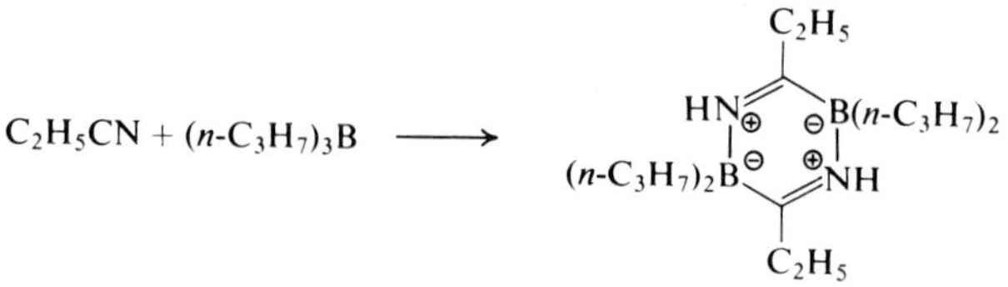

$$C_2H_5CN + (n\text{-}C_3H_7)_3B \longrightarrow \underset{(LIV)}{\overset{\overset{C_2H_5}{|}}{\underset{\underset{C_2H_5}{|}}{(n\text{-}C_3H_7)_2\overset{\ominus}{B}\overset{HN^{\oplus}}{\underset{\oplus NH}{\rightleftarrows}}\overset{\ominus B(n\text{-}C_3H_7)_2}{|}}}} \quad (21)$$

## IV. RELATED REACTIONS: THE REACTION OF CARBON MONOXIDE AND OTHER LEWIS BASES WITH ORGANOBORANES

The reaction of trialkylboranes with other Lewis bases have been studied, and these follow the pattern observed in the isonitrile–borane series. In particular, considerable effort has been devoted to the reaction of carbon monoxide with alkylboranes. Hillman[29,30] reported the formation (22) of a large series of diboradioxanes (LV) in excellent yields from carbon monoxide and alkylboranes in water (22). These compounds, which are formally analogous to VIII, would appear to arise from a series of B to C alkyl migrations starting with an adduct similar to V. In later work Hillman demonstrated the utility of this synthesis for the preparation of glycol esters[31] and 2-bora-1,3-dioxolanes.[32,33] Other workers have recently examined the reaction of alkyl-

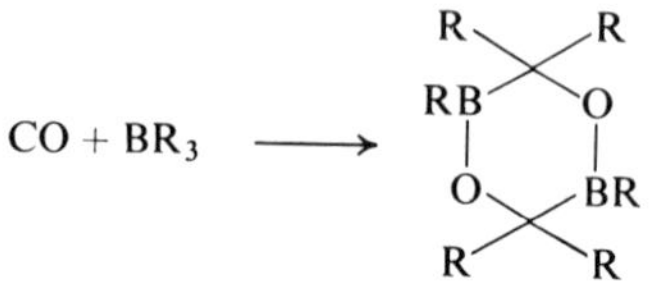

$$CO + BR_3 \longrightarrow \underset{(LV)}{\text{(structure)}} \quad (22)$$

boranes[50] and diborane[48,49] with carbon monoxide in water or alcohols. A novel application of the carbonylation of alkylboranes has been noted recently[59] and promises to be of special interest to the synthetic organic chemist. Trialkylboranes with carbon monoxide in the presence of sodium borohydride give high yields of alcohols,[60] probably via reduction of LVII, which is formally analogous to VI (23). If the initial reaction is carried out in water in the presence of alkaline hydrogen peroxide, ketones are formed in excellent yields.[8] These compounds could arise from the hydrolysis of LVII. When the borane and

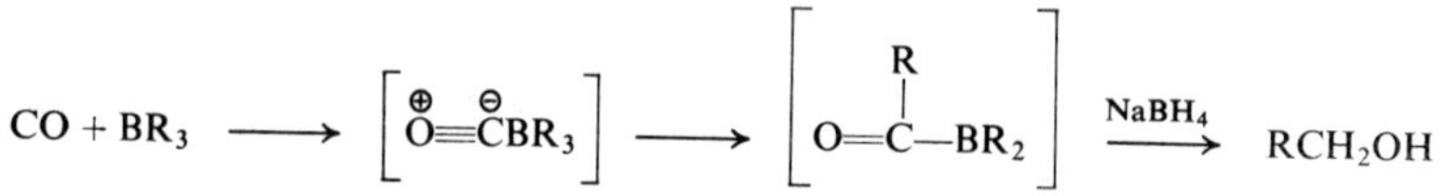

$$CO + BR_3 \longrightarrow [\overset{\oplus}{O}\equiv\overset{\ominus}{C}BR_3] \longrightarrow [O=\overset{\overset{R}{|}}{C}\text{—}BR_2] \xrightarrow{NaBH_4} RCH_2OH \quad (23)$$
$$\quad\quad\quad\quad\quad\quad (LVI) \quad\quad\quad (LVII)$$

## 6. Reaction of Isonitriles with Boranes

carbon monoxide are combined in a polar aprotic solvent (diglyme) and heated to 100–125°C, then oxidized with hydrogen peroxide, trialkylcarbinols are formed.[7] Again speculation is reasonable that in diglyme compounds such as LV are formed and at an elevated temperature undergo a third B to C alkyl migration, as was noted in the VIII→IX reaction sequence with isocyanide–borane adducts. Recently Brown[9] has reviewed the many synthetic possibilities of the carbon monoxide–organoborane reaction.

Other Lewis bases, though less frequently studied, nonetheless provide interesting parallel examples. The reaction of trialkylboranes with nitric oxide has been examined.[5,6,39,43] The variation in behavior of these reagents with temperature led Brois[5,6] to propose that NO behaves as an ambient nucleophile toward alkylboranes, and to dispute the structures proposed by Kuhn and Inatome[39,43] for this reaction. Yoshida[70] has reported on the reaction of nitrosyl chloride and nitrosylsulfuric acid with alkyl boranes. Inasmuch as the reactions were conducted at 50°C in sulfuric acid, it is not surprising that they followed a substantially different course. Low yields of oximes and lactams are reported to be the products.

Several other cases involving B to C alkyl group migrations in a Lewis base–borane adduct should be noted. Some of these represent potentially useful organic synthetic methods. Sulfoxonium and sulfonium ion stabilized carbanions have been found to undergo a facile reaction with trialkylboranes (24). The initial adduct (LIX) of trialkylboranes and dimethyloxosulfonium methylid (LVIII) undergo B to C alkyl group migration to form organoboranes which contain one more methylene group than the original borane (LX).[67] Interestingly, to a substantial extent, organoborane (LX) undergoes further B to C alkyl transfer reaction with the carbanion to give LXI. Oxidation of these compounds produces good yields of alcohols in which the carbon chain has been extended by one or two more carbons. The procedure has since been much improved[66] by the use of dimethylsulfonium methylid (LXII),

$$CH_3SO\overset{\ominus}{C}H_2 + BR_3 \longrightarrow [CH_3SOCH_2BR_3] \longrightarrow \begin{bmatrix} R \\ | \\ CH_2BR_2 \end{bmatrix} \xrightarrow{H_2O_2} RCH_2OH$$

(LVIII)　　　　　　(LIX)　　　　　　(LX)

$$\Bigg\downarrow CH_3SO\overset{\ominus}{C}H_2 \qquad (24)$$

$$\begin{bmatrix} R \\ | \\ CH_2CH_2BR_2 \end{bmatrix} \xrightarrow{H_2O_2} RCH_2CH_2OH$$

(LXI)

which produces after oxidation, 30–34% yield of the alcohol containing one more carbon atom. This suggests a quantitative conversion of the original reagents to LXIV via LXIII (25). The synthesis has since been extended to

$$(CH_3)_2\overset{\oplus}{S}\overset{\ominus}{C}H_2 + BR_3 \longrightarrow \left[(CH_3)_2\overset{\oplus}{S}CH_2\overset{\ominus}{B}R_3\right] \longrightarrow$$
(LXII) (LXIII)

$$(CH_3)_2S + \left[\begin{array}{c} R \\ | \\ CH_2BR_2 \end{array}\right] \xrightarrow{H_2O_2} RCH_2OH + 2ROH \quad (25)$$
(LXIV)

the preparation of homologous esters (LXVIII) (26) when dimethylsulfonium carbethyloxymethylid (LXV) is used as the Lewis base source. Surprisingly, the intermediate organoborane (LXVII) appears to be hydrolyzed faster than it is oxidized when subjected to the usual conditions for such a reaction.[71]

$$(CH_3)_2\overset{\oplus}{S}\overset{\ominus}{C}HCO_2R + BR_3 \longrightarrow \left[\begin{array}{c} (CH_3)_2\overset{\oplus}{S}CH\overset{\ominus}{B}R_3 \\ | \\ CO_2R \end{array}\right] \longrightarrow$$
(LXV) (LXVI)

$$(CH_3)_2S + \left[\begin{array}{c} R \\ | \\ CHBR_2 \\ | \\ CO_2R \end{array}\right] \xrightarrow{H_2O_2} RCH_2CO_2R \quad (26)$$
(LXVII) (LXVIII)

Musker and Stevens[52] have reported the extension of the reaction to include nitrogen ylids as a carbanion source. Hence they found that trimethylammonium methylid (LXIX) reacts with triphenylborane to produce a 1:1 adduct (LXX) which rearranges with displacement of trimethylamine to give the chain-extended organoborane (LXXI) (27).

$$(CH_3)_3\overset{\oplus}{N}\overset{\ominus}{C}H_2 + B(C_6H_5)_3 \longrightarrow \left[(CH_3)_3\overset{\oplus}{N}CH_2\overset{\ominus}{B}(C_6H_5)_3\right] \longrightarrow$$
(LXIX) (LXX)

$$(CH_3)_3N + C_6H_5CH_2B(C_6H_5)_2 \quad (27)$$
(LXXI)

A $^{11}$B chemical shift characteristic of alkyl borinate esters, and the absence of carbonyl IR absorption in the reaction product of diazoketones and trialkylboranes [cf. Hooz and Linke$^{35,36}$] has led Pasto$^{56}$ to propose that vinyloxyborinates (LXXII) are the first formed intermediates in this reaction.

$$\underset{R_2BO}{\overset{R'}{\diagdown}}C=C\underset{H}{\overset{R}{\diagup}} \qquad \underset{(i\text{-}C_3H_7)_2BO}{\overset{C_6H_5}{\diagdown}}C=C\underset{i\text{-}C_3H_7}{\overset{H}{\diagup}}$$

(LXXII)                           (LXXIII)

Pasto has also isolated LXXIII as the major product of the reaction of dimethylsulfonium phenacylid (LXXIV) with tri-$n$-propylborane. These results suggest that examination of the structure of intermediates in the reaction of sulfur

$$\underset{\ominus O}{\overset{C_6H_5}{\diagdown}}C=C\underset{\oplus S(CH_3)_2}{\overset{H}{\diagup}}$$

(LXXIV)

and nitrogen ylids, and of carbon monoxide with organoboranes may show that alkylborinate ester formation is more general than heretofore expected.

The introduction of a large variety of Lewis bases into the organoborane molecule to produce an adduct which undergoes sequential B to C alkyl or aryl group migration appears to be a general phenomenon, and presents a host of new and interesting synthetic possibilities. The opportunity to devise new synthetic applications so far is limited only by the ingenuity of the experimenter to conceive novel substrates for this reaction.

### REFERENCES

1. Allerhand, A., and Schleyer, P. von R., *J. Amer. Chem. Soc.* **85**, 866 (1963).
2. Bittner, G., Witte, H., and Hesse, G., *Justus Liebigs Ann. Chem.* **713**, 1 (1968).
3. Bresadola, S., Carraro, G., Pecile, C., and Turco, A., *Tetrahedron Lett.* p. 3185 (1964).
4. Bresadola, S., Rosetto, F., and Puosi, G., *Tetrahedron Lett.* p. 4775 (1965).
5. Brois, S. J., *144th Nat. Meeting, Amer. Chem. Soc., Los Angeles*, 1963 Paper No. 66, Org. Sect.
6. Brois, S. J., *Tetrahedron Lett.* p. 345 (1964).
7. Brown, H. C., and Rathke, M. W., *J. Amer. Chem. Soc.* **89**, 2737 (1967).
8. Brown, H. C., and Rathke, M. W., *J. Amer. Chem. Soc.* **89**, 2738 (1967).
9. Brown, H. C., *Accounts Chem. Res.* **2**, 65 (1969).
10. Bubnov, Yu. N., and Mikhailov, B. M., *Izv. Akad. Nauk SSSR* p. 472 (1967) (Russian); *Chem. Abstr.* **67**, 11523g (1967).
11. Casanova, J., unpublished results (1962).
12. Casanova, J., and Kiefer, R. H., *J. Org. Chem.* **34**, 2579 (1969).
13. Casanova, J., Kiefer, H. R., Kuwada, D., and Boulton, A. H., *Tetrahedron Lett.* p. 703 (1965).

14. Casanova, J., and Schuster, R. E., *Tetrahedron Lett.* p. 405 (1964).
15. Coates, G. E., Green, M. L. H., and Wade, K., *in* "Organometallic Compounds" (H. M. M. Shearer, and J. Willis, eds.), 3rd ed., Vol. 1, p. 264$^{427}$. Methuen, London, 1967.
16. Cotton, F. A., and Wilkinson, G., "Advanced Inorganic Chemistry," 2nd ed., pp. 744 and 745. Wiley, New York, 1966.
17. Dorokhov, V. A., and Lappert, M. F., *Chem. Commun.* p. 250 (1968).
18. Ferstandig, L. L., *J. Amer. Chem. Soc.* **84**, 1323 (1962).
19. Ferstandig, L. L., *J. Amer. Chem. Soc.* **84**, 3553 (1962).
20. Graybill, B. M., and Hawthorne, M. F., *J. Amer. Chem. Soc.* **83**, 2673 (1961).
21. Hertler, W. R., U.S. Pat. 3,429,923 (1965).
22. Hesse, G., and Witte, H., *Angew. Chem., Int. Ed. Engl.* **2**, 617 (1963).
23. Hesse, G., and Witte, H., *Angew. Chem., Int. Ed. Engl.* **4**, 355 (1965).
24. Hesse, G., and Witte, H., *Justus Liebigs Ann. Chem.* **687**, 1 (1965).
25. Hesse, G., Witte, H., and Haussleiter, H., *Angew. Chem., Int. Ed. Engl.* **5**, 723 (1966).
26. Hesse, G., Witte, H., and Bittner, G., *Justus Liebigs Ann. Chem.* **687**, 9 (1965).
27. Hesse, G., Witte, H., and Gulden, W., *Angew. Chem.* **77**, 591 (1965).
28. Hesse, G., Witte, H., and Gulden, W., *Tetrahedron Lett.* p. 2707 (1966).
29. Hillman, M. E. D., *J. Amer. Chem. Soc.* **84**, 4715 (1962).
30. Hillman, M. E. D., 142nd Nat. Meeting, *Amer. Chem. Soc.*, Atlantic City, 1962 Pap. No. 15, Org. Sect.
31. Hillman, M. E. D., *J. Amer. Chem. Soc.* **85**, 982 (1963).
32. Hillman, M. E. D., *J. Amer. Chem. Soc.* **85**, 1626 (1963).
33. Hillman, M. E. D., 144th Nat. Meeting, *Amer. Chem. Soc.*, Los Angeles, 1963 Pap. No. 13, Org. Sect.
34. Hoffmann, R., *J. Chem. Phys.* **40**, 2474 (1964).
35. Hooz, J., and Linke, S., *J. Amer. Chem. Soc.* **90**, 5936 (1968).
36. Hooz, J., and Linke, S., *J. Amer. Chem. Soc.* **90**, 6891 (1968).
37. Horn, E. M., U.S. Pat. 3,338,962 (1967); *Chem. Abstr.* **68**, 69111h (1968).
38. Hyatt, D. E., Owen, D. A., and Todd, L. J., *Inorg. Chem.* **5**, 1749 (1966).
39. Inatome, M., and Kuhn, L. P., *Advan. Chem. Ser.* **42**, 183 (1964).
40. Jennings, J. R., Lloyd, J. E., and Wade, K., *J. Chem. Soc.* p. 5083 (1965).
41. Jennings, J. R., Pattison, I., Summerford, C., Wade, K., and Wyatt, B. K., *Chem. Commun.* p. 250 (1968).
42. Klages, F., Monkemeyer, K., and Heinle, R., *Chem. Ber.* **85**, 109 (1952).
43. Kuhn, L. P., and Inatome, M., *J. Amer. Chem. Soc.* **85**, 1206 (1963).
44. Kuivila, H. G., *J. Amer. Chem. Soc.* **76**, 870 (1954).
45. Lappert, M. F., and Majumdar, M. K., *Proc. Chem. Soc.*, London p. 88 (1963).
46. Lappert, M. F., and Majumdar, M. K., *Advan. Chem. Ser.* **42**, 208 (1964).
47. Lloyd, J. E., and Wade, K., *J. Chem. Soc.* p. 1649 (1964).
48. Malone, L. J., *Inorg. Chem.* **7**, 1039 (1968).
49. Malone, L. J., and Manley, M. R., *Inorg. Chem.* **6**, 2260 (1967).
50. McCusker, P. A., and Bright, J. H., *J. Org. Chem.* 2093 (1964).
51. Miller, N. E., and Muetterties, E. L., *Inorg. Chem.* **3**, 1196 (1964).
52. Musker, W. K., and Stevens, R. R., *Tetrahedron Lett.* p. 995 (1967).
53. Nagasawa, K., *Inorg. Chem.* **5**, 442 (1966).
54. Niedenzu, K., *Angew. Chem., Int. Ed. Engl.* **3**, 86 (1964).
55. Paetzold, von P. I. *Z. Anorg. Allg. Chem.* **326**, 53 (1963).
56. Pasto, D. J., and Wojtkowski, P. W., private communication (1969).
57. Pasynkiewicz, S., Starowieski, K., and Rzepkowska, Z., *J. Organometal. Chem.* **10**, 527 (1967).

## 6. Reaction of Isonitriles with Boranes    131

58. Pattison, I., and Wade, K., *J. Chem. Soc.*, A p. 1098 (1967).
59. Rathke, M. W., *Diss. Abstr.* B **28**, 2355 (1967) [Univ. Microfilm 67-16, 701 (1968)].
60. Rathke, M. W., and Brown, H. C., *J. Amer. Chem. Soc.* **89**, 2740 (1967).
61. Reinheckel, H., and Jahnke, D., *Tenside* **2**, 249 (1965); *Chem. Abstr.* **63**, 9805a (1965).
62. Schaeffer, R., and Todd, L. J., *J. Amer. Chem. Soc.* **87**, 488 (1965).
63. Schleyer, P. von R., and Allerhand, A., *J. Amer. Chem. Soc.* **84**, 1322 (1962).
64. Tanaka, J., and Carter, J. C., *Tetrahedron Lett.* p. 329 (1965).
65. Tsai, C., and Streib, W., *Tetrahedron Lett.* p. 669 (1968).
66. Tufariello, J. J., *J. Amer. Chem. Soc.* **89**, 6804 (1967).
67. Tufariello, J. J., and Lee, L. T. C., *J. Amer. Chem. Soc.* **88**, 4757 (1966).
68. Tufariello, J. J., Wojtkowski, P. W., and Lee, L. T. C., *Chem. Commun.* p. 505 (1967).
69. Witte, H., *Tetrahedron Lett.* p. 1127 (1965).
70. Yoshida, Z., Ogushi, T., Manabe, O., and Hiyama, H., *Tetrahedron Lett.* p. 753 (1965).
71. Zweifel, G., and Brown, H. C., *Org. React.* **13**, 1 (1965).

# Chapter 7

# The Passerini Reaction and Related Reactions

D. Marquarding, G. Gokel, P. Hoffmann, and I. Ugi

| | |
|---|---|
| I. Introduction | 133 |
| II. α-Acyloxycarbonamides | 133 |
| III. The Mechanism of the Passerini Reaction | 136 |
| IV. Tetrazoles | 139 |
| V. α-Hydroxyamides | 140 |
| VI. α,γ-Diketoamides | 142 |
| References | 143 |

## I. INTRODUCTION

The chemistry of isonitriles differs remarkably from the rest of organic chemistry by the pronounced tendency of isonitriles to undergo multicomponent reactions by which more than two reactants selectively combine to form a single product (see Chapters 5–9). This unique feature may be attributed to the fact that isonitriles are the only stable organic compounds of formally divalent carbon and are capable of making the transition to compounds of formally tetravalent carbon by undergoing α-additions of suitable combinations of electrophiles and nucleophiles (see Chapter 4).

## II. α-ACYLOXYCARBONAMIDES

The first triple additions of isonitriles were reported in 1921 by Passerini[17,18,26] who observed the combination of carboxylic acids (I), carbonyl compounds (II) (aldehydes, ketones,* and acyl cyanides), and isonitriles (III) to form α-acyloxycarbonamides (IV).

$$R^1\text{—CO—OH} + R^2\text{—CO—}R^3 + R^4\text{—NC} \rightarrow R^1\text{—CO—O—}CR^2R^3\text{—CO—NH—}R^4 \quad (1)$$

(I)        (II)        (III)        (IV)

---

* Some α,β-unsaturated or sterically hindered ketones, like camphor, do not undergo the Passerini reaction.[2]

TABLE I

THE SYNTHESIS OF α-ACYLOXYCARBONAMIDES (IV)

| Example no. | $R^2$ | (II) $R^3$ | $R^1$(I) | $R^4$(III) | Yield (%) | Ref. |
|---|---|---|---|---|---|---|
| 1 | H— | H— | Phthalimido-$CH_2$— | $t$-$C_4H_9$—$O_2C$—$CH_2$— | 92 | 5 |
| 2 | H— | H— | Phthalimido-$CH_2$— | $CH_3$—$O_2C$—$CH(i$-$C_3H_7)$— | 94 | 5 |
| 3 | H— | H— | HCO—NH—CH($i$-$C_3H_7$)— | $t$-$C_4H_9$—$O_2C$—$CH_2$— | 18 | 5 |
| 4 | H— | H— | $N_3$—$CH_2$—$CO_2$—CH($i$-$C_3H_7$)—CO—NH—$CH_2$— | $t$-$C_4H_9$—$O_2C$—$CH_2$— | 56 | 5 |
| 5 | H— | $CH_3$— | $CH_3$— | $C_6H_5$— | 75 | 22 |
| 6 | H— | $CH_3$— | $N_3$—$CH_2$— | $t$-$C_4H_9$—$O_2C$—$CH_2$— | 76 | 5 |
| 7 | H— | $i$-$C_3H_7$— | H— | $c$-$C_6H_{11}$— | 20 | 11 |
| 8 | H— | $i$-$C_3H_7$— | $CH_3$— | $c$-$C_6H_{11}$— | 69 | 11 |
| 9 | H— | $i$-$C_3H_7$— | Cl—$CH_2$— | $t$-$C_4H_9$—$O_2C$—$CH_2$— | 87 | 5 |
| 10 | H— | $i$-$C_3H_7$— | $N_3$—$CH_2$— | $t$-$C_4H_9$—$O_2C$—$CH_2$— | 88 | 5 |
| 11 | H— | $i$-$C_3H_7$— | Cl—$CH_2$—CO—NH—$CH_2$— | $t$-$C_4H_9$—$O_2C$—$CH_2$— | 29 | 5 |
| 12 | H— | $i$-$C_3H_7$— | Cbo—NH—$CH_2$— | $t$-$C_4H_9$—$O_2C$—$CH_2$— | 80 | 5 |
| 13 | H— | $i$-$C_3H_7$— | L-Menthoxy-$CH_2$— | $C_6H_5$— | 98 | 3 |
| 14 | H— | $i$-$C_3H_7$— | Phthalimido-$CH_2$— | $C_2H_5$—$O_2C$—$CH_2$— | 79 | 5 |
| 15 | H— | $i$-$C_3H_7$— | Phthalimido-$CH_2$— | $t$-$C_4H_9$—$O_2C$—$CH_2$— | 85 | 5 |
| 16 | H— | $CH_3$—S—$(CH_2)_2$— | Phthalimido-$CH_2$— | $t$-$C_4H_9$—$O_2C$—$CH_2$— | 56 | 5 |
| 17 | H— | $C_6H_5$—$CH_2$—O—$CH_2$— | L-Menthoxy-$CH_2$— | $C_6H_5$— | — | 3 |
| 18 | H— | 3-$NO_2$—$C_6H_4$— | $C_6H_5$— | $C_6H_5$— | 57 | 22 |
| 19 | H— | $C_6H_5$— | 2-$C_4H_9$— | 4-$C_2H_5$—O—$C_6H_4$— | 72 | 1 |
| 20 | H— | $C_6H_5$— | $C_6H_5$— | $n$-$C_3H_7$— | 62 | 25 |
| 21 | H— | $C_6H_5$— | $C_6H_5$— | $C_6H_5$— | 35 | 18 |
| 22 | H— | $C_6H_5$— | $C_6H_5$— | 4-$C_6H_5$—N=N—$C_6H_4$— | 58 | 18 |
| 23 | H— | $C_6H_5$— | L-Menthoxy-$CH_2$— | $C_6H_5$— | 56 | 1 |
| 24 | H— | 3,4-$OCH_2O$—$C_6H_3$— | $C_6H_5$— | $C_6H_5$— | 34 | 22 |

# 7. Passerini Reaction

| No. | | | | | Yield | Ref. |
|---|---|---|---|---|---|---|
| 25 | NC— | CH$_3$— | H— | c-C$_6$H$_{11}$— | 58 | 16 |
| 26 | CF$_3$— | CF$_3$— | CH$_3$— | C$_2$H$_5$— | 71 | 12 |
| 27 | CF$_3$— | CF$_3$— | CH$_3$— | c-C$_6$H$_{11}$— | 75 | 12 |
| 28 | Cl—CH$_2$— | CH$_3$— | CH$_3$— | C$_6$H$_5$— | — | 22 |
| 29 | Cl—CH$_2$— | CH$_3$— | C$_6$H$_5$— | C$_6$H$_5$— | 29 | 22 |
| 30 | CH$_3$— | CH$_3$— | CH$_3$— | C$_6$H$_5$— | 45 | 18 |
| 31 | CH$_3$— | CH$_3$— | CH$_3$— | 4-C$_6$H$_5$—N=N—C$_6$H$_4$— | 14 | 17 |
| 32 | CH$_3$— | CH$_3$— | 2-HO—C$_6$H$_4$— | C$_6$H$_5$— | 51 | 22 |
| 33 | CH$_3$— | CH$_3$— | Phthalimido-CH$_2$— | t-C$_4$H$_9$—O$_2$C—CH$_2$— | 73 | 5 |
| 34 | CH$_3$— | C$_2$H$_5$— | CH$_3$— | C$_6$H$_5$— | 75 | 22 |
| 35 | CH$_3$— | C$_2$H$_5$— | L-Menthoxy-CH$_2$— | C$_6$H$_5$— | 20 | 3 |
| 36 | CH$_3$— | n-C$_3$H$_7$— | L-Menthoxy-CH$_2$— | C$_6$H$_5$— | 71 | 3 |
| 37 | CH$_3$— | C$_2$H$_5$—O$_2$C—CH$_2$— | CH$_3$— | 4-C$_6$H$_5$—N=N—C$_6$H$_4$— | 17–19 | 18 |
| 38 | CH$_3$— | n-C$_6$H$_{13}$— | L-Menthoxy-CH$_2$— | C$_6$H$_5$— | 76 | 1 |
| 39 | C$_6$H$_5$— | C$_6$H$_5$—CO— | C$_6$H$_5$— | C$_6$H$_5$— | 87 | 22 |
| 40 | —(CH$_2$)$_5$— | | C$_6$H$_5$— | C$_6$H$_5$— | 17 | 21 |
| 41 | —(CH$_2$)$_4$—CH(CH$_3$)— | | C$_6$H$_5$— | C$_6$H$_5$— | 27 | 21 |
| 42 | —(CH$_2$)$_3$—CH(CH$_3$)—CH$_2$— | | C$_6$H$_5$— | C$_6$H$_5$— | 95 | 21 |
| 43 | —(CH$_2$)$_3$—CH(CH$_3$)—CH$_2$— | | L-Menthoxy-CH$_2$— | C$_6$H$_5$— | 24 | 3 |
| 44 | —(CH$_2$)$_2$—CH(CH$_3$)—(CH$_2$)$_2$— | | C$_6$H$_5$— | C$_6$H$_5$— | 30 | 21 |
| 45 | —C$_{26}$H$_{20}$— (3-Cholestanone) | | C$_6$H$_5$— | C$_6$H$_5$— | 43 | 2 |
| 46 | —C$_{26}$H$_{20}$— (3-Cholestanone) | | C$_6$H$_5$— | 4-CH$_3$—C$_6$H$_4$— | 26 | 2 |
| 47 | —C$_{26}$H$_{20}$— (3-Cholestanone) | | C$_6$H$_5$— | 4-CH$_3$O—C$_6$H$_4$— | | 2 |

The Passerini reaction is usually effected by letting stand at 0–20°C, a mixture of the components or a concentrated solution of the components in an inert organic solvent for a time varying from hours to days. A wide variety of α-acyloxycarbonamides have been prepared by the Passerini reaction (1) (see Table I). As one can see from examples 1–4, 6, 9, 14–16 of Table I, the Passerini reaction is potentially useful for the synthesis of depsipeptides.[5] The results of Baker et al.[1–3] do not indicate clearly whether asymmetrically induced Passerini reactions (see Chapter 8, Section IV) are possible or not, because the observed stereoselectivities are within the accuracy limits of their experiments.

Lactones, e.g., IVa–IVc, are obtained by the Passerini reaction from suitable oxo acids (see also XIX).

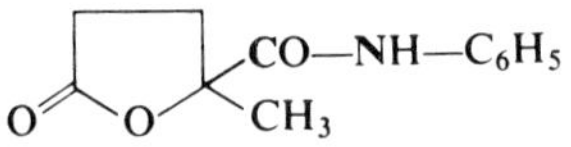

(IVa)[20] (66%)

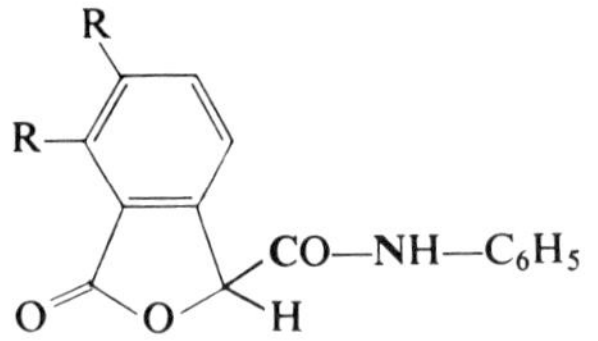

(IV)[27] b(20%): R = H—
c(20%): R = $CH_3O$—

## III. THE MECHANISM OF THE PASSERINI REACTION

A variety of different reaction mechanisms has been proposed for the Passerini reaction, but the experimental evidence which is presently available does not provide any basis upon which to decide between the alternatives. Isotopic labelling studies using $^{18}O$ carboxylic acids and carbonyl compounds as well as kinetic and stereoselectivity studies (see Chapter 8, Section IV) might lead to improved mechanistic insight.

Passerini postulated hemiacetallic adducts like V, as intermediates in the

$$I + II \rightleftharpoons R^1\text{—CO—O—}CR^2R^3\text{—OH} \xrightarrow{III} IV \qquad (2)$$
$$(V)$$

reaction (1). Baker and Stanonis[3] concluded from kinetic evidence (namely, the observed third order of the reaction) that the triple addition proceeds via the dipolar species (VI). The latter had been postulated by Dewar[4] to be an intermediate in this reaction.

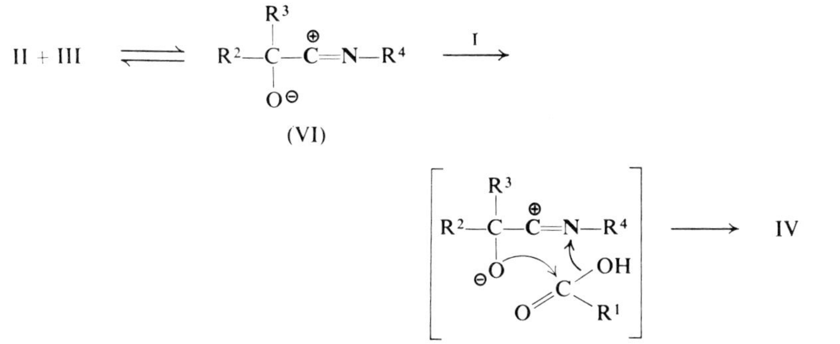

7. Passerini Reaction    137

$$\text{II + III} \rightleftharpoons \underset{(VI)}{\overset{R^3}{\underset{O^\ominus}{R^2-C-C=N-R^4}}} \xrightarrow{I}$$

$$\left[ \underset{\underset{O}{\overset{\ominus}{O}}\overset{R^3}{\underset{}{\overset{}{\underset{}{\overset{\oplus}{C}}}\overset{}{-C=N-R^4}}}}{\overset{}{\underset{}{\overset{}{\underset{}{}}}}} \underset{R^1}{\overset{OH}{\underset{}{\overset{}{C}}}} \right] \longrightarrow \text{IV} \quad (3)$$

Hagedorn and Eholzer[7] have suggested that N-protonated isonitriles (VII)

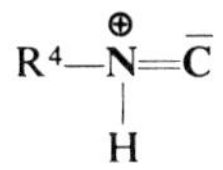

$$\underset{\underset{H}{|}}{R^4 - \overset{\oplus}{N} = \bar{C}}$$
(VII)

are the reactive species in the Passerini-type reactions; these authors do not suggest a process by which the C—C bond might be formed between the carbonyl carbon of II and the electrophilic carbon of VII.

Kagen and Lillien[8] observed the formation of X among other products (see Chapter 2) on reacting VIII with peracids. These authors pointed out

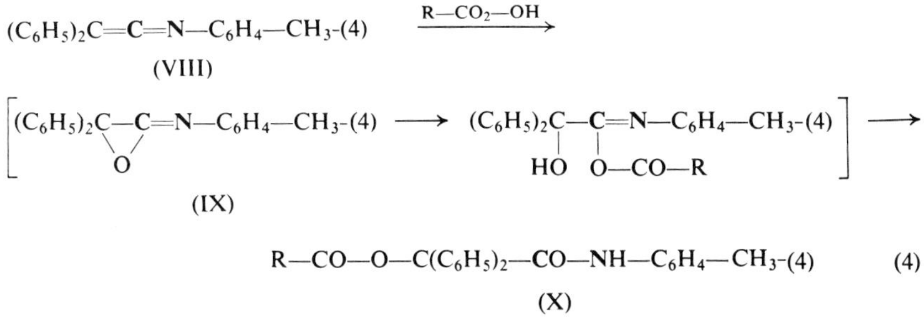

$$\underset{(VIII)}{(C_6H_5)_2C=C=N-C_6H_4-CH_3\text{-}(4)} \xrightarrow{R-CO_2-OH}$$

$$\left[ \underset{(IX)}{(C_6H_5)_2\underset{O}{\overset{}{\underset{\diagdown\diagup}{C-C}}}=N-C_6H_4-CH_3\text{-}(4)} \longrightarrow \underset{HO\phantom{xx}O-CO-R}{(C_6H_5)_2C-C=N-C_6H_4-CH_3\text{-}(4)} \right] \longrightarrow$$

$$\underset{(X)}{R-CO-O-C(C_6H_5)_2-CO-NH-C_6H_4-CH_3\text{-}(4)} \quad (4)$$

that intermediates like IX are probably also involved in the Passerini reaction. Saegusa, Taka-ishi, and Fujii[28] obtained small amounts of XII from acetone,

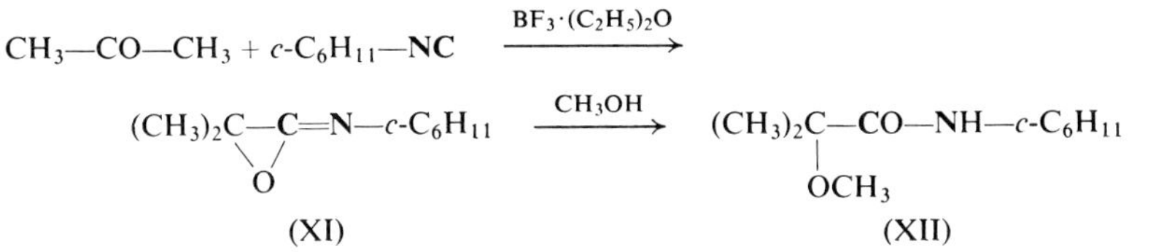

$$CH_3-CO-CH_3 + c\text{-}C_6H_{11}-NC \xrightarrow{BF_3\cdot(C_2H_5)_2O}$$

$$\underset{(XI)}{(CH_3)_2\underset{O}{\overset{}{\underset{\diagdown\diagup}{C-C}}}=N-c\text{-}C_6H_{11}} \xrightarrow{CH_3OH} \underset{(XII)}{(CH_3)_2\underset{OCH_3}{\overset{}{\underset{|}{C}}}-CO-NH-c\text{-}C_6H_{11}} \quad (5)$$

cyclohexyl isocyanide, and methanol in the presence of boron trifluoride etherate. They postulated XI as an intermediate. In addition, these authors found that some XV may be obtained from acetaldehyde and cyclohexyl isocyanide in the presence of boron trifluoride etherate and pyridine (6). The intermediates XIII and XIV are probable, but no firm evidence is available at this time. These observations support the assumption that iminooxiranes

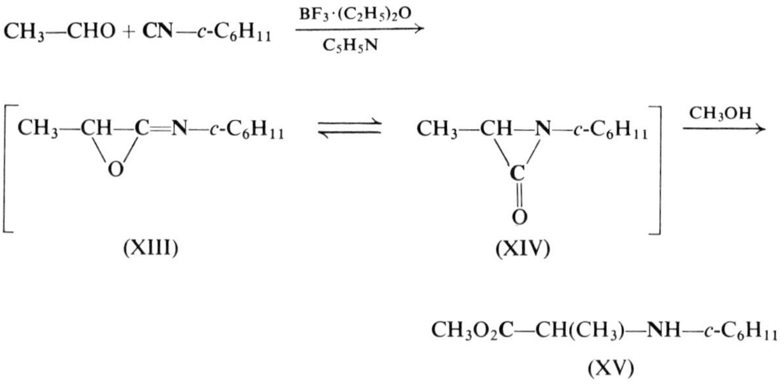

$$CH_3-CHO + CN-c\text{-}C_6H_{11} \xrightarrow[C_5H_5N]{BF_3\cdot(C_2H_5)_2O}$$

$$\left[ \begin{array}{cc} CH_3-CH-C=N-c\text{-}C_6H_{11} & CH_3-CH-N-c\text{-}C_6H_{11} \\ \diagdown\!\diagup & \diagdown\!\diagup \\ O & C \\ & \| \\ & O \\ (XIII) & (XIV) \end{array} \right] \xrightarrow{CH_3OH}$$

$$CH_3O_2C-CH(CH_3)-NH-c\text{-}C_6H_{11} \quad (6)$$
$$(XV)$$

like XI and XIII might also be the intermediates of the Passerini reaction.

The fact that polar solvents with a pronounced hydrogen bond affinity interfere with the Passerini reaction as well as other experimental observations like (8) can also be interpreted on the basis of mechanism (7).[29] This mechan-

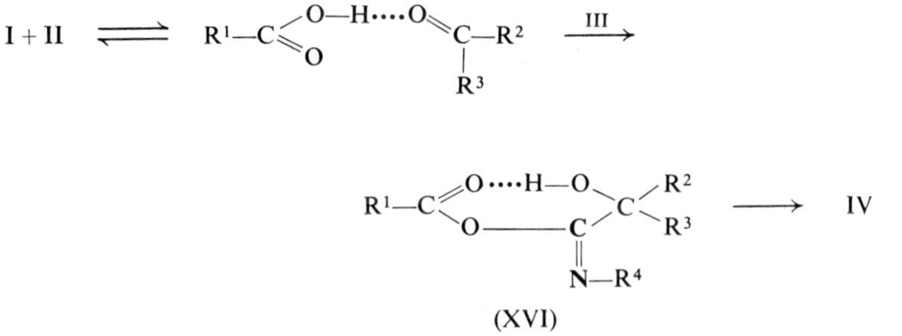

$$I + II \rightleftharpoons R^1-C\underset{O}{\overset{O-H\cdots O}{\diagup\!\diagdown}}C-R^2 \xrightarrow{III}$$
$$\phantom{I + II \rightleftharpoons R^1-C\underset{O}{\overset{O-H\cdots O}{\diagup\!\diagdown}}}\underset{R^3}{|}$$

$$R^1-C\underset{O\text{------}}{\overset{O\cdots H-O}{\diagup\!\diagdown}}\underset{\underset{N-R^4}{\|}}{C}\diagup\overset{R^2}{\diagdown}{R^3} \longrightarrow IV \quad (7)$$
$$(XVI)$$

ism, of course, complies with the well-known tendency of isonitriles to undergo α-additions.

No formation of the Passerini product (XIX)[10] was observed on reaction of 1-isocyanocyclohexylcarboxylic acid (XVII) and acetone, indicating that the cyclic adduct (XVIII) is more stable than the open-chain α-adducts (XVI).

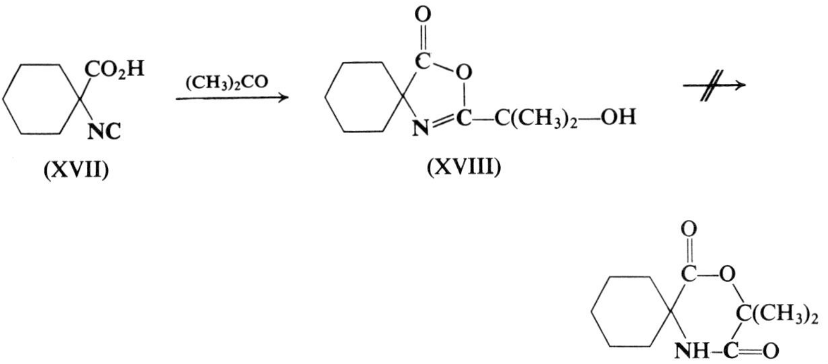

$$\underset{\text{(XVII)}}{\underset{\text{NC}}{\bigcirc}\text{CO}_2\text{H}} \xrightarrow{(CH_3)_2CO} \underset{\text{(XVIII)}}{\bigcirc\underset{N}{\overset{\overset{\overset{O}{\|}}{C-O}}{\underset{\diagdown}{\diagup}}}\!\!=\!\!C-C(CH_3)_2-OH} \xrightarrow{\;\;\;\not\!\!\longrightarrow\;\;\;}$$

$$\underset{\text{(XIX)}}{\bigcirc\underset{NH-C=O}{\overset{\overset{O}{\|}}{C-O}}\!\!\diagdown\!\!\underset{}{C(CH_3)_2}} \quad (8)$$

Reaction (8) can be considered confirming evidence for α-adducts (XVI) as intermediates in the Passerini reaction.

## IV. TETRAZOLES

With the knowledge of the Passerini reaction, the question arises whether the carboxylic acid component (I) of the Passerini reaction (1) can be replaced by other acids, and, if so, what the products will be. An investigation of this problem[27] showed that hydrazoic acid is the only noncarboxylic acid that is capable of undergoing triple additions with carbonyl compounds and

$$\underset{\text{(II)}}{R^2-CO-R^3} + \underset{\text{(III)}}{R^4-NC} + HN_3 \text{ (or Al(N}_3)_3) \longrightarrow$$

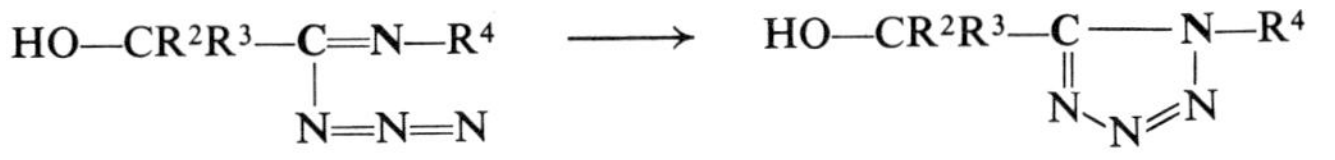

$$\underset{\underset{N=N=N}{|}}{HO-CR^2R^3-C=N-R^4} \longrightarrow \underset{\text{(XX)}}{HO-CR^2R^3-\underset{\underset{N\diagdown N\diagup}{\|}}{\overset{}{C}}\!\!-\!\!\underset{N}{\overset{}{N}}\!\!=\!\!N-R^4} \quad (9)$$

isonitriles (9). Only highly reactive carbonyl compounds produce appreciable yields of the tetrazole derivatives (XX) (Table II, Nos. 1–5), whereas with less reactive carbonyl components (Table II, Nos. 6–10) the competing formation of monosubstituted tetrazoles from $HN_3$ and isonitriles (see Chapter 4) becomes the predominant reaction. The tetrazoles (XX) can, however, be obtained in satisfactory yields by replacing the hydrazoic acid by aluminum azide (Table II, Nos. 6–10).

TABLE II

THE SYNTHESIS OF TETRAZOLE DERIVATIVES (XX)[29]

| Example no. | $R^2$ | (II) $R^3$ | $R^4$ | Yield with $HN_3$ | $Al(N_3)_3$ |
|---|---|---|---|---|---|
| 1  | H—    | H—                              | 2,6-$(CH_3)_2$—$C_6H_3$—        | 36 | —  |
| 2  | H—    | $CCl_3$—                        | $c$-$C_6H_{11}$—                | 69 | —  |
| 3  | H—    | $CCl_3$—                        | 2,6-$(CH_3)_2$—$C_6H_3$—        | 71 | —  |
| 4  | H—    | $CCl_3$—                        | 4$[(C_2H_5)_2N]$—$C_6H_4$—      | 95 | —  |
| 5  | H—    | $i$-$C_3H_7$—                   | $c$-$C_6H_{11}$—                | 85 | —  |
| 6  | $CH_3$— | $CH_3$—                       | $c$-$C_6H_{11}$—                | 32 | 80 |
| 7  | H—    | $C_6H_5$—                       | $c$-$C_6H_{11}$—                | 7  | 86 |
| 8  | H—    | $(CH_3)_2N$—$CH_2$—$C(CH_3)_2$— | $C_6H_5$—$CH_2$—                | —  | 82 |
| 9  | H—    | $(CH_3)_2N$—$CH_2$—$C(CH_3)_2$— | 2,6-$(CH_3)_2$—$C_6H_3$—        | —  | 81 |
| 10 | $CH_3$— | $C_6H_5$—                     | $c$-$C_6H_{11}$—                | —  | 68 |

## V. α-HYDROXYAMIDES

Boron trifluoride etherate and aqueous inorganic acids like hydrochloric, nitric, or phosphoric acid, but preferentially sulfuric acid, catalyze the addition of water and carbonyl compounds to isonitriles (10), to form α-hydroxyamides (XVII) (see Table III).

$$H_2O + R^2-CO-R^3 + R^4-NC \xrightarrow{H^\oplus} HO-CR^2R^3-CO-NH-R^4 \quad (10)$$
$$(XXI)$$

The hydrates of highly chlorinated aldehydes and ketones like chloral or α,α,β-trichloro-$i$-butyraldehyde or dichloroacetone form the α-hydroxyamides (Table III, Nos. 1–3, 5, 11) also in the absence of the catalytically active acids. Isodiazomethane is considered to be isocyanamide (XXII) because it reacts like an isonitrile (11)[13] in analogy to (10).

$$CN-NH_2 \xrightarrow{CCl_3-CHO} HO-CH(CCl_3)-CO-NH-N=CH-CCl_3 \quad (11)$$
$$(XXII)$$

An α-hydroxyamide is also formed XXIII from pernitrosomenthol and phenyl isocyanide,[24] whereas an unsaturated amide (XXIV) is formed from pernitrosocamphor.[23]

## TABLE III
### The Synthesis of α-Hydroxyamides (XXI)

| Example no. | $R^2$(II) | $R^3$ | $R^4$(III) | Yield (%) | Ref. |
|---|---|---|---|---|---|
| 1 | H— | CCl$_3$— | n-C$_3$H$_7$— | — | 25 |
| 2 | H— | CCl$_3$— | C$_6$H$_5$— | 50 | 19 |
| 3 | H— | CCl$_3$— | (C$_6$H$_5$—CH$_2$—)$_2$P(O)—CH$_2$— | 41 | 9 |
| 4 | H— | C$_2$H$_5$— | t-C$_4$H$_9$— | 47 | 14 |
| 5 | H— | CH$_3$—CHCl—CCl$_2$— | C$_6$H$_5$— | 37 | 19 |
| 6 | H— | n-C$_3$H$_7$— | t-C$_4$H$_9$— | 50 | 14 |
| 7 | H— | i-C$_3$H$_7$— | t-C$_4$H$_9$— | 70–80 | 31 |
| 8 | H— | i-C$_3$H$_7$— | c-C$_6$H$_{11}$— | 88 | 6,11 |
| 9 | H— | s-C$_4$H$_9$— | t-C$_4$H$_9$— | 55 | 31 |
| 10 | H— | C$_6$H$_5$— | 2,6-(CH$_3$)$_2$C$_6$H$_3$— | 62 | 6 |
| 11 | CHCl$_2$— | CH$_3$— | C$_6$H$_5$— | 12 | 22 |
| 12 | CH$_3$— | CH$_3$— | t-C$_4$H$_9$— | 72 | 6 |
| 13 | CH$_3$— | CH$_3$— | c-C$_6$H$_{11}$— | 71 | 6 |
| 14 | CH$_3$— | CH$_3$— | 2,6-(CH$_3$)$_2$C$_6$H$_3$— | 89 | 6 |
| 15 | CH$_3$— | CH$_3$— | β-C$_{10}$H$_7$— | 84 | 6 |
| 16 | CH$_3$— | C$_2$H$_5$— | t-C$_4$H$_9$— | 38 | 31 |
| 17 | CH$_3$— | C$_2$H$_5$— | 2,6-(CH$_3$)$_2$C$_6$H$_3$— | 63 | 6 |
| 18 | CH$_3$— | i-C$_3$H$_7$— | t-C$_4$H$_9$— | 22 | 31 |
| 19 | i-C$_3$H$_7$— | i-C$_3$H$_7$— | t-C$_4$H$_9$— | 1 | 31 |
| 20 | —(CH$_2$)$_4$— | | t-C$_4$H$_9$— | 70 | 14 |
| 21 | —(CH$_2$)$_5$— | | t-C$_4$H$_9$— | 75 | 14 |
| 22 | —(CH$_2$)$_5$— | | 2,6-(CH$_3$)$_2$C$_6$H$_3$— | 85 | 6 |
| 23 | —(CH$_2$)$_6$— | | t-C$_4$H$_9$— | 15–20 | 14 |
| 24 | —(CH$_2$)$_7$— | | t-C$_4$H$_9$— | 5–10 | 14 |
| 25 | —(CH$_2$)$_{11}$— | | t-C$_4$H$_9$— | 3–4 | 14 |

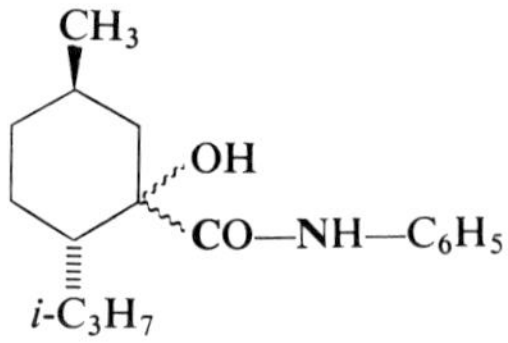

## VI. α,γ-DIKETOAMIDES

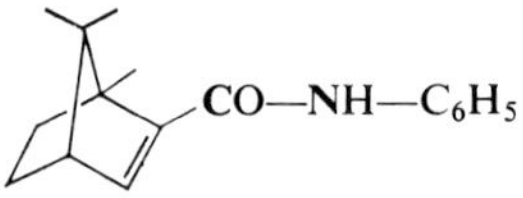

Diphenylketene reacts with carboxylic acids and isonitriles in ethereal solution at $-20°C$ to form α,γ-diketoamides (XXV)[30] (see Chapter 5 and Table IV). This reaction (12) probably involves the dipolar intermediate (XXV).

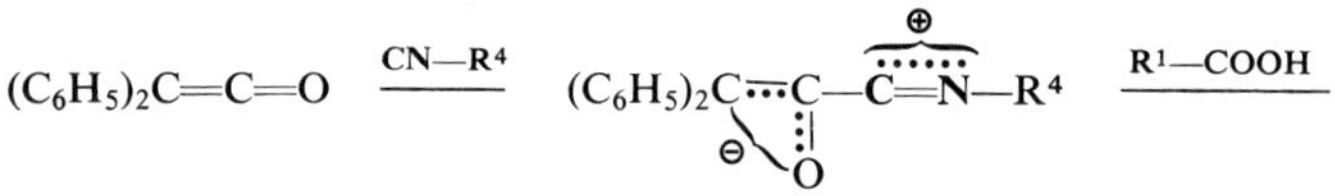

$$(C_6H_5)_2C=C=O \xrightarrow{CN-R^4} (C_6H_5)_2C\cdots C=N-R^4 \xrightarrow{R^1-COOH}$$
$$\qquad\qquad\qquad\qquad\qquad\qquad (XXV)$$

$$R^1-CO-C(C_6H_5)_2-CO-CO-NH-R^4 \qquad (12)$$
$$(XXVI)$$

TABLE IV

The Syntheses of the α,γ-Diketoamides (XXVI) According to (12)[9,30]

| Example No. | $R^1$— | $R^4$— | Yield (%) |
|---|---|---|---|
| 1 | $CH_2Cl$— | $c$-$C_6H_{11}$— | 69 |
| 2 | $CH_3$— | $c$-$C_6H_{11}$— | 79 |
| 3 | $CH_3$— | $(C_6H_5)_2P(O)$—$CH_2$— | 50[9] |
| 4 | $CH_3$—CO— | $c$-$C_6H_{11}$— | 52 |
| 5 | $t$-$C_4H_9$— | $c$-$C_6H_{11}$— | 67 |
| 6 | $C_6H_5$— | $c$-$C_6H_{11}$— | 78 |
| 7 | 2-HO—$C_6H_4$— | $c$-$C_6H_{11}$— | 79 |
| 8 | $(C_6H_5)CH$— | $c$-$C_6H_{11}$— | 74 |
| 9 | —$CH_2$— | $c$-$C_6H_{11}$— | 50 |

Acyl isocyanates and carboxylic acids react with isonitriles, e.g., (13), in analogy to (12) to form $N$-alkyl or $N$-aryl-$N'N'$-diacyloxamides (XXVIIa–XXVIIc).[15]

$$C_6H_5-COOH + R-CO-N=C=O + CN-R^4 \rightarrow$$

$$\begin{array}{c} C_6H_5-CO \\ R-CO \end{array} \!\!\!\!> N-CO-CO-NH-R^4 \qquad (13)$$

(XXVII)

a (89%): R = $C_6H_5-$
b (75%): R = $c\text{-}C_6H_{11}-$
c (87%): R = $1,3\text{-}c\text{-}C_6H_{10}-$

## REFERENCES

1. Baker, R. H., and Linn, L. E., *J. Amer. Chem. Soc.* **70**, 3721 (1948).
2. Baker, R. H., and Schlesinger, A. H., *J. Amer. Chem. Soc.* **67**, 1499 (1945).
3. Baker, R. H., and Stanonis, D., *J. Amer. Chem. Soc.* **73**, 699 (1951).
4. Dewar, M. J. S., "Electronic Theory of Organic Chemistry," p. 116. Oxford Univ. Press (Clarendon), London and New York, 1949.
5. Fetzer, U., and Ugi, I., *Justus Liebigs Ann. Chem.* **659**, 184 (1962).
6. Hagedorn, I., and Eholzer, U., *Chem. Ber.* **98**, 936 (1965); Ger. Pat. 1,173,082 (1964).
7. Hagedorn, I., Eholzer, U., and Winkelmann, H. D., *Angew. Chem.* **76**, 583 (1964); *Angew. Chem., Int. Ed. Engl.* **3**, 647 (1964).
8. Kagen, H., and Lillien, I., *J. Org. Chem.* **31**, 3728 (1966).
9. Kreutzkamp, N., and Lämmerhirt, K., *Angew. Chem.* **80**, 394 (1968).
10. Marquarding, D., unpublished results (1968).
11. McFarland, J. W., *J. Org. Chem.* **28**, 2179 (1963).
12. Middleton, W. J., England, D. C., and Krespan, C. G., *J. Org. Chem.* **32**, 948 (1967).
13. Müller, E., Kästner, P., Bentler, R., Rundel, W., Suhr, H., and Zeeh, B., *Justus Liebigs Ann. Chem.* **713**, 87 (1968).
14. Müller, E., and Zeeh, B., *Justus Liebigs Ann. Chem.* **696**, 72 (1966).
15. Neidlein, R., *Z. Naturforsch. B* **19**, 1159 (1964).
16. Neidlein, R., *Arch. Pharm.* **299**, 603 (1966).
17. Passerini, M., *Gazz. Chim. Ital.* **51**, II, 126 (1921).
18. Passerini, M., *Gazz. Chim. Ital.* **51**, II, 181 (1921).
19. Passerini, M., *Gazz. Chim. Ital.* **52**, I, 432 (1922).
20. Passerini, M., *Gazz. Chim. Ital.* **53**, 331 (1923).
21. Passerini, M., *Gazz. Chim. Ital.* **53**, 410 (1923).
22. Passerini, M., *Gazz. Chim. Ital.* **54**, 529 (1924).
23. Passerini, M., *Gazz. Chim. Ital.* **54**, 540 (1924).
24. Passerini, M., *Gazz. Chim. Ital.* **55**, 721 (1925).
25. Passerini, M., *Gazz. Chim. Ital.* **56**, 826 (1926).
26. Passerini, M., *Atti Reale Accad. Naz. Lincei, Mem. Cl. Sci. Fis., Mat. Natur.* [6] **2**, 377 (1927) (rev. art.).
27. Passerini, M., and Ragni, G., *Gazz. Chim. Ital.* **61**, 964 (1931).
28. Saegusa, T., Taka-ishi, N., and Fujii, H., *Tetrahedron* **24**, 3795 (1968).
29. Ugi, I., and Meyr, R., *Chem. Ber.* **94**, 2229 (1961).
30. Ugi, I., and Rosendahl, F. K., *Chem. Ber.* **94**, 2233 (1961).
31. Zeeh, B., and Müller, E., *Justus Liebigs Ann. Chem.* **715**, 47 (1968).

# Chapter 8

# Four-Component Condensations and Related Reactions

*G. Gokel, G. Lüdke, and I. Ugi*

I. Introduction and General Remarks . . . . . . . . 145
II. Hydantoin-4-imides and 2-Thiohydantoin-4-imides . . . . . 149
III. Acylated α-Amino- and α-Hydrazinocarbonamides . . . . . 155
  A. α-Acylaminocarbonamides . . . . . . . . . 155
  B. α-$N_\alpha$,$N_\beta$-Diacylhydrazino- and α-$N_\alpha$-Acyl-$N_\beta$-alkylidenehydrazinocarbonamides . . . . . . . . . . . . . 159
IV. Stereoselective Syntheses and the Reaction Mechanism of 4 C C . . . 161
V. β-Lactams and Penicillanic Acid Derivatives . . . . . . 181
  A. Simple β-Lactams . . . . . . . . . . 181
  B. Penicillanic Acid Derivatives . . . . . . . . 183
VI. Urethanes . . . . . . . . . . . . 185
VII. The Bucherer-Bergs Reaction . . . . . . . . . 186
VIII. N-Alkyl-3-acyl-4-hydroquinoline-4-carbonamides . . . . . 186
IX. Diacylimides, Amides, Thioamides, Selenoamides, and Amidines of α-Amino Acids . . . . . . . . . . . . . 189
X. Tetrazoles . . . . . . . . . . . . 193
    References . . . . . . . . . . . . 197

## I. INTRODUCTION AND GENERAL REMARKS

Amines (I) and carbonyl compounds (II), such as aldehydes or ketones,* react with isonitriles (III) and suitable acids (IV) to form unstable α-adducts (V) which are converted by spontaneous secondary reactions into stable α-amino acid derivatives (VI).[15,19,41,53,59,60,62,67,76] The term four-component condensation† is used for multicomponent reactions according to (1).

---

\* Instead of the amines and the carbonyl compounds, it is also possible and often advisable to use their condensation products, such as aminals, Schiff bases, and enamines.

† In this article we will use the abbreviation "4 C C." The 4 C C is also occasionally called "α-amino alkylation of isonitriles and acids" or "α-addition of immonium ions and anions to isonitriles, coupled with secondary reactions." See also footnote, p. 95.

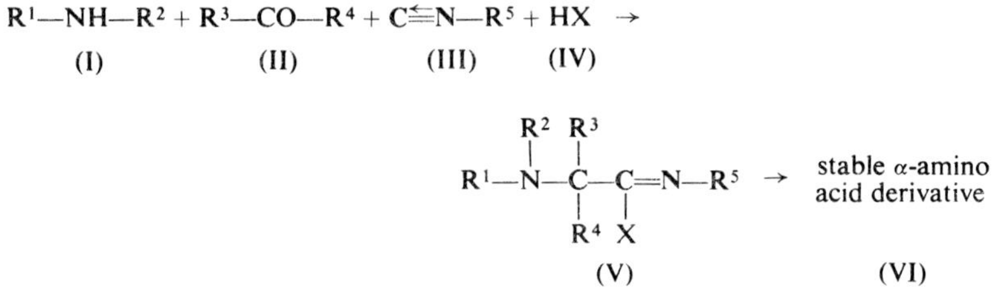

146  Gokel, Lüdke, and Ugi

$$R^1-NH-R^2 + R^3-CO-R^4 + C\equiv N-R^5 + HX \rightarrow$$
   (I)           (II)          (III)        (IV)

$$R^1-\underset{\underset{R^4}{|}}{\overset{\overset{R^2}{|}}{N}}-\underset{\underset{X}{|}}{\overset{\overset{R^3}{|}}{C}}-C=N-R^5 \rightarrow \text{stable } \alpha\text{-amino acid derivative} \quad (1)$$

            (V)                         (VI)

Any known type of C-isonitrile* (see Chapter 2) can be used as the isonitrile component, and there are only a few restrictions with respect to the selection of the carbonyl component. Except for sterically hindered diaryl ketones, aldehydes and ketones can generally be used as carbonyl components (II).

Ammonia, primary and secondary amines, as well as hydrazine derivatives can be used as amine components (I).

Acids which do not form stable α-aminoalkylation products[16] (see footnote, p. 148) but which do form α-adducts (V) are suitable acid components (IV) provided they can lead to stable α-amino acid derivatives (VI) via secondary reactions of V. The nature of the secondary reaction is largely dependent on

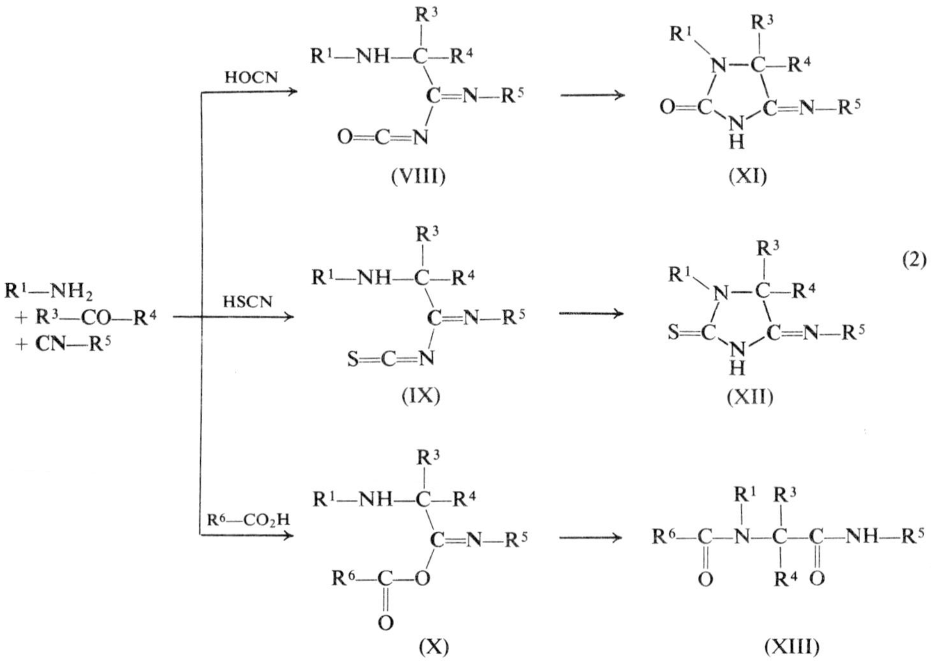

(2)

$R^1-NH_2$
$+ R^3-CO-R^4$
$+ CN-R^5$

via HOCN → (VIII) → (XI)

via HSCN → (IX) → (XII)

via $R^6-CO_2H$ → (X) → (XIII)

---

* In these the isocyano group is attached to a carbon atom.

the type of acid component used. It might be noted here that the reaction of $N$-ethyl-benzisoxazolium cation with a suitable anion involves similar secondary reactions.[22-24,83,84]

There are some acids, such as hydrogen cyanate, hydrogen thiocyanate, and carboxylic acids, which behave differently toward primary amines than toward secondary amines in the 4 C C. The α-adducts (VIII–X) of these contain acylating moieties and generally can be transformed by secondary reactions into stable α-amino acid derivatives (XI–XIII) only if the α-adduct contains an acylatable $\rangle$NH group (2). In the absence of acylatable compounds, the 4 C C of carboxylic acids and secondary amines leads to the diacyl imides (XIX) (see Section IX). If the α-adduct (XIV) has some alternative by which it may transfer its acyl group to some nucleophile, HY (such as an alcohol or amine), an α-amino acid amide (XX) is formed (3).

In contrast to the above acids, water, hydrogen thiosulfate, hydrogen selenide, and hydrazoic acid are acid components which undergo the same

$$R^1\text{—}NH\text{—}R^2 + R^3\text{—}CO\text{—}R^4 + CN\text{—}R^5$$

$\xrightarrow{R^6\text{—}CO_2H}$ 

$R^1\text{—}N\text{—}C\text{—}C\text{=}N\text{—}R^5$ with $R^2$, $R^4$, $O\text{—}CO\text{—}R^6$ below (XIV) $\longrightarrow$ $R^1\text{—}N\text{—}C\text{—}C\text{—}N\text{—}R^5$ with $R^2$, $R^4$, $O$, $CO\text{—}R^6$ (XIX)

$\xrightarrow{H_2O}$ $R^1\text{—}N\text{—}C\text{—}C\text{=}N\text{—}R^5$ with $R^2$, $R^4$, $OH$ (XV) $\xrightarrow{HY}$ $R^1\text{—}N\text{—}C\text{—}C\text{—}NH\text{—}R^5$ with $R^2$, $R^4$, $O$ (XX)

$\xrightarrow{H_2Se}$ $R^1\text{—}N\text{—}C\text{—}C\text{=}N\text{—}R^5$ with $R^2$, $R^4$, $SeH$ (XVI) $\longrightarrow$ $R^1\text{—}N\text{—}C\text{—}C\text{—}NH\text{—}R^5$ with $R^2$, $R^4$, $Se$ (XXI)     (3)

$\xrightarrow{H_2S_2O_3}$ $R^1\text{—}N\text{—}C\text{—}C\text{=}N\text{—}R^5$ with $R^2$, $R^4$, $S\text{—}SO_3^{\ominus}$ (XVII) $\xrightarrow{HO}$ $R^1\text{—}N\text{—}C\text{—}C\text{—}NH\text{—}R^5$ with $R^2$, $R^4$, $S$ (XXII)

$\xrightarrow{HN_3}$ $R^1\text{—}N\text{—}C\text{—}C\text{=}N\text{—}R^5$ with $R^2$, $R^4$, $N_3$ (XVIII) $\longrightarrow$ $R^1\text{—}N\text{—}C\text{—}C\text{—}N\text{—}R^5$ with $R^2$, $R^4$, and triazole ring $N\text{—}N\text{=}N$ (XXIII)

type of 4 C C with primary and secondary amines, because the nitrogen of the amine component does not take part in the secondary reactions (3) of the α-adducts (XIV–XVIII).

Reaction scheme (4) represents the reaction mechanism of the 4 C C in simplified form.[60]*

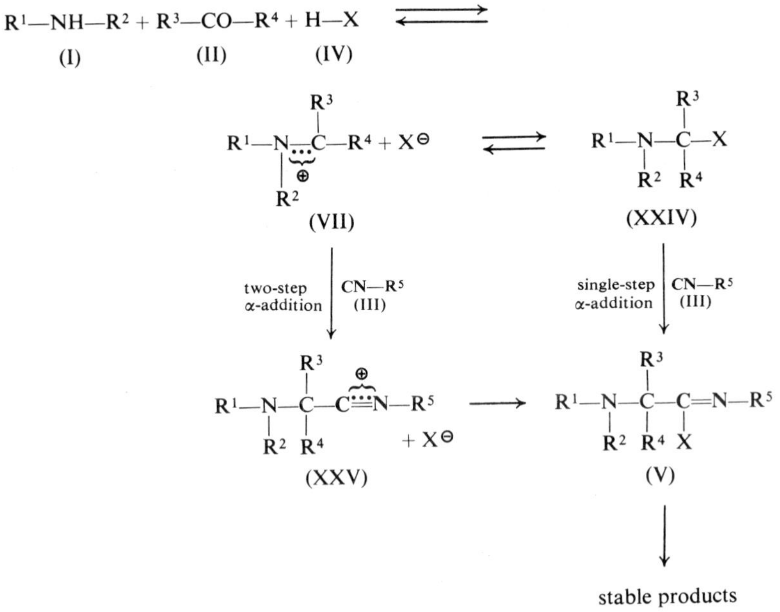

$$R^1-NH-R^2 + R^3-CO-R^4 + H-X \rightleftarrows$$
(I)        (II)        (IV)

$$\begin{matrix} R^3 \\ | \\ R^1-N\cdots C-R^4 + X^\ominus \\ | \quad \oplus \\ R^2 \\ (VII) \end{matrix} \rightleftarrows \begin{matrix} R^3 \\ | \\ R^1-N-C-X \\ | \quad | \\ R^2 \quad R^4 \\ (XXIV) \end{matrix}$$

two-step α-addition | CN—R⁵ (III)      single-step α-addition | CN—R⁵ (III)     (4)

$$\begin{matrix} R^3 \quad \oplus \\ | \\ R^1-N-C-C\equiv N-R^5 \\ | \quad | \\ R^2 \quad R^4 \quad + X^\ominus \\ (XXV) \end{matrix} \longrightarrow \begin{matrix} R^3 \\ | \\ R^1-N-C-C=N-R^5 \\ | \quad | \quad | \\ R^2 \quad R^4 \quad X \\ (V) \end{matrix}$$

↓

stable products

The reactive intermediates VII and XXIV are in mobile equilibrium† with the starting materials. The formation of the α-adduct (V) may take place by two different pathways which differ with regard to the formal kinetic reaction order. From the immonium ion (VII), which is separated from the counterion, $X^\ominus$, by solvent molecules, V is formed in a two-step reaction via the nitrilium ion (XXV), while it is formed directly from XXIV. It is not known, however, whether III inserts itself in the C—X bond of XXIV or whether XXIV reacts with III as an oriented ion pair.[66]

4 C C can easily be carried out. As a rule, the isonitrile component is added to the solution of the other three components, while stirring and cooling‡

---

* The detailed reaction mechanism (19)[66] will be discussed in Section IV.
† Therefore, hydrogen sulfide or hydrogen cyanide might be examples of unsuitable acid components.
‡ Generally, 4 C C are fast, strongly exothermic reactions.

8. Four-Component Condensations    149

and, after the reaction is over, the product is isolated. Frequently, the 4 C C product crystallizes out. The yields of 4 C C are generally high (80–100%). The formation of uniform products from four different reactants is accounted for by the fact that most of the potential side reactions are reversible, while the main reaction is not (see Section IV). Since four different starting materials take part in the 4 C C, the reaction is extremely versatile. If, for example, 40 each of the different components are reacted with one another, the result is $40^4 = 2,560,000$ reaction products, which is quite a high figure considering that it is of the same order of magnitude as the total number of chemical compounds described to date.

## II. HYDANTOIN-4-IMIDES AND 2-THIOHYDANTOIN-4-IMIDES

Formally, the 4 C C syntheses of the hydantoin-4-imides (XI) and the 2-thiohydantoin-4-imides (XII),[24,27] according to scheme (5), are very similar.

$$R^1-NH_2 \cdot HCl + R^3-CO-R^4(R^1-N=\overset{\overset{\displaystyle R^3}{|}}{C}-R^4 + Py \cdot HCl) + CN-R^5 + KYCN \longrightarrow$$

$$Y=\underset{\underset{H}{N}}{\overset{\overset{R^1}{\underset{|}{N}}\underset{5}{\overset{R^3}{\diagdown}}R^4}{\underset{3}{\overset{}{\diagup}}\underset{4}{\overset{}{C}}=N-R^5}} \qquad (5)$$

(XI: Y = O; XII: Y = S)

Nevertheless, there are remarkable differences between the syntheses of the two classes of compounds. When using an aldehyde as the carbonyl component (II), the synthesis of the hydantoin-4-imides takes place smoothly, whereas it is possible only in exceptional cases to react ketones to yield hydantoin-4-imides. On the contrary, for unknown reasons, 2-thiohydantoin-4-imides are obtained in high yields from ketones,* but not from aldehydes.† Since the formation of the monosubstituted urea (XXVI) gets the better of the main reaction in the hydantoin-4-imide synthesis from relatively unreactive carbonyl

$$R^1-NH-CO-NH_2$$
(XXVI)

components,[42] it is advisable to first prepare the Schiff bases from the carbonyl and amine components, and to react this Schiff base with isonitriles and potassium cyanate in the presence of pyridine hydrochloride (see Table I).

---
\* Certain ketones like acetophenone do not undergo this type of 4 C C.
† Except when combined with an acyl hydrazine.[4]

TABLE I

THE PREPARATION OF 1,5-DI- AND 1,5,5-TRISUBSTITUTED HYDANTOIN-4-IMIDES (XI) BY REACTING CARBONYL COMPOUNDS AND HYDROCHLORIDES OF PRIMARY AMINES WITH ISONITRILES AND POTASSIUM CYANATE IN AQUEOUS METHANOL (METHOD A) OR BY REACTING SCHIFF BASES AND PYRIDINE HYDROCHLORIDE WITH ISONITRILES AND POTASSIUM CYANATE IN ANHYDROUS METHANOL/ETHYLENE GLYCOL (METHOD B) AT 0–20°C[4,42,69]

| Designation no. (XI) | $R^3$ (II) | $R^4$ (II) | $R^1$ (I) | $R^5$ (III) | Yield (%) (by method) |
|---|---|---|---|---|---|
| a | H— | H— | $C_6H_5$—$CH_2$— | $C_6H_5$—$CH_2$— | 74 A |
| b | H— | H— | $C_2H_5O_2C$—$(CH_2)_2$— | $C_6H_5$— | 58 A |
| c | H— | $CH_3$— | 3,4,6-$Cl_3C_6H_2$— | $i$-$C_3H_7$ | 96 A |
| d | H— | $C_2H_5$— | $\beta$-$C_{10}H_7$— | $C_2H_5$— | 80 A |
| e | H— | $CH_3$—S—$(CH_2)_2$— | 2-$CH_3$—$C_6H_4$— | 4-$CH_3$—$C_6H_4$— | 47 A |
| f | H— | $CH_3$—$CH(C_6H_5)$— | 2-OH—$C_6H_4$— | 2,6-$(CH_3)_2$—$C_6H_3$— | 45 A |
| g | H— | $n$-$C_3H_7$— | $CH_3$— | $t$-$C_4H_9$— | 63 A |
| h | H— | $i$-$C_3H_7$— | H— | $t$-$C_4H_9$— | 25 A |
| i | H— | $i$-$C_3H_7$— | $i$-$C_3H_7$— | $c$-$C_6H_{11}$ | 81 A |
| j | H— | $i$-$C_3H_7$— | $n$-$C_4H_9$— | $c$-$C_6H_{11}$ | 84 A |

# 8. Four-Component Condensations

| | | | | |
|---|---|---|---|---|
| k | H— | i-C$_3$H$_7$— | t-C$_4$H$_9$— | t-C$_4$H$_9$— | 76 A |
| l | H— | i-C$_3$H$_7$— | C$_6$H$_5$— | t-C$_4$H$_9$— | 69 A |
| m | H— | i-C$_3$H$_7$— | 2,6-(CH$_3$)$_2$—C$_6$H$_3$— | t-C$_4$H$_9$— | 62 A |
| n | H— | i-C$_4$H$_9$— | 4-NO$_2$—C$_6$H$_4$— | C$_6$H$_5$—CH$_2$— | 18 A |
| o | H— | i-C$_4$H$_9$— | —(CH$_2$)$_2$— | t-C$_4$H$_9$— | 61 A |
| p | H— | —(CH$_2$)$_3$— | C$_6$H$_5$— | t-C$_4$H$_9$— | 66 A |
| q | H— | c-C$_6$H$_{11}$— | n-C$_{12}$H$_{25}$— | 4-CH$_3$—C$_6$H$_4$— | 78 A |
| r | H— | C$_6$H$_5$— | n-C$_4$H$_9$— | c-C$_6$H$_{11}$— | 60 B |
| s | H— | C$_6$H$_5$— | C$_6$H$_5$— | c-C$_6$H$_{11}$— | 87 A |
| t | H— | 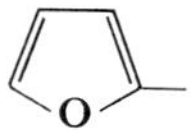 \<furyl\> | C$_6$H$_5$— | c-C$_6$H$_{11}$— | 53 A |
| u | CH$_3$— | CH$_3$— | c-C$_6$H$_{11}$— | c-C$_6$H$_{11}$— | 16 B |
| v | CH$_3$— | C$_6$H$_5$— | C$_6$H$_5$—CH$_2$— | c-C$_6$H$_{11}$— | 45 B |
| w | —(CH$_2$)$_5$— | | i-C$_3$H$_7$— | c-C$_6$H$_{11}$— | 67 B |
| x | H— | i-C$_3$H$_7$— | C$_6$H$_5$—CO—NH— | c-C$_6$H$_{11}$— | 41 B |
| y | H— | i-C$_3$H$_7$— | C$_6$H$_5$—CO—NH— | C$_6$H$_5$—CH$_2$— | 67 B |
| z | CH$_3$— | CH$_3$— | C$_6$H$_5$—CO—NH— | t-C$_4$H$_9$— | 75 B |

The hydantoin-4-imide synthesis only takes place in strongly polar solvents, such as methanol–water or methanol–ethylene glycol, and in moderately concentrated solution. This is probably due to the fact that under the aforementioned conditions, undissociated cyanic acid, as well as its α-aminoalkylation product, XXVII, occur only in small concentrations. Both are capable of undergoing irreversible side reactions, like the formation of the

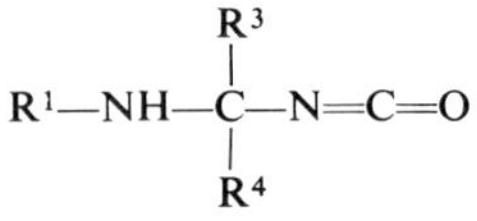

$$R^1-NH-\underset{R^4}{\overset{R^3}{\underset{|}{\overset{|}{C}}}}-N=C=O$$

(XXVII)

corresponding disubstituted urea resulting from addition of a molecule of primary amine ($R^1$—$NH_2$) at the cyanate carbon atom of XXVII.

As bifunctional components, both ethylenediamine and pentane-1,5-dial react bilaterally to form XIo and XIp.

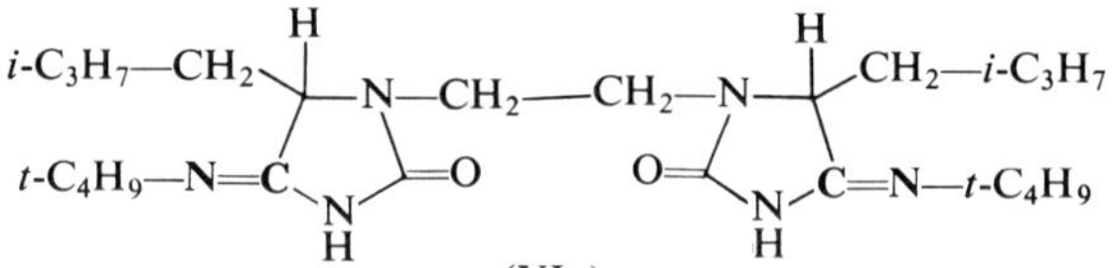

(XIo)

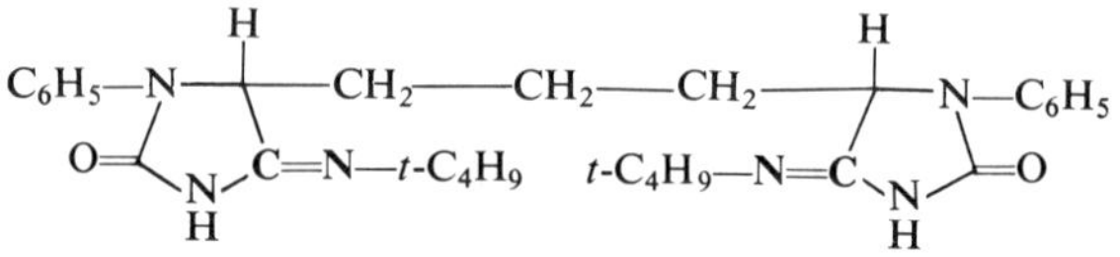

(XIp)

The union of 2,2,5,5-tetramethyl-$\Delta^3$-thiazoline with cyclohexyl isocyanide, and cyanic acid produces XXVIII, which can be desulfurized by means of Raney nickel to XIi.

The 4 C C synthesis of 2-thiohydantoin-4-imides proceeds well in concentrated methanol and chloroform solutions (see Table II).

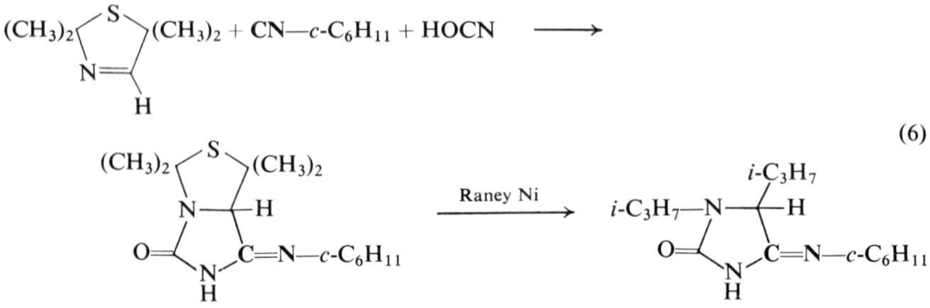

$$(CH_3)_2\underset{N=\underset{H}{\ }}{\overset{S}{\diagup}}(CH_3)_2 + CN-c\text{-}C_6H_{11} + HOCN \longrightarrow$$

(6)

(XXVIII) $\xrightarrow{\text{Raney Ni}}$ (XIi)

TABLE II

THE PREPARATION OF 2-THIOHYDANTOIN-4-IMIDES (XIIa–r) PREPARED BY CONDENSATION OF PRIMARY AMINES AND KETONES OR ACYL HYDRAZONES WITH ISONITRILES AND THIOCYANIC ACID IN METHANOL AT 0–20°C[4,48,75]

| Designation no. (XII) | $R^3$ (II) | $R^4$ (II) | $R^1$ (I) | $R^5$ (III) | Yield (%) |
|---|---|---|---|---|---|
| a | $CH_3$— | $CH_3$— | $i$-$C_3H_7$— | $c$-$C_6H_{11}$ | 61 |
| b | $CH_3$— | $CH_3$— | $C_6H_5$—$CH_2$— | $C_6H_5$—$CH_2$ | 71 |
| c | $CH_3$— | $CH_3$— | $C_6H_5$—$CH_2$— | 2,6-$(CH_3)_2C_6H_3$ | 73 |
| d | $CH_3$— | $CH_3$— | $C_6H_5$— | $c$-$C_6H_{11}$ | 65 |
| e | $CH_3$— | $C_2H_5$— | $C_6H_5$— | $t$-$C_4H_9$ | 79 |
| f | $C_2H_5$— | $C_2H_5$— | $C_6H_5$—$CH_2$— | $i$-$C_3H_7$ | 74 |
| g | —$(CH_2)_5$— | —$(CH_2)_5$— | $i$-$C_3H_7$— | $c$-$C_6H_{11}$ | 74[a] |
| h | —$(CH_2)_5$— | —$(CH_2)_5$— | $n$-$C_4H_9$— | $c$-$C_6H_{11}$ | 55 |
| i | —$(CH_2)_5$— | —$(CH_2)_5$— | $C_6H_5$—$CH_2$— | $n$-$C_4H_9$ | 90 |
| j | —$(CH_2)_5$— | —$(CH_2)_5$— | $C_6H_5$—$CH_2$— | $t$-$C_4H_9$ | 80 |
| k | —$(CH_2)_5$— | —$(CH_2)_5$— | $C_6H_5$— | $c$-$C_6H_{11}$ | 63 |
| l | —$(CH_2)_5$— | —$(CH_2)_5$— | 4-Cl—$C_6H_4$— | $i$-$C_3H_7$ | 61 |
| m | —$(CH_2)_5$— | —$(CH_2)_5$— | —$(CH_2)_2$— | $i$-$C_3H_7$ | 89 |
| n | $CH_3$— | $C_6H_5$—$CH_2$— | $C_6H_5$— | $c$-$C_6H_{11}$ | 61 |
| o | $CH_3$— | 2,4-$(CH_3O)_2$—$C_6H_3$—$CH_2$— | $C_6H_5$—$CH_2$— | $c$-$C_6H_{11}$ | 71 |
| p | H— | $i$-$C_3H_7$— | $C_6H_5$—CO—NH— | $t$-$C_4H_9$ | 83 |
| q | $C_2H_5$— | $C_2H_5$— | $C_6H_5$—$SO_2$—NH— | $c$-$C_6H_{11}$ | 78 |
| r | —$(CH_2)_5$— | | OHC—NH— | $t$-$C_4H_9$ | 92 |

[a] Yield in chloroform: 81%.

Structural proofs for the hydantoin-4-imides and the corresponding thiohydantoin-4-imides were carried out according to (7). The combination of the infrared spectra, the empirical formulas, and the alkaline hydrolysis products confirmed structures XI and XII. The hydantoin-4-imides and the 2-thiohydantoin-4-imides are extremely resistant to thermal decomposition and to both acidic and alkaline hydrolysis. Drastic degradation procedures had to be employed for them. Reductive desulfurization was particularly useful for the elucidation of the structure of the thio compounds.

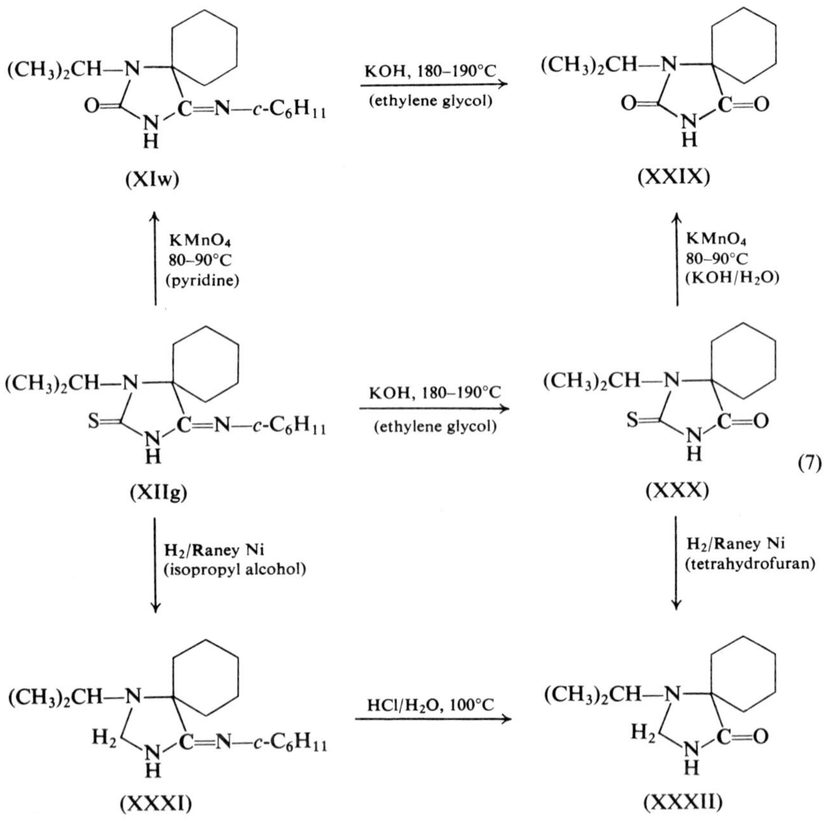

(7)

Oxidative desulfurization of XIIg to XIw is possible with potassium permanganate in pyridine at 80–90°C; XIIg can be reductively desulfurized to XXXI by heating with Raney nickel in isopropyl alcohol.

The product XIw of oxidative desulfurization of XIIg is identical with the condensation product of isopropyl amine hydrochloride, cyclohexanone, cyclohexyl isocyanide, and potassium cyanate; it can be hydrolyzed by means of KOH in ethylene glycol at 180–190°C to yield XXIX and cyclohexyl

8. Four-Component Condensations    155

amine. XXIX is also obtained by oxidative desulfurization of XXX with alkaline $KMnO_4$ solution at 80–90°C. The constitution of XXIX was proved by synthesis from 1-isopropylamino-1-cyanocyclohexane and KOCN according to the method of Long.[28]

The product (XXXI) of the reductive desulfurization of XIIg is cleaved by heating with aqueous HCl to form 1-isopropyl-5,5-pentamethyleneimidazolidone-4 (XXXII). The constitution of the latter corresponds to its formation from XXX by reductive desulfurization by means of Raney-Ni in tetrahydrofuran.

*p*-Nitrophenyl cyanamide condenses in 93% yield with cyclohexanone-*n*-butyl imide, and *t*-butyl isocyanide to form a product whose empirical formula is $C_{22}H_{33}N_5O_2$.[42] It must be assumed that in this case *p*-nitrophenyl cyanamide reacts in a fashion analogous to that for cyanic acid. The exact structure of this compound, however, was not elucidated.

### III. ACYLATED α-AMINO- AND α-HYDRAZINOCARBONAMIDES

#### A. α-Acylaminocarbonamides

The combination of ammonia or primary amines (I) and aldehydes or ketones (II) reacts with carboxylic acids (XXXIII) and isonitriles (III) to form the intermediate α-adducts (X). These undergo spontaneous O → N acyl transfer (2) by a cyclic mechanism yielding α-acylaminocarbonamides (XIII) (see Table III and Chapter 9). The formation of the two carbonamide

$$R^1\text{—}NH_2 + R^3\text{—}CO\text{—}R^4 + R^6\text{—}CO_2H + CN\text{—}R^5 \rightarrow$$
(I)          (II)              (XXXIII)          (III)

$$\begin{array}{c} R^4 \\ | \\ R^3\text{—}C\text{——}C\text{=}N\text{—}R^5 \\ | \\ R^1\text{—}NH \quad O \\ \quad \diagup \\ R^6\text{—}C \\ \quad \diagdown O \end{array} \rightarrow \begin{array}{c} R^1 \quad R^3 \\ | \quad\; | \\ R^6\text{—}CO\text{—}N\text{—}C\text{—}CO\text{—}NH\text{—}R^5 \\ | \\ R^4 \end{array} \quad (8)$$

(X)                                    (XIII)

groups in the final product (XIII) provides the reaction with a strong driving force. Because of its cyclic mechanism (see also Chapter 9) the O → N acyl transfer competes very effectively with possible side reactions of the α-adducts (X). Even α-aminocarbonamides which are so loaded with bulky groups that serious steric repulsion of some parts of the molecules cannot be avoided are often formed smoothly, e.g., XIIIk, in 87% yield. Yet, there is a limit, beyond

## TABLE III

### THE PREPARATION OF α-ACYLAMINOCARBONAMIDES (XIII) BY 4 C C OF PRIMARY AMINES AND CARBONYL COMPOUNDS (OR SCHIFF BASES) WITH CARBOXYLIC ACIDS AND ISONITRILES

| Designation no. (XIII) | $R^3$ (II) | $R^4$ (II) | $R^1$ (I) | $R^6$ (XXXIII) | $R^5$ (III) | Yield (%) | Ref. |
|---|---|---|---|---|---|---|---|
| a | H— | H— | $i$-$C_3H_7$— | $C_6H_5$—$CH_2$— | $c$-$C_6H_{11}$— | 72 | 78 |
| $b_1$ | H— | $i$-$C_3H_7$— | H— | H— | $c$-$C_6H_{11}$— | 54 | 78 |
| $b_2$ | H— | $i$-$C_3H_7$— | H— | $CH_3$— | $c$-$C_6H_{11}$— | 52 | 78 |
| $b_3$ | H— | $i$-$C_3H_7$— | H— | $CH_3$—CO—NH—$CH_2$— | $c$-$C_6H_{11}$— | 40 | 78 |
| $b_4$ | H— | $i$-$C_3H_7$— | H— | Phthalimido-$CH_2$— | $t$-$C_4H_9$— | 72 | 78 |
| c | H— | $i$-$C_3H_7$— | $n$-$C_3H_7$— | $C_6H_5$— | $c$-$C_6H_{11}$— | 80 | 78 |
| $d_1$ | H— | $i$-$C_3H_7$— | $C_6H_5$—$CH_2$— | $C_2H_5$— | $c$-$C_6H_{11}$— | 78 | 78 |
| $d_2$ | H— | $i$-$C_3H_7$— | $C_6H_5$—$CH_2$— | $CH_3CO$—NH—$CH_2$— | $i$-$C_3H_7$— | 92 | 78 |
| $d_3$ | H— | $i$-$C_3H_7$— | $C_6H_5$—$CH_2$— | $CH_3CO$—NH—$CH_2$— | $n$-$C_4H_9$— | 85 | 78 |
| $d_4$ | H— | $i$-$C_3H_7$— | $C_6H_5$—$CH_2$— | $CH_3CO$—NH—$CH_2$— | $t$-$C_4H_9$— | 94 | 78 |
| $d_5$ | H— | $i$-$C_3H_7$— | $C_6H_5$—$CH_2$— | $CH_3CO$—NH—$CH_2$— | $c$-$C_6H_{11}$— | 88 | 78 |
| $d_6$ | H— | $i$-$C_3H_7$— | $C_6H_5$—$CH_2$— | $CH_3CO$—NH—$CH_2$— | $C_6H_5$—$CH_2$— | 95 | 78 |
| e | H— | $CH_3$— | $C_6H_5$—$O_2C$—$CH_2$—$CH_2$— | Phthalimido-$CH_2$— | $t$-$C_4H_9$— | 92 | 78 |
| $f_1$ | H— | $i$-$C_3H_7$— | $C_2H_5O_2C$—$CH_2$—$CH(CH_3)$— | $CH_3$— | $C_2H_5$— | 89 | 71 |
| $f_2$ | H— | $i$-$C_3H_7$— | NC—$CH_2$—$CH(CH_3)$— | $C_6H_5$— | $c$-$C_6H_{11}$— | 98 | 71 |
| $f_3$ | H— | $i$-$C_3H_7$— | $C_2H_5O_2C$—$CH_2$—$CH(CO_2C_2H_5)$— | $C_6H_5$— | $c$-$C_6H_{11}$— | 98 | 71 |
| $f_4$ | H— | $i$-$C_3H_7$— | $C_2H_5O_2C$—$CH_2$—$CH(CO_2C_2H_5)$— | $C_6H_5$— | $t$-$C_4H_9$— | 92 | 71 |
| $f_5$ | H— | $i$-$C_3H_7$— | $C_2H_5O_2C$—$CH_2$—$CH(C_6H_5)$— | $C_6H_5$— | $C_6H_5$— | 87 | 71 |
| $f_6$ | H— | $i$-$C_3H_7$— | $C_2H_5O_2C$—$CH_2$—$CH(C_6H_5)$— | $C_6H_5$— | $t$-$C_4H_9$— | 84 | 71 |
| $f_7$ | H— | $i$-$C_3H_7$— | $C_2H_5O_2C$—$CH_2$—$CH(C_6H_5)$— | $C_6H_5$— | $c$-$C_4H_9$— | 92 | 71 |
| $f_8$ | H— | $H_3C$—S—$(CH_2)_2$— | $C_2H_5O_2C$—$CH_2$—$CH(C_6H_5)$— | Phthalimido-$CH_2$— | $(CH_3)_3CO_2C$—$CH_2$— | 96 | 71 |
| $f_9$ | H— | $C_6H_5$— | $C_2H_5O_2C$—$CH_2$—$CH(C_6H_5)$— | $C_6H_5$— | $C_6H_5$— | 94 | 71 |
| $f_{10}$ | H— | $i$-$C_3H_7$— | $C_2H_5O_2C$—$CH_2$=$C(CH_3)$— | $C_6H_5$— | $2,6$-$(CH_3)$—$C_6H_{11}$— | 39 | 71 |
| $g_1$ | H— | $i$-$C_3H_7$— | $CH_3$—CO—CH=$C(CH_3)$— | Phthalimido-$CH_2$— | $t$-$C_4H_9$—$O_2C$—$CH_2$— | 89 | 71 |
| $g_2$ | H— | $i$-$C_3H_7$— | NC—CH=$C(CH_3)$— | Phthalimido-$CH_2$— | $t$-$C_4H_9$—$O_2C$—$CH_2$— | 47 | 71 |
| $g_3$ | H— | $i$-$C_3H_7$— | $CH_3$—CO—CH=$C(CH_3)$— | Phthalimido-$CH_2$— | $t$-$C_4H_9$—$O_2C$—$CH_2$— | 17 | 71 |
| $g_4$ | H— | $i$-$C_3H_7$— | $2$-$C_2H_5O_2C$—$\Delta^1$-$c$-$C_5H_6$— | Phthalimido-$CH_2$— | $t$-$C_4H_9$—$O_2C$—$CH_2$— | 19 | 71 |
| $g_5$ | H— | $i$-$C_3H_7$— | $2$-$C_2H_5O_2C$—$\Delta^1$-$c$-$C_6H_8$— | Phthalimido-$CH_2$— | $t$-$C_4H_9$—$O_2C$—$CH_2$— | 45 | 71 |
| $h_1$ | H— | $i$-$C_3H_7$— | $C_6H_5$—$CH_2$— | $CH_3CO$—NH—$CH_2$— | $t$-$C_4H_9$— | 62 | 71 |
|  |  |  |  |  |  | 88 | 78 |

## 8. Four-Component Condensations

| | | | | | | |
|---|---|---|---|---|---|---|
| h₂ | H— | $i$-C₃H₇— | C₆H₅—CH₂— | CF₃CO—NH—CH₂— | $t$-C₄H₉— | 87 | 78 |
| i₁ | H— | H— | C₆H₅—CH₂— | Phthalimido-CH₂— | $t$-C₄H₉— | 83 | 78 |
| i₂ | H— | CH₃— | C₆H₅—CH₂— | Phthalimido-CH₂— | $t$-C₄H₉— | 73 | 78 |
| i₃ | H— | △O | C₆H₅—CH₂— | Phthalimido-CH₂— | $t$-C₄H₉— | 57 | 78 |
| i₄ | H— | CH₃—S—(CH₂)₂— | C₆H₅—CH₂— | Phthalimido-CH₂— | $t$-C₄H₉— | 69 | 78 |
| i₅ | H— | $i$-C₃H₇— | C₆H₅—CH₂— | Phthalimido-CH₂— | $t$-C₄H₉— | 92 | 78 |
| i₆ | H— | $i$-C₄H₉— | C₆H₅—CH₂— | Phthalimido-CH₂— | $t$-C₄H₉— | 78 | 78 |
| i₇ | H— | C₆H₅—CH(CH₃)— | C₆H₅—CH₂— | Phthalimido-CH₂— | $t$-C₄H₉— | 74 | 78 |
| i₈ | H— | C₆H₅— | C₆H₅—CH₂— | Phthalimido-CH₂— | $t$-C₄H₉— | 91 | 78 |
| i₉ | H— | 2-C₄H₃S— | C₆H₅—CH₂— | Phthalimido-CH₂— | $t$-C₄H₉— | 61 | 78 |
| i₁₀ | —(CH₂)₅— | | C₆H₅—CH₂— | Phthalimido-CH₂— | $t$-C₄H₉— | 90 | 78 |
| i₁₁ | CH₃— | C₆H₅— | C₆H₅—CH₂— | Phthalimido-CH₂— | $t$-C₄H₉— | 91 | 78 |
| j | H— | $i$-C₃H₇— | C₆H₅—CH₂— | Phthalimido-CH(CH₃)— | $t$-C₄H₉— | 76 | 16 |
| k | H— | $i$-C₃H₇— | C₆H₅—CH₂— | CH₃—CO—(CH₂)₂— | $c$-C₆H₁₁— | 40 | 16 |
| l₁ | H— | $i$-C₃H₇— | C₆H₅—CH₂— | (CH₃O)₂CH— | $c$-C₆H₁₁— | — | 16 |
| l₂ | —(CH₂)₅— | | C₆H₅—CH₂— | Cl₂CH— | $c$-C₆H₁₁— | — | 16 |
| l₃ | —(CH₂)₅— | | C₆H₅—CH₂— | (CH₃O)₂CH— | $c$-C₆H₁₁— | — | 16 |
| l₄ | —(CH₂)₅— | | C₆H₅—CH₂— | CH₃— | $c$-C₆H₁₁— | — | 16 |
| m | H— | C₆H₅— | $c$-C₆H₁₁— | | $c$-C₆H₁₁— | 78 | 78 |
| n | —(CH₂)₅— | N—CO— | $n$-C₄H₉— | C₆H₅— | $c$-C₆H₁₁— | 55 | 16 |
| o | —(CH₂)₄— | | $i$-C₃H₇— | H— | —1,4-$c$-C₆H₁₀— | 29 | 39 |
| p₁ | —(CH₂)₅— | | $i$-C₃H₇— | H— | 3-C₅H₄N—CH₂ | 29 | 40 |
| p₂ | —(CH₂)₅— | | $i$-C₃H₇— | H— | 2-C₄H₃O—CH₂— | 30 | 40 |
| q₁ | —(CH₂)₅— | | $n$-C₄H₉— | H— | —1,4-$c$-C₆H₁₀—C₆H₁₃— | 29 | 39 |
| q₂ | —(CH₂)₅— | | $n$-C₄H₉— | ClCH— | $c$-C₆H₁₁— | 93 | 78 |
| q₃ | —(CH₂)₅— | | $n$-C₄H₉— | CF₃— | 2,6-(CH₃)₂—C₆H₃— | 84 | 78 |
| q₄ | —(CH₂)₅— | | $n$-C₄H₉— | OCH—NH—CH₂— | $t$-C₄H₉— | 75 | 78 |
| q₅ | —(CH₂)₅— | | $n$-C₄H₉— | OCH—NH—CH₂— | $c$-C₆H₁₁— | 95 | 78 |
| q₆ | —(CH₂)₅— | | $n$-C₄H₉— | CH₃CO—NH—CH₂— | $c$-C₆H₁₁— | 91 | 78 |
| r | —(CH₂)₅— | | C₆H₅—CH₂— | H— | Δ¹-$c$-C₆H₉— | 69 | 48, 74 |

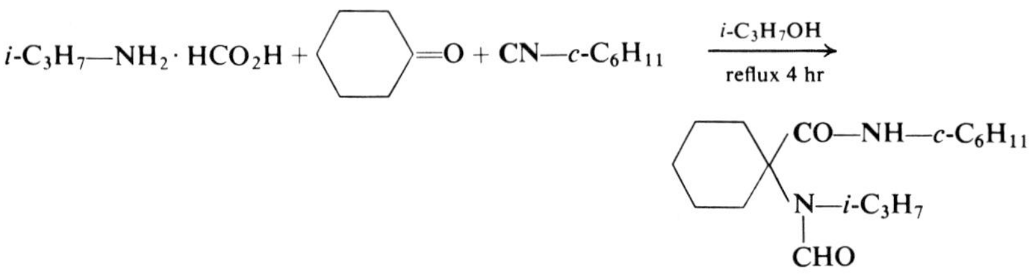

$i\text{-}C_3H_7\text{—}NH_2 \cdot HCO_2H +$ [cyclohexanone] $=O + CN\text{—}c\text{-}C_6H_{11}$ $\xrightarrow[\text{reflux 4 hr}]{i\text{-}C_3H_7OH}$

[cyclohexane with] CO—NH—$c$-$C_6H_{11}$ / N—$i$-$C_3H_7$ / CHO

(XIIIk)

(9)

which the secondary reaction fails to occur.

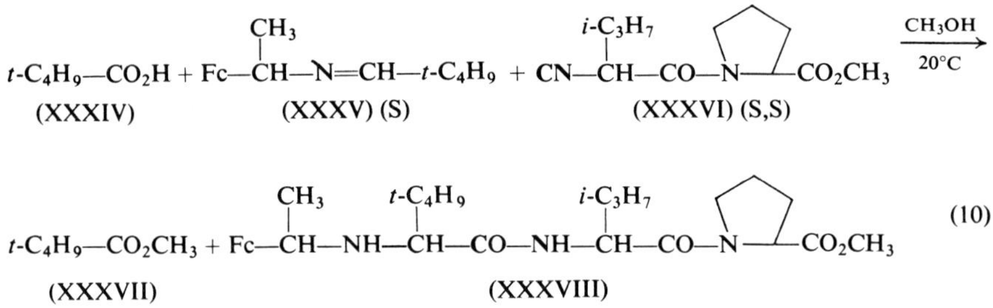

$t\text{-}C_4H_9\text{—}CO_2H + Fc\text{—}\underset{\underset{CH_3}{|}}{CH}\text{—}N=CH\text{—}t\text{-}C_4H_9 + CN\text{—}\underset{\underset{i\text{-}C_3H_7}{|}}{CH}\text{—}CO\text{—}N\text{—}[\text{cyclopentane}]\text{—}CO_2CH_3 \xrightarrow[20°C]{CH_3OH}$

(XXXIV)  (XXXV) (S)  (XXXVI) (S,S)

$t\text{-}C_4H_9\text{—}CO_2CH_3 + Fc\text{—}\underset{\underset{CH_3}{|}}{CH}\text{—}NH\text{—}\underset{\underset{t\text{-}C_4H_9}{|}}{CH}\text{—}CO\text{—}NH\text{—}\underset{\underset{i\text{-}C_3H_7}{|}}{CH}\text{—}CO\text{—}N\text{—}[\text{cyclopentane}]\text{—}CO_2CH_3$

(XXXVII)    (XXXVIII)

(10)

The components XXXIV–XXXVI of reaction (10) undergo a 4 C C type of reaction only in solvents that are acylatable by the α-adduct; in methanol, XXXVII and XXXVIII are formed.[20] If the "isocyano peptide" (XXXVI) of reaction (10) is replaced by $t$-butyl isocyanide, the components do not react at all even in methanol.

An analogous solvent participation is also observed when α-amino acids are reacted with aldehydes and isonitriles, e.g. (11), as the six-membered cyclic α-adduct (XXXIX) is not capable of undergoing transannular O → N acyl transfer to form an α-lactam (see Section V).[57]

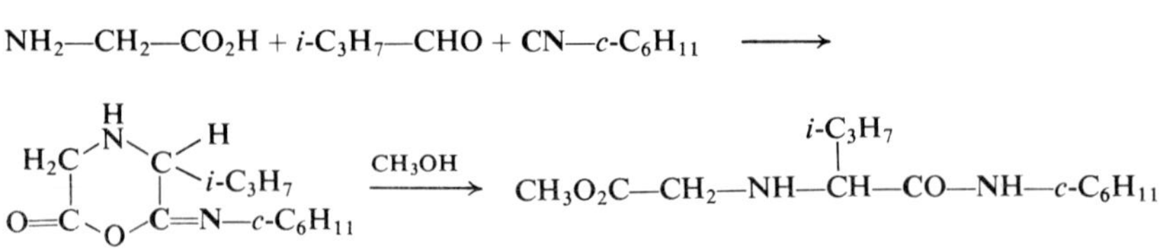

$NH_2\text{—}CH_2\text{—}CO_2H + i\text{-}C_3H_7\text{—}CHO + CN\text{—}c\text{-}C_6H_{11} \longrightarrow$

[six-membered ring: $H_2C$—N(H)—C(H)($i$-$C_3H_7$) / O=C—O—C=N—$c$-$C_6H_{11}$]

(XXXIX)

$\xrightarrow{CH_3OH}$ $CH_3O_2C\text{—}CH_2\text{—}NH\text{—}\underset{\underset{i\text{-}C_3H_7}{|}}{CH}\text{—}CO\text{—}NH\text{—}c\text{-}C_6H_{11}$

(11)

The reaction [e.g., (12)] of Schiff bases with α-isocyano acids leads to the formation of α-aminoalkyl azlactones, remarkably stable cyclic 4 C C α-adducts.[29]

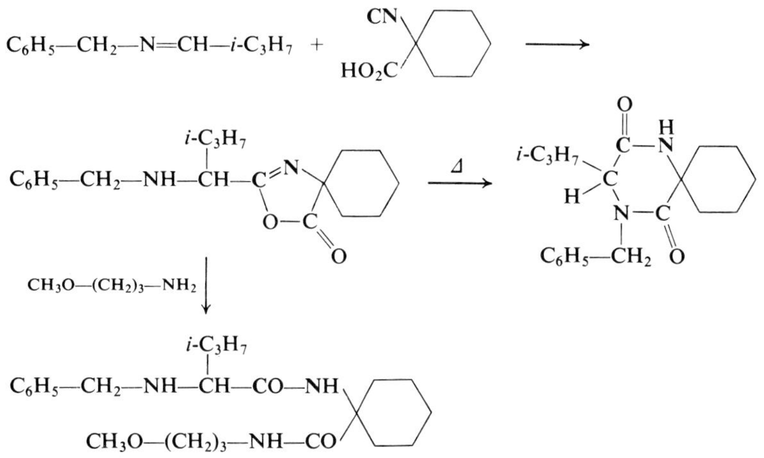

## 8. Four-Component Condensations

$C_6H_5-CH_2-N=CH-i-C_3H_7$ + [cyclohexane with CN and $HO_2C$ substituents] →

[structure with $i-C_3H_7$, $C_6H_5-CH_2-NH-CH-C$ bonded to N and O-C=O, spiro cyclohexane]  $\xrightarrow{\Delta}$  [bicyclic product with $i-C_3H_7$, $C_6H_5-CH_2$, N, C=O groups]   (12)

$CH_3O-(CH_2)_3-NH_2 \downarrow$

$C_6H_5-CH_2-NH-\underset{\underset{\displaystyle i\text{-}C_3H_7}{|}}{CH}-CO-NH$ [cyclohexyl]

$CH_3O-(CH_2)_3-NH-CO$

**B. α-$N_\alpha$,$N_\beta$-Diacylhydrazino- and α-$N_\alpha$-Acyl-$N_\beta$-alkylidenehydrazinocarbonamides**

The union of acyl hydrazones (XL) with carboxylic acids and isonitriles produces α-$N_\alpha$,$N_\beta$-diacylhydrazinocarbonamides (XLI) (see Table IV).[63]

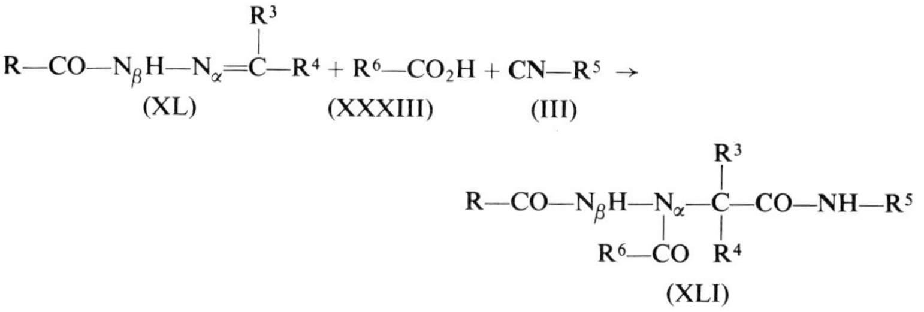

$R-CO-N_\beta H-N_\alpha=\underset{\underset{\displaystyle R^4}{|}}{\overset{\overset{\displaystyle R^3}{|}}{C}}-R^4$ + $R^6-CO_2H$ + $CN-R^5$ →
(XL)                  (XXXIII)           (III)

$R-CO-N_\beta H-N_\alpha-\underset{\underset{\displaystyle R^6-CO}{}}{\overset{\overset{\displaystyle R^3}{|}}{C}}-CO-NH-R^5$   (13)
                                    $R^4$
(XLI)

The structure of these products follows from their IR spectra and their reductive cleavage (14), e.g., of XLIc$_1$.[4]

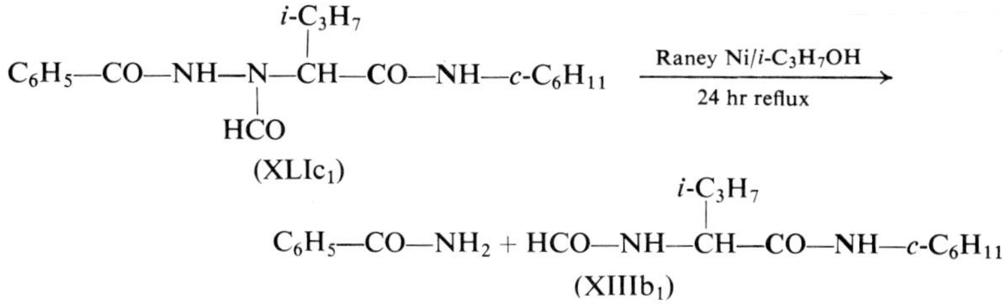

$C_6H_5-CO-NH-\underset{\underset{\displaystyle HCO}{|}}{N}-\overset{\overset{\displaystyle i\text{-}C_3H_7}{|}}{CH}-CO-NH-c\text{-}C_6H_{11}$ $\xrightarrow[\text{24 hr reflux}]{\text{Raney Ni}/i\text{-}C_3H_7OH}$
(XLIc$_1$)

$C_6H_5-CO-NH_2$ + $HCO-NH-\overset{\overset{\displaystyle i\text{-}C_3H_7}{|}}{CH}-CO-NH-c\text{-}C_6H_{11}$   (14)
                              (XIIIb$_1$)

TABLE IV

THE PREPARATION OF $\alpha$-$N_\alpha,N_\beta$-DIACYLHYDRAZINOCARBONAMIDES (XLI) BY 4 C C FROM ACYLHYDRAZONES (XL), CARBOXYLIC ACIDS, AND ISONITRILES[4,63]

| Designation no. (XLI) | $R^3$ (XL) | $R^4$ (XL) | R (XL) | $R^6$ (XXXIII) | $R^5$ (III) | Yield (%) |
|---|---|---|---|---|---|---|
| a | H— | $i$-$C_3H_7$— | $C_2H_5$—O— | $CHCl_2$— | $t$-$C_4H_9$— | 80 |
| $b_1$ | H— | $i$-$C_3H_7$— | $C_6H_5$— | H— | $t$-$C_4H_9$— | 76 |
| $b_2$ | H— | $i$-$C_3H_7$— | $C_6H_5$— | $CH_2Cl$— | $t$-$C_4H_9$— | 79 |
| $c_1$ | H— | $i$-$C_3H_7$— | $C_6H_5$— | H— | $c$-$C_6H_{11}$— | 80 |
| $c_2$ | H— | $i$-$C_3H_7$— | $C_6H_5$— | $CH_3$— | $c$-$C_6H_{11}$— | 26 |
| d | H— | $C_6H_5$—$CH_2$— | $C_6H_5$— | $C_6H_5$—$CH_2$— | $t$-$C_4H_9$— | 31 |
| e | $CH_3$— | $CH_3$— | H— | $CH_2Cl$— | $t$-$C_4H_9$— | 65 |
| f | $CH_3$— | $CH_3$— | $H_2N$— | $CH_2Cl$— | $t$-$C_4H_9$— | 80 |
| $g_1$ | $CH_3$— | $CH_3$— | $C_6H_5$— | $CH_3$— | $t$-$C_4H_9$— | 61 |
| $g_2$ | $CH_3$— | $CH_3$— | $C_6H_5$— | $CH_2Cl$— | 2,6-$(CH_3)_2$—$C_6H_3$— | 25 |
| $g_3$ | $CH_3$— | $CH_3$— | $C_6H_5$— | NC—$CH_2$— | $t$-$C_4H_9$— | 45 |
| $g_4$ | $CH_3$— | $CH_3$— | $C_6H_5$— | $ClCH_2$—$CH_2$— | $t$-$C_4H_9$— | 48 |
| $g_5$ | $CH_3$— | $CH_3$— | $C_6H_5$— | $t$-$C_4H_9$— | $t$-$C_4H_9$— | 33 |
| $g_6$ | $CH_3$— | $CH_3$— | $C_6H_5$— | $C_6H_5$—O—$CH_2$— | $t$-$C_4H_9$— | 52 |
| $g_7$ | $CH_3$— | $CH_3$— | $C_6H_5$— | $C_6H_5$— | $t$-$C_4H_9$— | 56 |
| $g_8$ | $CH_3$— | $CH_3$— | $C_6H_5$— | Phthalimido-$CH_2$— | $t$-$C_4H_9$— | 61 |
| $g_9$ | $CH_3$— | $CH_3$— | $C_6H_5$— | 1-acetamino-2-[indolyl-(3)]-ethyl— | $t$-$C_4H_9$— | 46 |
| h | —$(CH_2)_5$— | | $t$-$C_4H_9$— | $CH_2Cl$— | $t$-$C_4H_9$— | 60 |

8. Four-Component Condensations    161

Diethyl ketone sulfonylhydrazone reacts, in analogy to (13), with formic acid and $t$-butyl isocyanide to yield 58% of XLII.

The combination of the azines of isobutyraldehyde or acetone (XLIIIa or XLIIIb) with acetic acid and $t$-butyl isocyanide results in 51 or 50%, respectively, of XLIVa and XLIVb.[63]

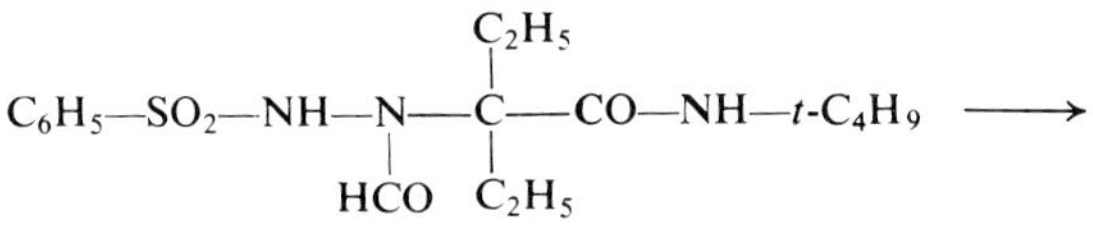

$$C_6H_5-SO_2-NH-N-\underset{\underset{C_2H_5}{|}}{\overset{\overset{C_2H_5}{|}}{C}}-CO-NH-t-C_4H_9 \longrightarrow$$
$$\phantom{C_6H_5-SO_2-NH-N-}HCO$$
(XLII)

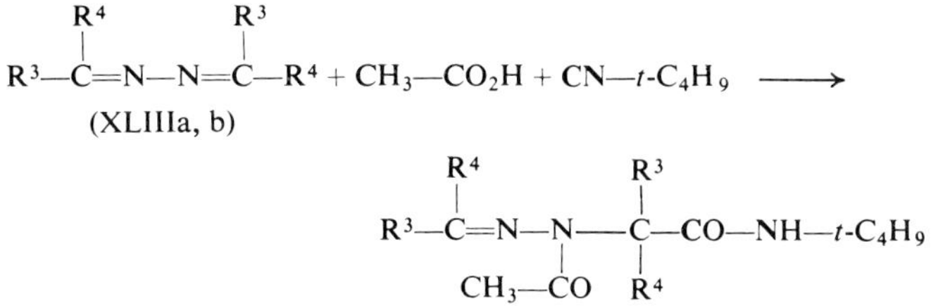

$$R^3-\underset{\underset{}{}}{\overset{\overset{R^4}{|}}{C}}=N-N=\underset{}{\overset{\overset{R^3}{|}}{C}}-R^4 + CH_3-CO_2H + CN-t-C_4H_9 \longrightarrow$$
(XLIIIa, b)

$$R^3-\underset{\underset{CH_3-CO}{|}}{\overset{\overset{R^4}{|}}{C}}=N-N-\underset{\underset{R^4}{|}}{\overset{\overset{R^3}{|}}{C}}-CO-NH-t-C_4H_9 \quad (15)$$
(XLIV)

a: $R^3 = H$, $R^4 = i\text{-}C_3H_7-$
b: $R^3 = R^4 = CH_3-$

It is interesting to note that (XLIIIa and XLIIIb) do not undergo bilateral 4 C C with excess acetic acid and $t$-butyl isocyanide although the products of (15) are acylhydrazones [see (13)].

## IV. STEREOSELECTIVE SYNTHESES AND THE REACTION MECHANISM OF 4 C C

4 C C which involve unsymmetrically substituted carbonyl compounds ($R^3$—CO—$R^4$, where $R^3 \neq R^4$) lead to the formation of products with new centers of chirality.[21] Because of this, the 4 C C in which chiral components take part lead to mixtures of diastereoisomers.

In the absence of asymmetric induction,*[5,13,25,31,34,35,46,47,58] the diastereoisomers are formed at equal rates, i.e., in equal amounts. The stereoselectivity of asymmetrically induced syntheses depends on the "chemical chirality"[50] of the chiral reference system, i.e., the asymmetric inducing power of an element of chirality, which may be described as its "chirality

---
* That is, the formation of new elements of chirality in the presence of chiral reference systems under conditions under which stereoisomers are produced in different amounts.

product."[49,50,61] The chemical chirality of the reference system is determined by three parameters, viz., its distance from the newly formed element of chirality, the chemical properties of the reactants, and the reaction conditions.[12,59,60,62]

The dependence of the ratio $Q_{pn}$ of the $p$- and $n$-products* of stereoselective 4 C C upon all these factors has to be investigated in order to obtain the information which is needed for synthesizing the desired stereoisomers of 4 C C products in optimum yields. Tables V and VI contain experimental data on some model reactions carried out according to (16) and (18).

The structure and configuration of the stereoisomers (XLVI) which were isolated follow from their hydrolyses by hydrochloric acid to the optically active amino acids, $NH_2$—$CHR^3$—$CO_2H$. These hydrolyses also remove the $N$-aralkyl groups. The diastereoisomer ratios $Q$ of the products (XLVI) were either obtained from the optical rotation of the diastereoisomer mixtures or by chromatographic analysis.[17,51,73]

$$R^3\text{---CHO} \quad + \quad C_6H_5\text{---}CO_2H \quad + \quad CN\text{---}t\text{-}C_4H_9$$

```
  R'—CH₂                                    R
    |                                       |
  R—C—NH₂                           R'—CH₂—C—NH₂
    |                                       |
    H                                       H
  (XLV) (S)  ↓                         (XLV) (R)  ↓

  R'—CH₂   R³                          R       R³
    |      |                           |       |
  R—C—N———C—CO—NH—t-C₄H₉       R'—CH₂—C—N———C—CO—NH—t-C₄H₉
    |  |   |                           |   |   |
    H  CO  H                           H   CO  H
       |                                   |
       C₆H₅                                C₆H₅
   (XLVI) (S) (S)                      (XLVI) (R) (S)            (16)

        +                                   +

  R'—CH₂   CO—NH—t-C₄H₉                R       CO—NH—t-C₄H₉
    |      |                           |       |
  R—C—N———C—R³                   R'—CH₂—C—N———C—R³
    |  |   |                           |   |   |
    H  CO  H                           H   CO  H
       |                                   |
       C₆H₅                                C₆H₅
   (XLVI) (S) (R)                      (XLVI) (R) (R)
```

* If there are two centers of chirality within a molecule, the molecule is termed "positive" if the configurations according to the (R)(S) nomenclature[11] are the same [(R)(R) or (S)(S)] and "negative" if different [(S)(R) or (R)(S)].[12,49,50,61] The "$p$–$n$" nomenclature is less ambiguous than the "threo–erythro" and is advantageous for the theoretical treatment of stereochemistry.

## TABLE V

Preparation of $\alpha$-$N$-Benzoylamino acid-$N'$-$t$-butylamides (XLVI) by Stereoselective 4 C C of Benzoic Acid, Aldehydes, Optically Active Primary Amines, and $t$-Butyl Isocyanide According to (16)

| $R^3$ (II) | R (XLV) | R' (XLV) | Solvent[a] | Initial concentration (moles/kg)[b] | Temp. (°C) | Configuration of the main product | | Stereo-selectivity ($Q_{pn}$) | Ref. |
|---|---|---|---|---|---|---|---|---|---|
| $n$-C$_3$H$_7$— | C$_6$H$_5$— | H— | M | 1.0 | −70 | (S) | (R) | 0.43 | 18, 51 |
| $n$-C$_3$H$_7$— | C$_6$H$_5$— | H— | M | 0.1 | 0 | (S) | (S) | 2.50 | 18, 51 |
| $i$-C$_3$H$_7$— | C$_6$H$_5$— | H— | E | 1.6 | −70 | (R) | (S)[c] | 0.39 | 17 |
| $i$-C$_3$H$_7$— | C$_6$H$_5$— | H— | M | 0.29 | 0 | (R) | (R)[c] | 2.50 | 17 |
| $i$-C$_3$H$_7$— | C$_6$H$_5$— | H— | M | 0.05 | −40 | (S) | (S)[c] | 3.35 | 17 |
| $i$-C$_3$H$_7$— | C$_6$H$_5$— | H— | M | 1.0 | −78 | (S) | (R)[c] | 0.33 | 17 |
| $i$-C$_3$H$_7$— | C$_6$H$_5$— | C$_2$H$_5$O$_2$C— | E | 0.88 | 0 | (R) | (S) | 0.56 | 70 |
| $i$-C$_3$H$_7$— | C$_2$H$_5$O$_2$C— | C$_2$H$_5$O$_2$C— | E | 0.89 | 0 | (S) | (R) | 0.47 | 70 |
| $i$-C$_3$H$_7$— | C$_{10}$H$_9$Fe— | H— | M | 1.00 | −60 | (S) | (R) | 0.59 | 20 |
| $i$-C$_3$H$_7$— | C$_{10}$H$_9$Fe— | H— | M | 0.0375 | 0 | (S) | (S) | 3.76 | 20 |
| C$_2$H$_5$CH(CH$_3$)—[d] | C$_6$H$_5$— | H— | E | 1.9 | −70 | (S) | (R) (S) | 0.40 | 17 |
| C$_2$H$_5$CH(CH$_3$)—[d] | C$_6$H$_5$— | H— | M | 0.38 | +10 | (S) | (R) (R) | 2.69 | 17 |
| C$_2$H$_5$CH(CH$_3$)—[d] | C$_6$H$_5$— | H— | M | 0.38 | +10 | (S) | (S) (S) | 2.69 | 17 |
| C$_2$H$_5$CH(CH$_3$)—[d] | C$_6$H$_5$— | H— | E | 2.1 | −70 | (S) | (S) (R) | 0.47 | 17 |

[a] E = ethanol; M = methanol.
[b] Initial concentration of the equimolar starting materials.
[c] Reaction according to (17).
[d] (S)-2-Methyl butanal.[2]

The four diastereoisomeric 4 C C products (XLVIII) [(S)(S)(S) + (S)(S)(R) + (S)(R)(S) + (S)(R)(R)] are formed (17) from the components in methanol at $C_0 = 0.10$ moles/kg and 0°C in a ratio of 60:16:9:15, and at $C_0 = 1.00$ moles/kg in a ratio of 38:13:14:36.[30]

$C_6H_5$—CH($CH_3$)—$NH_2$ + $C_6H_5$—CH($CH_3$)—CHO + $C_6H_5$—$CO_2H$ + CN—$t$-$C_4H_9$ →
(XLV) (S)                          (XLVII), racemic

$$\underset{\text{(XLVIII) [(S) (S) (S) + (S) (S) (R) + (S) (R) (S) + (S) (R) (R)]}}{C_6H_5-CH(CH_3)-\underset{\underset{C_6H_5-CO}{|}}{N}-\overset{\overset{CH_3-CH-C_6H_5}{|}}{CH}-CO-NH-t\text{-}C_4H_9} \quad (17)$$

This result shows that the antipodes of XLVII are rapidly interconverted and that simultaneous asymmetric induction occurs at two centers of chirality.

The 4 C C of isobutyraldehyde-(S)-α-phenylethylimine (18), **A**, [or isobutyraldehyde and (S)-α-phenylethylamine (16)], with benzoic acid, **B**, and $t$-butyl isocyanide, **C**, in methanol at 0°C to form the diastereoisomer mixture of the (S)(S)- and (R)(S)-valine derivatives, $Y_p$ and $Y_n$, served as a model for the elucidation of the reaction mechanism of stereoselective 4 C C.[66,68,71–73]

$$\underset{\text{A}}{\overset{\overset{C_6H_5}{|}}{\underset{\underset{H}{|}}{CH_3-C}}-N=CH-i\text{-}C_3H_7} + \underset{\text{B}}{C_6H_5-CO_2H} + \underset{\text{C}}{CN-t\text{-}C_4H_9} \rightarrow$$

$$\underset{Y_p = \text{(XLVI) (S) (S)}}{\overset{\overset{C_6H_5 \quad i\text{-}C_3H_7}{| \quad |}}{\underset{\underset{H \quad CO \quad H}{| \quad | \quad |}}{CH_3-C-N-C}}-CO-NH-t\text{-}C_4H_9} + \underset{Y_n = \text{(XLVI) (S) (R)}}{\overset{\overset{C_6H_5 \quad CO-NH-t\text{-}C_4H_9}{| \quad |}}{\underset{\underset{H \quad CO \quad H}{| \quad | \quad |}}{CH_3-C-N-C}}-i\text{-}C_3H_7} \quad (18)$$

[R = $C_6H_5$, R′ = H, $R^3$ = $i$-$C_3H_7$ (16)]

It became apparent during preliminary experiments[73] that, depending on the reaction conditions, one or the other of the diastereoisomers was formed preferentially (see Table VI), and that (18) takes place by a complex reaction mechanism, presumably (19).

The concept of "pairs of corresponding reactions"*[49] is very useful for the analysis of stereoselective reactions. If the products of a stereoselective

---

* The abbreviation PCR is used in this book. The four PCR of (19) are symbolized by ⇉.

TABLE VI
STEREOSELECTIVE SYNTHESES (16) AND (18) OF $N$-BENZOYL-$N$-$\alpha$-PHENYLETHYLVALINE-$N'$-$t$-BUTYLAMIDE, Y (XLVI)[73]

| Exp. no. | Solvent | Procedure | Initial concentration $a_0 = b_0$ of A and B (moles/kg)[a] | Temperature (°C) | Added components (moles/kg) | $Q_{pn}$ |
|---|---|---|---|---|---|---|
| 1 | Chloroform | 16 | 0.10 | 0 | — | 2.264 |
| 2 | Chloroform | 16 | 0.63 | 0 | — | 1.704 |
| 3 | Chloroform | 16 | 1.00 | 0 | — | 1.762 |
| 4 | Chloroform | 18 | 0.20 | 0 | — | 2.486 |
| 5 | Chloroform | 18 | 0.54 | 0 | — | 2.317 |
| 6 | Chloroform | 18[b] | 0.54 | 0 | — | 2.260 |
| 7 | Chloroform | 18[c] | 0.54 | 0 | — | 2.175 |
| 8 | Chloroform | 18[d] | 0.54 | 0 | — | 2.128 |
| 9 | Chloroform | 18 | 1.00 | 0 | — | 2.065 |
| 10 | Methanol | 16 | 0.15 | 0 | — | 3.089 |
| 11 | Methanol | 16 | 0.95 | −40 | — | 0.782 |
| 12 | Methanol | 16 | 0.95 | −20 | — | 0.973 |
| 13 | Methanol | 16 | 0.95 | 0 | — | 1.243 |
| 14 | Methanol | 16 | 0.95 | +20 | — | 1.503 |
| 15 | Methanol | 16 | 0.95 | +40 | — | 1.794 |
| 16 | Methanol | 16 | 0.95 | +60 | — | 1.978 |
| 17 | Methanol | 18 | 0.01 | 0 | — | 3.490 |
| 18 | Methanol | 18 | 0.03 | −42 | — | 3.327 |
| 19 | Methanol | 18 | 0.03 | −23 | — | 3.673 |
| 20 | Methanol | 18 | 0.03 | 0 | — | 3.428 |
| 21 | Methanol | 18 | 0.03 | 27 | — | 3.007 |
| 22 | Methanol | 18 | 0.03 | 60 | — | 2.313 |
| 23 | Methanol | 18 | 0.10 | −42 | — | 2.371 |

[Table continued overleaf]

## TABLE VI (Continued)

| Exp. no. | Solvent | Procedure | Initial concentration $a_0 = b_0$ of **A** and **B** (moles/kg)[a] | Temperature (°C) | Added components (moles/kg) | $Q_{pn}$ |
|---|---|---|---|---|---|---|
| 24 | Methanol | 18[e] | 0.10 | −40 | — | 3.076 |
| 25 | Methanol | 18 | 0.10 | 0 | — | 2.950 |
| 26 | Methanol | 18 | 0.10 | 0 | 0.05 Tetraethyl-ammonium-benzoate | 2.110 |
| 27 | Methanol | 18 | 0.10 | 0 | 0.75 Tetraethyl-ammonium-benzoate | 1.078 |
| 28 | Methanol | 18 | 0.10 | 0 | 1.50 Tetraethly-ammonium-benzoate | 0.818 |
| 29 | Methanol | 18 | 0.30 | 0 | — | 2.446 |
| 30 | Methanol | 18 | 0.30 | 0 | 1.12 Tetraethyl-ammonium-benzoate | 0.822 |
| 31 | Methanol | 18 | 0.50 | 0 | — | 1.805 |
| 32 | Methanol | 18 | 1.00 | −78 | — | 0.338 |
| 33 | Methanol | 18 | 1.00 | −78 | 0.25 Schiff base 0.70 Triethyl-ammonium-benzoate | 0.282 |
| 34 | Methanol | 18 | 1.00 | −78 | 0.50 Triethyl-ammonium-benzoate | 0.307 |
| 35 | Methanol | 18 | 1.00 | −60 | — | 0.497 |

8. *Four-Component Condensations*    167

| | | | | | | |
|---|---|---|---|---|---|---|
| 36 | Methanol | 18 | 1.00 | — | −23 | 1.119 |
| 37 | Methanol | 18 | 1.00 | — | −23 | 1.225 |
| 38 | Methanol | 18 | 1.00 | 1.00 Benzoic acid | −23 | 0.759 |
| 39 | Methanol | 18 | 1.00 | 1.00 Schiff base | 0 | 1.484 |
| 40 | Methanol | 18 | 1.00 | 1.50 LiCl | 0 | 2.728 |
| 41 | Methanol | 18 | 1.00 | 0.50 Tetraethylammoniumbenzoate | 0 | 0.807 |
| 42 | Methanol | 18 | 1.00 | 1.50 Tetraethylammoniumbenzoate | 0 | 0.576 |
| 43 | Methanol | 18 | 1.00 | 2.00 Schiff base | 0 | 0.847 |
| 44 | Methanol | 18 | 1.00 | — | 60 | 1.855 |
| 45 | Methanol | 18 | 2.00 | — | 0 | 0.723 |
| 46 | Methanol | 18 | 2.50 | — | 0 | 0.770 |
| 47 | $CH_3OD^f$ | 18 | 0.10 | — | 0 | 2.581 |
| 48 | $CH_3OD$ | 18 | 0.94 | — | 0 | 0.803 |
| 49 | $CH_3OD$ | 18 | 2.00 | — | 0 | 1.000 |
| 50 | Acetonitrile | 16 | 0.10 | — | 0 | 2.164 |
| 51 | Acetonitrile | 16 | 1.00 | — | 0 | 1.576 |
| 52 | Acetonitrile | 18 | 0.10 | — | 0 | 2.093 |
| 53 | Acetonitrile | 18 | 0.86 | — | 0 | 1.937 |
| 54 | Acetonitrile | $18^b$ | 0.86 | — | 0 | 1.148 |
| 55 | Acetonitrile | $18^c$ | 0.86 | — | 0 | 1.072 |
| 56 | Acetonitrile | $18^d$ | 0.86 | — | 0 | 1.028 |
| 57 | Ethanol | 16 | 0.30 | — | 0 | 0.733 |
| 58 | Ethanol | 16 | 0.96 | — | 0 | 0.636 |
| 59 | Ethanol | 18 | 0.01 | — | 0 | 1.965 |
| 60 | Ethanol | 18 | 1.00 | 0.50 Triethylammoniumbenzoate | −78 | 0.271 |
| 61 | Ethanol | 18 | 1.00 | — | 0 | 0.642 |
| 62 | $C_2H_5OD$ | 18 | 2.00 | — | 0 | 0.490 |

[*Table continued overleaf*]

TABLE VI (Continued)

| Exp. no. | Solvent | Procedure | Initial concentration $a_0 = b_0$ of **A** and **B** (moles/kg)[a] | Temperature (°C) | Added components (moles/kg) | $Q_{pn}$ |
|---|---|---|---|---|---|---|
| 63 | Ethylene glycol | 18 | 0.10 | 0 | — | 2.497 |
| 64 | Ethylene glycol | 18 | 0.60 | 0 | — | 2.072 |
| 65 | Dimethylformamide | 16 | 0.10 | 0 | — | 2.220 |
| 66 | Dimethylformamide | 18 | 2.00 | 0 | — | 0.710 |
| 67 | Diethyl ether | 18 | 0.92 | 0 | — | 2.338 |
| 68 | Diethyl ether | 18[g] | 0.92 | 0 | — | 2.149 |
| 69 | Diethyl ether | 18[h] | 0.92 | 0 | — | 2.213 |
| 70 | Diethyl ether | 18[i] | 0.92 | 0 | — | 2.163 |
| 71 | n-Butanol | 16 | 0.30 | 0 | — | 0.627 |
| 72 | n-Butanol | 18 | 2.00 | 0 | — | 0.867 |
| 73 | Pyridine | 16 | 0.84 | 0 | — | 0.866 |
| 74 | Tetramethylurea | 16 | 0.84 | 0 | — | 0.608 |
| 75 | Toluene | 16 | 0.91 | 0 | — | 1.659 |

[a] The concentration of any N is indicated throughout this paper as $n$; the initial concentration of N is, therefore, $n_0$.
[b] **C** was added after 24 hr.
[c] **C** was added after 48 hr.
[d] **C** was added after 168 hr.
[e] **A** + **B** was added to a solution of **C**.
[f] The products of exp. nos. 47–49 do not contain any C—D bonds.
[g] **A** + **B** was added to a solution of **C** after 24 hr.
[h] **A** + **B** was added to a solution of **C** after 48 hr.
[i] **A** + **B** was added to a solution of **C** after 168 hr.

## 8. Four-Component Condensations

reaction are pairs of stereoisomers in thermodynamic equilibrium, or are formed by kinetically controlled reactions, via a pair of corresponding stereoisomeric transition complexes[14] which are in thermodynamic equilibrium, then the reaction is a PCR. Complex kinetically controlled stereoselective reactions which produce pairs of stereoisomeric products can be treated as systems of competing PCR.

$$C_6H_5-CH(CH_3)-\overset{\oplus}{NH}\cdots CH-i\text{-}C_3H_7 \underset{K_1}{\overset{H^\oplus}{\rightleftharpoons}} C_6H_5-CH(CH_3)-N=CH-i\text{-}C_3H_7$$
$$(X_1) \hspace{4cm} (A)$$

$$C_6H_5CO_2^{\ominus} \Big\updownarrow K_2 \hspace{2cm} (C) \searrow^{k_1}$$
$$(X_5)$$

$$\begin{array}{cc} C_6H_5-CH(CH_3)-NH-CH-i\text{-}C_3H_7 & C_6H_5-CH(CH_3)-NH-CH-i\text{-}C_3H_7 \\ | & | \\ O-CO-C_6H_5 & \overset{\oplus}{C}\cdots N-t\text{-}C_4H_9 \\ (X_2, p+n) & (N, p+n) \end{array}$$

$$H^\oplus \Big\updownarrow K_3 \hspace{2cm} (C) \searrow^{k_2} \hspace{2cm} \Big\downarrow (X_5)$$

(19)[66]

$$\begin{array}{cc} C_6H_5-CH(CH_3)-\overset{\oplus}{NH}_2-CH-i\text{-}C_3H_7 & \xrightarrow[(C)]{k_3} & C_6H_5-CH(CH_3)-NH-CH-i\text{-}C_3H_7 \\ | & & | \\ O-CO-C_6H_5 & & C=N-t\text{-}C_4H_9 \\ (X_3, p+n) & & | \\ & & C_6H_5-CO-O \\ & & (Z, p+n) \end{array}$$

$$(X_5) \Big\updownarrow K_4 \hspace{2cm} {}^{k_4}\!\!\nearrow(C)$$

$$\begin{array}{cc} C_6H_5-CH(CH_3)-NH-CH-i\text{-}C_3H_7 \\ \vdots \hspace{1cm} | \\ C_6H_5-CO-OH \hspace{0.5cm} O-CO-C_6H_5 \\ (X_4, p+n) \end{array} \hspace{1cm} \begin{array}{c} C_6H_5-CH(CH_3)-N-CH-i\text{-}C_3H_7 \\ | \hspace{2cm} | \\ C_6H_5-CO \hspace{0.5cm} CO-NH-t\text{-}C_4H_9 \\ (Y, p+n) \end{array}$$

If the equilibrium $A \rightleftharpoons \ldots \rightleftharpoons X_4$ of (19) is established considerably faster than the $t$-butyl isocyanide (C) adds to the intermediates $X_1$ to $X_4$, the Curtin-Hammett[66] principle can be applied to (19). The ratio $Q_{pn}$ of the diastereomeric products $Y_p$ and $Y_n$ of (19) is then determined by the competition of the four PCR in (19)

$$\begin{array}{ccccc}
& N_p & \xleftarrow{k_{1p}} & X_1 & \xrightarrow{k_{1n}} & N_n \\
Z_p & \xleftarrow{k_{2p}} & \{X_{2p} & \rightleftharpoons & X_{2n}\} & \xrightarrow{k_{2n}} & Z_n \\
Z_p & \xleftarrow{k_{3p}} & \{X_{3p} & \rightleftharpoons & X_{3n}\} & \xrightarrow{k_{3n}} & Z_n \\
Z_p & \xleftarrow{k_{4p}} & \{X_{4p} & \rightleftharpoons & X_{4n}\} & \xrightarrow{k_{4n}} & Z_n
\end{array} \qquad (20)$$

(i.e., the equilibrium concentrations $x_1$ to $x_4$ of $\mathbf{X}_1$ to $\mathbf{X}_4$ and the rate constants $k = k_{v,p} + k_{v,n}$ of the PCR) and by the internal stereoselectivities, $Q_{v,pn} = k_{v,p}/k_{v,n}$ of the individual PCR [see (60)].

The assumption that in methanol solution benzoic acid (**B**) and isobutyraldehyde-(S)-α-phenylethylimine (**A**) are in mobile equilibrium with their adducts ($\mathbf{X}_1$–$\mathbf{X}_4$) could be confirmed by comparing the electrical conductivities of the solutions of **A** + **B** and trimethyl-α-phenylethylammoniumbenzoate* (19).[66]

These adducts, ($\mathbf{X}_2$–$\mathbf{X}_4$), occur in the mobile equilibrium system in both diastereoisomeric (*p*- and *n*-) forms. Because the ratio of these diastereoisomers is independent of the concentrations of **A** and **B**, the sum of their concentrations can be used in the pertinent equilibrium equations (1).

The present investigations[66] yield information only on the stoichiometry of the adducts, ($\mathbf{X}_1$–$\mathbf{X}_4$), but not on their structure.

$$\mathbf{A} \xrightleftharpoons{H^{\oplus}} \mathbf{X}_1 \xrightleftharpoons{X_5} \mathbf{X}_2 \xrightleftharpoons{H^{\oplus}} \mathbf{X}_3 \xrightleftharpoons{X_5} \mathbf{X}_4 \quad (19a)$$

The equilibrium system, (19a) reacts with *t*-butyl isocyanide (**C**) via the intermediates (**N**) and (**Z**) to form **Y**† (19). The equilibrium system of (19) can be approximately described by eqs. 1–4.

$$\frac{x_1 \cdot x_5}{a \cdot b} = K_1 \tag{1}$$

$$\frac{x_1 \cdot x_5}{x_2} = K_2 \tag{2}$$

$$\frac{x_3 \cdot x_5}{x_2 \cdot b} = K_3 \tag{3}$$

$$\frac{x_4}{x_3 \cdot x_5} = K_4 \tag{4}$$

where $a$ and $b$ are the concentrations of **A** and **B**, and $x_1$ to $x_5$ are the concentrations of $\mathbf{X}_1$ to $\mathbf{X}_5$. If **A** and **B** are consumed from the equilibrium system $\mathbf{A} \rightleftharpoons \ldots \mathbf{X}_4$ by reaction with **C** according to (19), then $a_\Sigma$ and $b_\Sigma$ will become

---

* It is assumed that the ion mobilities of $\mathbf{X}_1$, $\mathbf{X}_3$, and trimethyl-α-phenylethyl ammonium cation are approximately equal. Those concentrations of trimethyl-α-phenylethylammoniumbenzoate were determined which have the same electrical conductivity as the chosen concentrations of **A** + **B**.

† Mutual conversion of the *n*- and *p*-isomers of the intermediates **N** and **Z** by proton shifts does not take place. See Table VI, exp. nos. 47–49.

## 8. Four-Component Condensations

$a_\Sigma = a_0 - y$ and $b_\Sigma = b_0 - y$, with $a_0$ and $b_0$ as the initial concentrations of **A** and **B**; $a_\Sigma$ and $b_\Sigma$ refer to the total amounts of the reactants **A** and **B** present in the equilibrium system; $y$ is the concentration of the product **Y**.

There are no further equilibrium relations between **A**, **B**, $X_1$, ..., $X_5$ which are independent of **1**–**4**; for the rank of the stoichiometrical matrix*[66]

$$M = \begin{pmatrix} 17 & 6 & 18 & 23 & 24 & 29 & 5 \\ 12 & 7 & 12 & 19 & 19 & 26 & 7 \\ 1 & 0 & 1 & 1 & 1 & 1 & 0 \\ 0 & 2 & 0 & 2 & 2 & 4 & 2 \end{pmatrix}$$

$$= \begin{pmatrix} -1 & 0 & 18 & 3 \\ 0 & 0 & 0 & \tfrac{1}{2} \\ 1 & 0 & -17 & -3 \\ 0 & 1 & -12 & -\tfrac{7}{2} \end{pmatrix}^{-1} \begin{pmatrix} 1 & 0 & 0 & 1 & 0 & 1 & 1 \\ 0 & 1 & 0 & 1 & 1 & 2 & 1 \\ 0 & 0 & 1 & 0 & 1 & 0 & -1 \\ \hline 0 & 0 & 0 & 0 & 0 & 0 & 0 \end{pmatrix} = T^{-1}N \quad (5)$$

in the representation $m = aM$ of $m = (A, B, X_1, ..., X_5)$ by $a = (H, C, N, O)$ equals the rank of the factor $N$, i.e., 3. Consequently the number of linearly independent stoichiometric relations between **A**, **B**, $X_1$, ..., $X_5$ equals $7 - 3 = 4$. With the submatrix $Q$ (**6**)

$$Q = \begin{pmatrix} 1 & 0 & 1 & 1 \\ 1 & 1 & 2 & 1 \\ 0 & 1 & 0 & -1 \end{pmatrix} \quad (6) \qquad R_1 = \begin{pmatrix} 0 & 1 & -1 & 0 \\ 0 & 0 & 1 & -1 \\ 0 & 0 & 0 & 0 \\ 1 & -1 & 1 & -1 \end{pmatrix} \quad (7)$$

of $N$, the (4,4) unit matrix $E$ and an arbitrary (4,4) matrix $R_1$ (with det $R_1 \neq 0$), e.g., **7**, it is possible to state the quadruple of independent stoichiometric relations between **A**, **B**, $X_1$, ..., $X_5$ in the form (**8**) ≡ (**9**) ≡ (**10**) + (**11**)

$$mS = m\begin{pmatrix} -Q \\ E \end{pmatrix} R_1 = m\begin{pmatrix} -Q & R_1 \\ R_1 & \end{pmatrix} = 0 \quad (8)$$

$$(A, B, X_1, X_2, X_3, X_4, X_5) \cdot \begin{pmatrix} -1 & 0 & 0 & 0 \\ -1 & 0 & -1 & 0 \\ 1 & -1 & 0 & 0 \\ \hline 0 & 1 & -1 & 0 \\ 0 & 0 & 1 & -1 \\ 0 & 0 & 0 & 1 \\ 1 & -1 & 1 & -1 \end{pmatrix} = 0 \quad (9)$$

* The "element coefficients" of the empirical formulas of the equilibrium species $H_{17}C_{12}N_1O_0$ (**A**), $H_6C_7N_0O_2$ (**B**), $H_{18}C_{12}N_1O_0$ ($X_1$), $H_{23}C_{19}N_1O_2$ ($X_2$), $H_{24}C_{19}N_1O_2$ ($X_3$), $H_{29}C_{26}N_1O_4$ ($X_4$), and $H_5C_7N_0O_2$ ($X_5$) are the matrix elements of $M$. This type of mathematical treatment is generally applicable to the analysis of complex reactions.

$$A + B = X_1 + X_5 = X_2 \tag{10}$$

$$X_2 + B = X_3 + X_5 = X_4 \tag{11}$$

Because rank $M = 7 -$ rank $S = 3$, there are precisely three linearly independent conservation of matter equations, **17–19**, between $A, B, X_1, \ldots, X_5$; the most general triplet is **12**

$$P(C - C_0) = 0 \tag{12}$$

with the (3,7)-matrix $P = R_2 (E, Q) = (R_2, R_2 Q)$, in which $E$ now means the (3,3) unit matrix, $R_2$ an arbitrary (3,3) matrix (with det $R_2 \neq 0$), and

$$C = \begin{pmatrix} a \\ b \\ x_1 \\ \vdots \\ x_5 \end{pmatrix} \tag{13} \qquad C_0 = \begin{pmatrix} a_\Sigma \\ b_\Sigma \\ 0 \\ \vdots \\ 0 \end{pmatrix} \tag{14}$$

The special case **15** for which **12** is given by **16** yields the evident conservation of matter equations **17–19**.

$$R_2 = \begin{pmatrix} 1 & 0 & 1 \\ 0 & 1 & 0 \\ 0 & 0 & 1 \end{pmatrix} \tag{15}$$

$$\begin{pmatrix} 1 & 0 & 1 & 1 & 1 & 1 & 0 \\ 0 & 1 & 0 & 1 & 1 & 2 & 1 \\ 0 & 0 & 1 & 0 & 1 & 0 & -1 \end{pmatrix} \cdot \begin{pmatrix} a - a_\Sigma \\ b - b_\Sigma \\ x_1 \\ \vdots \\ x_5 \end{pmatrix} = 0 \tag{16}$$

$$a = a_\Sigma - x_1 - x_2 - x_3 - x_4 \tag{17}$$

$$b = b_\Sigma - x_2 - x_3 - 2x_4 - x_5 \tag{18}$$

$$x_5 = x_1 + x_3 \tag{19}$$

Elimination of $x_1$ to $x_4$ from **1–4** and **17–19*** yields the equations **20–22**.

$$a = a_\Sigma - x_5[1 + K_4 x_5 + (1 - K_2 K_4) G(x_5)] \tag{20}$$

$$b = b_\Sigma - 2x_5 - 2K_4 x_5^2 + [K_2 + (2K_2 K_4 - 1)x_5]G(x_5) \tag{21}$$

$$\frac{K_2 \cdot x \cdot G(x_5)}{a \cdot b} - K_1 = 0 \tag{22}$$

---

* If in the equilibrium system (18) concentration $x_5$ of the benzoate anions $X_5$ is increased, for example by adding tetraethylammoniumbenzoate (S), then all the equilibrium concentrations of $A \ldots X_5$ will change. The effect of tetraethylammoniumbenzoate (S) on the concentration $s_0$ is computationally accounted for by replacing $x_5$ by $x_5 - s_0$ in the equations **17–19**.

with

$$G(x_5) = -\frac{K_2 + K_3(b_\Sigma - 2x_5 - 2K_4x_5^2) - \sqrt{4x_5[K_2K_3 + (2K_2K_4 - 1)K_3x_5] + [K_2 + K_3(b_\Sigma - 2x_5 - 2K_4x_5^2)]^2}}{2[K_2K_3 + (2K_2K_4 - 1)K_3x_5]}$$
(23)

By varying $K_1$ to $K_4$ these equilibrium constants can be determined from measured values* of $x_5$ by minimizing the deviation square sum (24), which is based upon 22. In 24, $\mu$ is the serial number of the individual measurement and $m$ ($= 41$) the total number of the measurements.

$$Z = \sum_{\mu=1}^{m} \left( \frac{K_2 \cdot x_{5,\mu} \cdot G(x_{5,\mu})}{a_\mu \cdot b_\mu} - K_1 \right)$$
(24)

The equilibrium constants

$K_1 = 1.4$

$K_2 = 0.023$(mole/liter); $\quad [K_2 = 0.019$(mole/kg)$]$

$K_3 = 3.3$

$K_4 = 0$

are meaningful in a chemical sense.

The composition of the product $\mathbf{Y}_{p,n}$ of (19) is given by 25.

$$\gamma_p = \sum_{\nu=1}^{4} \gamma_\nu \cdot \gamma_{\nu,p}$$
(25)

with

$$\gamma_p = y_p/(y_p + y_n) = Q_{pn}/(1 + Q_{pn})$$
(26)

$$\gamma_\nu = \frac{y_{\nu,t=\infty}}{y_{t=\infty}} = \int_0^\infty \left(\frac{dy_\nu}{dt}\right) dt \bigg/ \sum_{\nu=1}^{\nu=4} \int_0^\infty \left(\frac{dy_\nu}{dt}\right) dt$$
(27)

$$\gamma_{\nu,p} = k_{\nu,p}/(k_{\nu,p} + k_{\nu,n}) = k_{\nu,p}/k_\nu$$
(28)

The observed mole fraction of $\mathbf{Y}_p$ in the diastereoisomer mixture $\mathbf{Y}_{p+n}$ is $\gamma_p$; $\gamma_\nu$ is the mole fraction of the product $\mathbf{Y}_{\nu,p+n}$ which is formed by the $\nu$th PCR, and $\gamma_{\nu,p}$ is the mole fraction of $\mathbf{Y}_{\nu,p}$, i.e., that part of $\mathbf{Y}_p$ which arises from the $\nu$th PCR.

The $\gamma_\nu$ of 25 are, according to 27, determined by the relative values of the four PCR of (19), which are described by the differential equations 29 and

* See footnote, p. 172.

**30.** These are based upon the solution $X_1$ to $X_4$ of the equilibrium equations **1–4** and the rate constants $k_1$–$k_4$ of the PCR of (19).

$$dy_\nu/dt = k_\nu x_\nu c \qquad (\nu = 1, 2, 3, 4) \tag{29}$$

$$dy_4/dt = k_4 \cdot x_4 = k'' \cdot x_3 \cdot x_5 \tag{30}$$

**30** is introduced in order to account for the fact that a very small $K_4$ can be compensated by a large $k'_4$, despite the fact that $K_4 = 0$ requires $x_4 = K_4 \cdot x_3 \cdot x_5 = 0$. As $x_4 \ll a, b$, and $x_\nu$ ($\nu = 1, 2, 3$), it is possible to neglect $x_4$ in equations **17** and **18**. Formally, $k''_4$ may then be equated to $k_4$ and $x_4$ to $x_3 x_5$.

By transformation (**31**)

$$t \to t' = k_1 \int_0^t c\, dt \tag{31}$$

one obtains differential equation **32** from **29**.

$$\frac{dy_\nu}{dt'} = k'_\nu x_\nu, \qquad k'_\nu = \frac{k_\nu}{k_1} \qquad (\nu = 1, 2, 3, 4) \tag{32}*$$

The variable

$$y = \sum_{\lambda=1}^{4} y_\lambda \begin{cases} 0 \geqslant y \geqslant \beta \\ \beta = \min(a_0, b_0) \end{cases} \tag{33}$$

runs monotonically from 0 to $y_{t=\infty} = \beta$: $t'$ is eliminated from **32** and **34**

$$\frac{dy}{dt'} = \sum_{\lambda=1}^{4} k'_\lambda x_\lambda \tag{34}$$

$$\frac{dy_\nu}{dy} = k'_\nu x_\nu \Big/ \sum_{\lambda=1}^{4} k'_\lambda x_\lambda \qquad (\nu = 1, 2, 3, 4) \tag{35}$$

The mole fraction $\gamma_\nu$ which is finally observed of Eqs. **25** and **27** is given by **36**.

$$\gamma_\nu = \frac{1}{\beta} \int_0^\beta \frac{k'_\nu \dfrac{x_\nu}{x_1}\, dy}{1 + k'_2 \dfrac{x_2}{x_1} + k'_3 \dfrac{x_3}{x_1} + k'_4 \dfrac{x_4}{x_1}} \qquad (\nu = 1, 2, 3, 4) \tag{36}†$$

The $\gamma_\nu$ have to be obtained by numerical integration. The dependence of the $x_1$ to $x_4$ upon $y$ is implicitly given by **1–4** and **17–19**; since $x_1$ to $x_4$

---

* $x_\nu = x_{\nu p} + x_{\nu n}$ ($\nu = 2, 3, 4$).

† Since no specific time or rate scale is relevant for the solution of this essentially time-independent problem, we use $k_1 = 1$.

can be represented by explicit functions of $b$ (**37–40**), the reaction coordinate $y$ is replaced by $b$ according to **46**.

$$x_1 = K'' \cdot F \tag{37}$$

$$x_2 = K' \cdot a \cdot b \tag{38}$$

$$x_3 = F \cdot b \tag{39}$$

$$x_4 = K \cdot a \cdot b^2 \tag{40}$$

with

$$a = a_0 - b_0 + b(1 + F) \tag{41}$$

$$F = \frac{1}{2(b + K'')} \{Kb^2 - s_0 \pm [(Kb^2 - s_0)^2 + 4Kb(a_0 - b_0 + b)(b + K)]^{1/2}\} \tag{42}*$$

$$K = K_1 K_3 / K_2 \tag{43}$$

$$K' = K_1 / K_2 \tag{44}$$

$$K'' = K_2 / K_3 \tag{45}$$

$$\gamma_\nu = \frac{1}{\beta} \int_0^\beta f_\nu(x_1, x_2, x_3, x_4)\, dy = \frac{1}{\beta} \int_{b'}^{b''} f_\nu(b) y'(b)\, db \tag{46}$$

The new integration limits $b'$ and $b''$ are determined by **47** and **48**

$$y(b') = 0 \tag{47}$$

$$b'' = \max(0, b_0 - a_0) \tag{48}$$

The monotony which is necessary for $b$ as a function of $y$ is not granted under all experimental conditions. Inside the wedge-shaped areas of Fig. 1, there are experimental conditions $(a_0, b_0, s_0)$, for which $b$ is not a (monotonic) reaction coordinate. For determining the values of $k'_\nu$ and $\alpha_{\nu b}$ (Table 7), the experiments were correspondingly selected at best on the (permitted) wedge rims (Fig. 1) but never within it. Figures 2 and 3 show the equilibrium system for a set of experimental conditions outside and one set of experimental conditions located inside the "wedge."

The overall stereoselectivity of (19) is determined by measuring the optical rotation, $\alpha$, of the product mixture Y, $p + n$. Equation **49** holds because optical rotations are additive:

$$\alpha = \gamma_p \cdot \alpha_p + \gamma_n \cdot \alpha_n \tag{49}$$

$$\gamma_p = (\alpha - \alpha_p)/(\alpha_p - \alpha_n) \tag{49a}$$

* (+) or (−) for $a_0 - b_0 \leq (s_0/K)^{1/2}$ or $a_0 - b > (s_0/K)^{1/2}$.

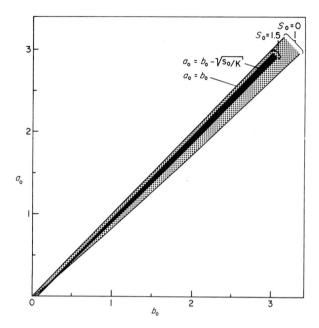

**Fig. 1** The permitted reaction conditions for (19). (For experimental conditions $a_0$, $b_0$, $s_0$ (moles/kg) inside the wedge b runs through a minimum during the reaction. The 62 experiments of Table VII lie outside the wedge or on a wedge rim.)

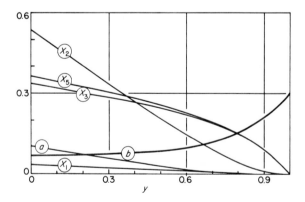

**Fig. 2** The equilibrium concentrations $a$, $b$, $x_1$, ... $x_5$ (moles/kg) as functions of the product concentration $y$ (moles/kg) (for an experiment outside the "wedge" of Fig. 1.)

8. Four-Component Condensations  177

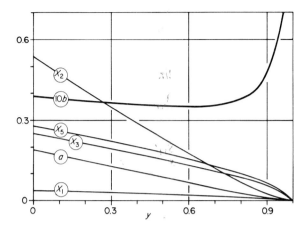

**Fig. 3** The equilibrium concentrations $a, b, x_1, \ldots x_5$ (moles/kg) of $A, B, X_1$ to $X_2$ as functions of the product concentration $y$ (moles/kg) (for an experiment inside the "wedge" of Fig. 1.)

Equation **50** follows from **25** and **49**.

$$\alpha = \sum_{\nu=1}^{4} \gamma_\nu \alpha_\nu \qquad (50)*$$

with

$$\alpha_\nu = \gamma_\nu^{(p)} \alpha_p + \gamma_\nu^{(n)} \alpha_n \qquad (51)$$

Since the $\gamma_\nu$ can be calculated by **46**, eq. **50** provides a basis for determining the relative rate constants $K'_\nu$ of the four PCR and their internal stereoselectivities (represented by the $\alpha_\nu$) from the concentration dependence of those measured optical rotations ($\alpha_{\text{exp}}$) of the diastereomer mixtures $Y_{p+n}$ (see Table VII). The values of $K'_\nu$ and $\alpha_\nu$ can be determined for which the sum **52** goes to a minimum. The $\alpha_{\mu,\text{calc}}$ represent calculated values according to **46** and **50**.

$$Z' = \sum_{\mu=1}^{m} (\alpha_{\mu,\text{exp}} - \alpha_{\mu,\text{calc}})^2 \qquad (52)$$

The problem can be solved in two stages, because the

$$\alpha_{\mu,\text{calc}} = \sum_{\nu=1}^{4} \gamma_{\nu\mu} \alpha_\nu \qquad (\mu = 1, \ldots m) \qquad (50a)$$

depend only upon the $\alpha_\nu$ ($\gamma_{\nu\mu}$ is determined by the quantities $a_{0\mu}$, $b_{0\mu}$, and $s_{0,\mu}$ which are specified for the reaction, and by $k'_\nu$ which can be adjusted by

---

* $[\alpha_p]^{20} = +36°$ and $[\alpha_n]^{20} = -200°$ are the optical rotations of $Y_p$ and $Y_n$. $\alpha_\nu$ is the optical rotation of a product $Y_n$ which is exclusively formed by the $\nu$th PCR.

the variation technique). For arbitrary, but firmly selected $k'_2, k'_3, k'_4$ there are optimal values $\alpha_1$ to $\alpha_4$ which impart the minimum value to $Z'$ (52a).

$$Z'(k'_2, k'_3, k'_4; \alpha_1, \alpha_2, \alpha_3, \alpha_4) = Z'(\alpha_1, \alpha_2, \alpha_3, \alpha_4) \qquad (52a)$$

The $\alpha_\nu$ are obtained by solving the linear equation system (53)

$$\frac{\partial Z'(\alpha_1, \alpha_2, \alpha_3, \alpha_4)}{\partial \alpha_\nu} = 0 \qquad (\nu = 1, 2, 3, 4) \qquad (53)$$

i.e.,

$$\begin{pmatrix} \alpha_1 - \alpha_4 \\ \alpha_2 - \alpha_4 \\ \alpha_3 - \alpha_4 \\ \alpha_4 \end{pmatrix} = \begin{pmatrix} S_{11} S_{12} S_{13} \bar{\gamma}_1 \\ S_{12} S_{22} S_{23} \bar{\gamma}_2 \\ S_{13} S_{23} S_{33} \bar{\gamma}_3 \\ \bar{\gamma}_1 \; \bar{\gamma}_2 \; \bar{\gamma}_3 \; 1 \end{pmatrix}^{-1} \begin{pmatrix} S_1 \\ S_2 \\ S_3 \\ \bar{\alpha}_{\exp} \end{pmatrix} \qquad (54)$$

with the mean values $\gamma_\nu$ that have been calculated from $\alpha_{\mu,\exp}$, $a_{0\mu}$, $b_{0\mu}$, $s_{0\mu}$, as well as $k'_2, k'_3, k'_4$ according to **46**, and with

$$S_{ij} = \frac{1}{m} \sum_{\mu=1}^{m} \gamma_{i\mu} \gamma_{j\mu}, \qquad S_i = \frac{1}{m} \sum_{\mu=1}^{m} \gamma_{i\mu} \alpha_{\mu,\exp} \qquad (55)$$

$$\bar{\gamma}_i = \frac{1}{m} \sum_{\mu=1}^{m} \gamma_{i\mu}, \qquad \bar{\alpha}_{\exp} = \frac{1}{m} \sum_{\mu=1}^{m} \alpha_{\mu,\exp} \qquad (56)$$

The data $a_{0\mu}, b_{0\mu}, s_{0\mu}$, and $\alpha_{\mu,\exp}$ ($\mu = 1, 2, \ldots, 62$) from 62 experiments (see Fig. 1 and Table VII) were processed by a digital computer IBM 7074 in order to establish the numerical values of $k'_2, k'_3, k'_4$ which minimize

$$Z' = Z'(k'_i; \alpha'_j) = Z'[k'_i; \alpha'_j(k'_i)] = Z'(k'_i) \qquad (57)$$

A FORTRAN program which coordinates the mole fractions $\gamma_{\nu\mu}$ ($\nu = 1, 2, 3, 4$; $\mu = 1, 2, \ldots, 62$) and optimum angles of rotation $\alpha_1, \alpha_2, \alpha_3, \alpha_4$, (equations 37–40, 46, and 54) to $k'_2, k'_3, k'_4$ and thereby determines the mean square deviation

$$\Delta = [(1/m) Z']^{1/2} \qquad (m = 62)$$

requires for one case

$$k'_2, k'_3, k'_4 \rightarrow \alpha_1, \alpha_2, \alpha_3, \alpha_4; \Delta$$

a calculating time of approximately 3 min. The final result of several hours of computer calculation, in which $k'_2, k'_3, k'_4$ were automatically altered in accordance with a minimum determining strategy, was $\Delta = 10.73°$ and

$$k'_1 = 1; \qquad k'_2 = 0.63; \qquad k'_3 = 1.15; \qquad k'_4 = 1.20; \qquad (58)$$

$$\alpha_1 = 17°; \qquad \alpha_2 = -107°; \qquad \alpha_3 = -17°; \qquad \alpha_4 = -160° \qquad (59)$$

Table VII contains, in addition to the experimental data $a_{0\mu}, b_{0\mu}, s_{0\mu}$, and $\alpha_{\mu,\exp}$, the calculated values $\alpha_{\mu,\text{calc}}$ belonging to the "best" $k'_\nu$, (58).

## 8. Four-Component Condensations

### TABLE VII[66]

REACTION (19) AT 0°C IN METHANOL, PARTLY IN THE PRESENCE OF TETRA-ETHYLAMMONIUMBENZOATE S. INITIAL CONCENTRATIONS: $a_0$, $b_0$, $s_0$; MEASURED OPTICAL ROTATION (AT 578 M$\mu$) OF $Y_{,p+n}$: $\alpha_{exp}$; ANGLE OF ROTATION CALCULATED: $\alpha_{calc}$ WITH THE OPTIMIZED VALUES OF $K_\nu$, $k_\nu$, AND $\alpha_\nu$

| $a_0$ (mole/kg) | $b_0$ (mole/kg) | $s_0$ (mole/kg) | $\alpha_{exp}$ | $\alpha_{calc}$ |
|---|---|---|---|---|
| 0.030 | 0.030 | 0.00 | −17.30 | −8.46 |
| 0.080 | 0.300 | 0.00 | −23.20 | −23.89 |
| 0.100 | 0.100 | 0.00 | −22.20 | −23.96 |
| 0.150 | 0.300 | 0.00 | −30.60 | −30.28 |
| 0.300 | 0.075 | 0.00 | −26.16 | −24.91 |
| 0.300 | 0.150 | 0.00 | −30.18 | −35.09 |
| 0.300 | 0.300 | 0.00 | −32.20 | −39.71 |
| 0.300 | 0.600 | 0.00 | −35.64 | −37.24 |
| 0.300 | 1.200 | 0.00 | −35.84 | −36.00 |
| 0.333 | 1.000 | 0.00 | −39.14 | −37.73 |
| 0.500 | 1.000 | 0.00 | −42.85 | −44.50 |
| 0.600 | 0.300 | 0.00 | −38.30 | −47.05 |
| 0.666 | 1.000 | 0.00 | −49.40 | −49.63 |
| 0.750 | 1.000 | 0.00 | −53.14 | −51.98 |
| 0.754 | 0.376 | 0.00 | −41.00 | −50.89 |
| 1.000 | 0.333 | 0.00 | −45.84 | −50.45 |
| 1.000 | 0.500 | 0.00 | −53.20 | −55.60 |
| 1.000 | 0.666 | 0.00 | −55.10 | −58.24 |
| 1.000 | 1.000 | 0.00 | −58.10 | −57.59 |
| 1.000 | 1.250 | 0.00 | −60.05 | −56.72 |
| 1.000 | 1.500 | 0.00 | −60.36 | −56.66 |
| 1.000 | 2.000 | 0.00 | −58.92 | −58.41 |
| 1.050 | 1.000 | 0.00 | −61.90 | −59.21 |
| 1.100 | 1.000 | 0.00 | −62.83 | −60.35 |
| 1.250 | 1.000 | 0.00 | −66.95 | −62.53 |
| 1.270 | 0.320 | 0.00 | −44.60 | −50.49 |
| 1.500 | 1.000 | 0.00 | −73.44 | −64.49 |
| 2.000 | 1.000 | 0.00 | −83.12 | −66.35 |
| 2.500 | 2.500 | 0.00 | −97.30 | −71.01 |
| 3.000 | 1.000 | 0.00 | −91.80 | −67.83 |
| 0.100 | 0.100 | 0.05 | −39.90 | −50.77 |
| 0.100 | 0.100 | 0.10 | −43.20 | −63.72 |
| 0.100 | 0.100 | 0.25 | −56.60 | −81.73 |
| 0.100 | 0.100 | 0.50 | −68.20 | −93.16 |
| 0.100 | 0.100 | 0.75 | −77.60 | −98.35 |
| 0.100 | 0.100 | 1.00 | −78.10 | −101.35 |
| 0.100 | 0.100 | 1.25 | −82.70 | −103.31 |
| 0.100 | 0.100 | 1.50 | −93.80 | −104.68 |
| 0.300 | 0.300 | 1.00 | −93.50 | −100.36 |
| 0.300 | 1.200 | 1.00 | −97.60 | −94.46 |
| 1.000 | 1.000 | 0.05 | −70.00 | −65.22 |

[*Table continued overleaf*]

TABLE VII (Continued)

| $a_0$ (mole/kg) | $b_0$ (mole/kg) | $s_0$ (mole/kg) | $\alpha_{exp}$ | $\alpha_{calc}$ |
|---|---|---|---|---|
| 1.000 | 1.000 | 0.10 | −77.60 | −70.48 |
| 1.000 | 1.000 | 0.25 | −87.50 | −80.74 |
| 1.000 | 1.000 | 0.50 | −94.60 | −90.32 |
| 1.000 | 1.000 | 0.75 | −102.10 | −96.06 |
| 1.000 | 1.000 | 1.00 | −103.60 | −99.97 |
| 1.000 | 1.000 | 1.25 | −108.50 | −102.82 |
| 1.000 | 1.000 | 1.50 | −118.50 | −105.02 |
| 1.000 | 2.000 | 0.05 | −62.60 | −61.97 |
| 1.000 | 2.000 | 0.25 | −75.70 | −73.80 |
| 1.000 | 2.000 | 0.50 | −82.30 | −84.81 |
| 1.000 | 2.000 | 0.75 | −97.00 | −93.10 |
| 1.000 | 2.000 | 1.00 | −104.80 | −99.59 |
| 1.000 | 3.000 | 0.50 | −69.10 | −86.37 |
| 1.200 | 0.300 | 1.00 | −98.80 | −103.25 |
| 2.000 | 1.000 | 0.05 | −78.30 | −75.86 |
| 2.000 | 1.000 | 0.25 | −98.80 | −97.26 |
| 2.000 | 1.000 | 0.50 | −106.70 | −97.93 |
| 2.000 | 1.000 | 0.75 | −115.50 | −101.03 |
| 2.000 | 1.000 | 1.00 | −119.20 | −102.81 |
| 2.000 | 2.000 | 0.50 | −93.30 | −91.67 |
| 3.000 | 1.000 | 0.50 | −115.80 | −98.97 |

The quality of the "best" constants, **58** and **59**, for which $\Delta = 10.73°$, is illustrated by Table VIII which contains "wrong" sets of $k'_\nu$ and $\alpha_\nu$ with a mean square deviation $\Delta > 10.73°$; not only the case of equal $k_\nu$ but also the cases in which one of the $k_\nu$ is ten times larger than the other $k_\nu$ ($k_\nu = k'_\nu k_1$), are included.

TABLE VIII[66]

THE INFLUENCE OF THE PARAMETERS $k'_\nu$ AND $\alpha_\nu$ ON $\Delta$

| $k'_1$ | $k'_2$ | $k'_3$ | $k'_4$ | $\alpha_1$ | $\alpha_2$ | $\alpha_3$ | $\alpha_4$ | $\Delta$ |
|---|---|---|---|---|---|---|---|---|
| 1 | 1 | 1 | 1 | 42 | −102 | −12 | −169 | 11.12 |
| 1 | 0.1 | 0.1 | 0.1 | −29 | −120 | −13 | −161 | 12.07 |
| 1 | 10 | 1 | 1 | 315 | −86 | 26 | −227 | 16.70 |
| 1 | 1 | 10 | 1 | 47 | −116 | −37 | −608 | 12.28 |
| 1 | 1 | 1 | 10 | 33 | −105 | 29 | −99 | 11.91 |

In view of the fact that the $\alpha_{exp}$ are determined with an accuracy of $\pm 3°$ to $5°$ ($\hat{=} \Delta\gamma_p \cdot 100 = \pm 1.3$ to $2.1$ mole %) and the rather crude approximation in the description of the equilibria $A \rightleftharpoons \ldots \rightleftharpoons X_{4,p+n}$ by **1–4**,* the mean square deviation found, $\Delta = 10.73°$, ($\hat{=} \Delta_{\gamma p} \cdot 100 = \pm 4.5$ mole %) indicates excellent coincidence of the mathematical model with the experimental data observed. It follows from the calculated $\alpha_v$ that the stereoselectivity coefficients $Q_{v,pn}$ of the four PCR of (19) are:

$$Q_{1,pn} = 11.4 \qquad Q_{2,pn} = 0.655$$
$$Q_{3,pn} = 3.45 \qquad Q_{4,pn} = 0.204 \qquad (60)$$

This shows that under reaction conditions for which $\gamma_1$ or $\gamma_3$ or both acquire a high value $Y_p$ is preferentially obtained as the reaction product, whereas reaction conditions with high values of $\gamma_2$ or $\gamma_4$ or $\gamma_2$ and $\gamma_4$ lead to the preferential formation of $Y_n$.

Thus, the elucidation of the present system (19) of four competing PCR allows the selection of reaction conditions under which the reaction $A + B + C \to Y$ provides maximum yields of $Y_p$ or $Y_n$, respectively. It may be assumed that (19) can also be transferred by analogy to other 4 C C and is potentially useful for planning of peptide syntheses.

It should be noted that the available results do not permit any statements concerning the details of the reaction mechanism, such as the answer to the question about the precise structure of the intermediates $X_1$–$X_4$ or the mechanism of the formation of the $\alpha$-adducts $(Z, p+n)$.

## V. β-LACTAMS AND PENICILLANIC ACID DERIVATIVES

### A. Simple β-Lactams

The synthesis of $\beta$-lactams by cyclization of $\beta$-amino acids is generally difficult to achieve by standard methods.[52] It proceeds remarkably smoothly by 4 C C on condensing $\beta$-amino acids with isonitriles and carbonyl compounds (21). Examples are listed in Table IX. In (21) the $\beta$-amino acid

* The determined $\alpha_v$ (59) are only slightly dependent upon the $K_v$. If $K_1$ to $K_3$ and $k'_2$ to $k'_4$ are varied simultaneously, the $K_v$ have changed more drastically than the $k'_v$ and the $\alpha_v$ remain in the original order of magnitude;

$K_1 = 4.6;$   $K_2 = 0.056;$   $K_3 = 11.7;$   $k'_1 = 1;$   $k'_2 = 1.04;$   $k'_3 = 0.91;$   $k'_4 = 1.21;$

$\alpha_1 = 11°;$   $\alpha_2 = -112°;$   $\alpha_3 = -13°;$   $\alpha_4 = 142°;$   $\Delta = -9.82$   (cf. p. 178 and **58** and **59**).

The $\alpha_v$ are stereochemically the most relevant data as they are simple functions of the internal stereoselectivities $Q_{v,pn}$ of the four PCR of (19).

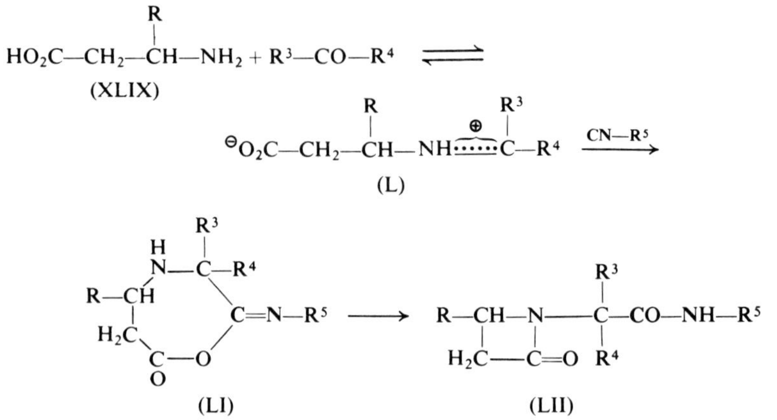

182   Gokel, Lüdke, and Ugi

$$\text{HO}_2\text{C—CH}_2\text{—CH(R)—NH}_2 + \text{R}^3\text{—CO—R}^4 \rightleftarrows$$
(XLIX)

$$^{\ominus}\text{O}_2\text{C—CH}_2\text{—CH(R)—NH}\overset{\oplus}{\cdots}\text{C(R}^3\text{)—R}^4 \xrightarrow{\text{CN—R}^5}$$
(L)

(21)

$$\begin{array}{c}\text{R—CH} \\ \text{H}_2\text{C} \end{array} \begin{array}{c} \text{H} \quad \text{R}^3 \\ \text{N—C—R}^4 \\ \text{C=N—R}^5 \\ \text{C—O} \\ \text{O} \end{array} \longrightarrow \text{R—CH—N——C(R}^3\text{)—CO—NH—R}^5$$
$$\qquad\qquad\qquad\qquad\qquad\qquad\quad \text{H}_2\text{C——C=O  R}^4$$
(LI)                    (LII)

(XLIX) functions simultaneously as the amine and acid component. The immonium betaine (L) carries opposite charges on the reactive centers. This favors the formation of the cyclic α-adduct (LI), which undergoes spontaneous transannular reacylation to the β-lactam (LII).

TABLE IX

THE PREPARATION OF β-LACTAMS (LII) FROM β-AMINO ACIDS (XLIX), CARBONYL COMPOUNDS (II), AND ISONITRILES (III) (21)

| Designation no. (LII) | $R^3$ (II) | $R^4$ (II) | $R^5$ (III) | R (XLIX) | Yield (%) | Ref. |
|---|---|---|---|---|---|---|
| a | $i\text{-C}_3\text{H}_7$— | H— | $c\text{-C}_6\text{H}_{11}$— | H— | 80 | 56 |
| b | $i\text{-C}_3\text{H}_7$— | H— | $c\text{-C}_6\text{H}_{11}$— | $\text{C}_6\text{H}_5$— | 84 | 56 |
| c | —(CH$_2$)$_5$— | —(CH$_2$)$_5$— | $c\text{-C}_6\text{H}_{11}$— | H— | 85 | 56 |
| d | —(CH$_2$)$_5$— | —(CH$_2$)$_5$— | $t\text{-C}_4\text{H}_9$— | $\text{C}_6\text{H}_5$— | 89 | 56 |
| e | CH$_3$— | CH$_3$— | —(CH$_2$)$_2$— | H— | — | 39 |
| f | CH$_3$— | CH$_3$— | —(CH$_2$)$_6$— | H— | — | 39 |
| g | —(CH$_2$)$_5$— | —(CH$_2$)$_5$— | 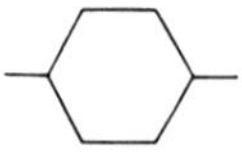 —⟨⟩— | H— | — | 39 |

To date, no investigations have been carried out to determine whether it is also possible to synthesize, in an analogous fashion, lactams of other ring sizes by 4 C C.* We assume that the lactams with 5–8-membered rings will

*The syntheses of N-n-butyl- and N-benzyl-5-methylpyrrolidone 5-[N'-cyclohexylcarbonamides] in 70% yield[16] from levulinic acid, n-butyl or benzylamine and cyclohexyl isocyanide are examples of a different type of 4 C C lactam synthesis.

8. Four-Component Condensations    183

not be produced as easily by 4 C C because the corresponding cyclic (8–11-membered) α-adducts would probably be too strained.

**B. Penicillanic Acid Derivatives**

$\Delta^3$-Thiazolines are readily available by the Asinger condensation of α-mercapto aldehydes or ketones with ammonia and carbonyl compounds.[1]

Like the Schiff bases, $\Delta^3$-thiazolines undergo 4 C C with isonitriles and carboxylic acids. The cysteine derivatives and penicillamine derivatives (LIV) are conveniently accessible from 2,2-dimethyl-$\Delta^3$-thiazoline or 2,2,5,5-tetramethyl-$\Delta^3$-thiazoline (LIII), respectively.[42,79]

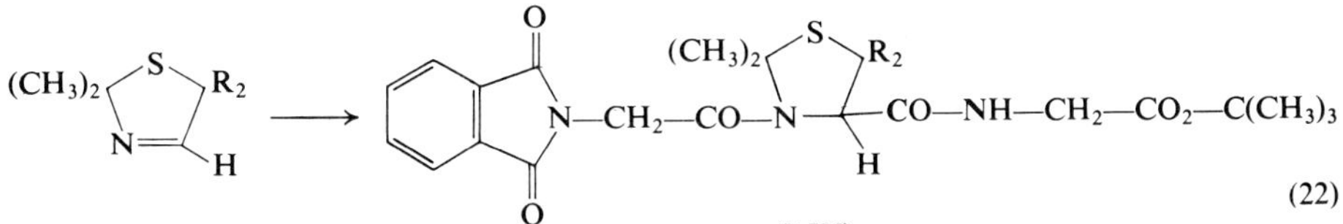

(LIII) ⟶ (LIV)    (22)

R = H or CH$_3$

The Asinger condensation of α-mercaptoisobutyraldehyde which is easily accessible from α-bromoisobutyraldehyde and trimethylammonium hydrogen sulfide in methylene chloride and ethyl β-aminocrotonate or ethyl-α-formylpropionate and ammonia, respectively, followed by alkaline hydrolysis leads to (LVa or b).[54,81]

In the two-phase mixture of water and petroleum ether, (LVa and b) react with isonitriles to form 5- or 6-methyl penicillanic amides* (LVIII)[79,81] (see Table X).[79] The reaction with the isonitrile (23) presumably takes place

TABLE X

The 4 C C Synthesis (23) of 5- and 6-Methyl cis-Penicillanic Amides (LVIIIa and b)[79,81]

| | | Yield of LVIII (%) | |
|---|---|---|---|
| Exp. no. | R$^5$ | a | b |
| 1 | C$_2$H$_5$ | 89 | 56 |
| 2 | i-C$_3$H$_7$ | 94 | 57 |
| 3 | t-C$_4$H$_9$ | 88 | 48 |
| 4 | c-C$_6$H$_{11}$ | 36 | 17 |

* Surprisingly these can be purified by sublimation *in vacuo*.

in the aqueous phase and "under dilution conditions." If the reaction is carried out in a single organic phase, however, the formation of resins predominates.

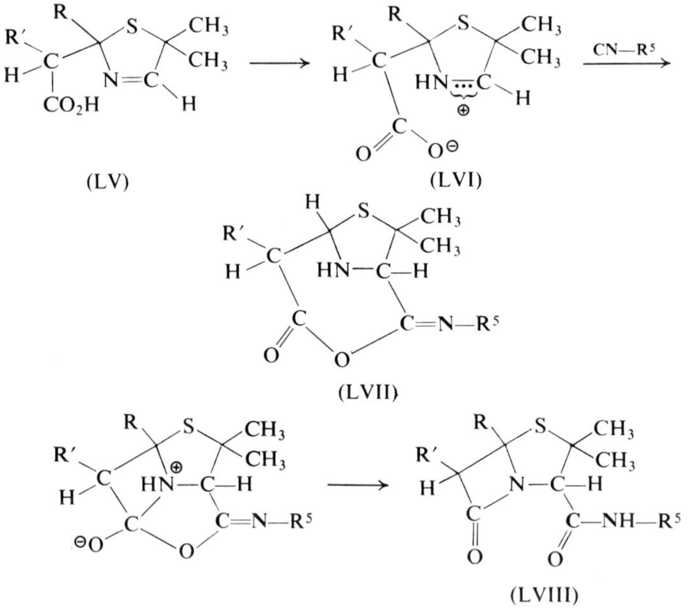

[Structures (LV) → (LVI) → (LVII) → (LVIII), with reagent CN—R⁵, labeled as equation (23)]

a: $R = CH_3$, $R' = H$
b: $R = H$, $R' = CH_3$

In aqueous solution LV is in equilibrium with the zwitterion (LVI), which adds to the isonitrile to form the bicyclic α-adduct (LVII). The latter is converted into LVIII by transannular acyl migration. It follows necessarily from this mechanism that reaction (23) is stereospecific and that the carbonamide group and the β-lactam ring are in *cis* relation.

Oxidation of (LVIIIa$_3$) with potassium permanganate in acetic acid leads to the sulfone (LIX); N-acylthiazolidines can be oxidized to the corresponding sulfones, whereas in contrast, thiazolidines which carry no acyl group at the ring nitrogen are subject to oxidative ring cleavage.

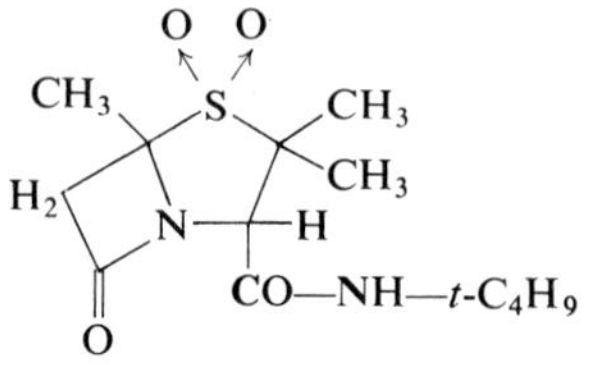

[Structure (LIX): thiazolidine sulfone with CO—NH—$t$-C$_4$H$_9$ group]

(LIX)

Sjöberg[54] recently succeeded in synthesizing *cis*-6-phenylacetamidopenicillanic acid-*t*-butylamide (LX) by the following procedure:

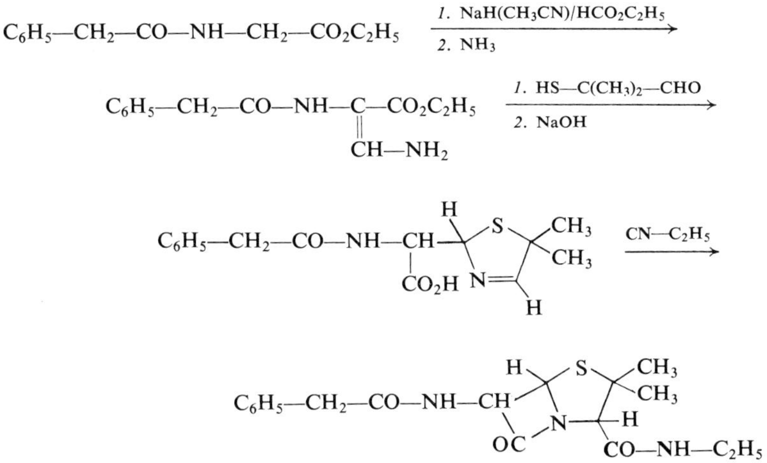

$$C_6H_5-CH_2-CO-NH-CH_2-CO_2C_2H_5 \xrightarrow[\text{2. NH}_3]{\text{1. NaH(CH}_3\text{CN)/HCO}_2\text{C}_2\text{H}_5}$$

$$C_6H_5-CH_2-CO-NH-\underset{\underset{CH-NH_2}{\parallel}}{C}-CO_2C_2H_5 \xrightarrow[\text{2. NaOH}]{\text{1. HS--C(CH}_3\text{)}_2\text{--CHO}}$$

C$_6$H$_5$—CH$_2$—CO—NH—CH(CO$_2$H)—[thiazolidine ring with CH$_3$, CH$_3$] $\xrightarrow{\text{CN—C}_2\text{H}_5}$

C$_6$H$_5$—CH$_2$—CO—NH—CH—[β-lactam/thiazolidine bicyclic: S, CH$_3$, CH$_3$, N, H, OC, CO—NH—C$_2$H$_5$]   (24)

(LX)

## VI. URETHANES

Monomethyl carbonate (CH$_3$O—CO—OH) behaves toward *n*-butylamine, isobutyraldehyde, and cyclohexyl isonitrile like a carboxylic acid. If isobutyraldehyde and cyclohexyl isocyanide are added to a solution of *n*-butylamine in methanol which is saturated with carbon dioxide, and the solution is then allowed to stand for 20 hr at about 20°C, the urethane (LXI) can be isolated in almost quantitative yield.[78]

An analogous 4 C C is observed with benzylamine, *n*-butyraldehyde and cyclohexyl isocyanide.[16]

The unique, almost quantitative combination of a total of five reactants to yield a uniform product (25) is due to the fact that all side reactions are reversible and the main reaction consists of a chain of coupled equilibria, which lead to an irreversible final step.

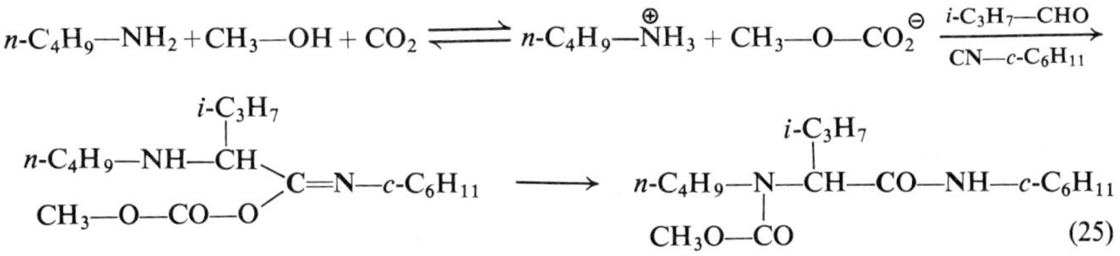

$$n\text{-}C_4H_9-NH_2 + CH_3-OH + CO_2 \rightleftharpoons n\text{-}C_4H_9-\overset{\oplus}{N}H_3 + CH_3-O-CO_2^{\ominus} \xrightarrow[\text{CN}-c\text{-}C_6H_{11}]{i\text{-}C_3H_7-\text{CHO}}$$

$$\underset{CH_3-O-CO-O}{n\text{-}C_4H_9-NH-CH} \overset{i\text{-}C_3H_7}{\underset{}{>}}C=N-c\text{-}C_6H_{11} \longrightarrow n\text{-}C_4H_9-\underset{CH_3O-CO}{N}-\overset{i\text{-}C_3H_7}{CH}-CO-NH-c\text{-}C_6H_{11} \quad (25)$$

(LXI)

## VII. THE BUCHERER-BERGS REACTION

Beside the 4 C C of isonitriles, to our knowledge, the hydantoin synthesis according to Bergs[3] and Bucherer[6-10] (26) is the only reaction in which the union of four different reactants to a uniform product occurs.[80] Probably, the formation of hydantoins (LXVI) from ammonia, carbonyl compounds, hydrogen cyanide, and carbon dioxide[55] occurs via the intermediates LXIII–LXV. The fact that primary amino nitriles (LXII) and $CO_2$ react to yield the hydantoins,[10] whereas secondary amino nitriles (LXII, R ≠ H) do not undergo an analogous cyclization to form $N$-substituted hydantoins, supports the relevance of LXV as an intermediate.

Thus, not only with regard to the components but even with regard to the mechanism, there is an analogy with the 4 C C, and (26) could be considered a 4 C C of hydrogen isocyanide.

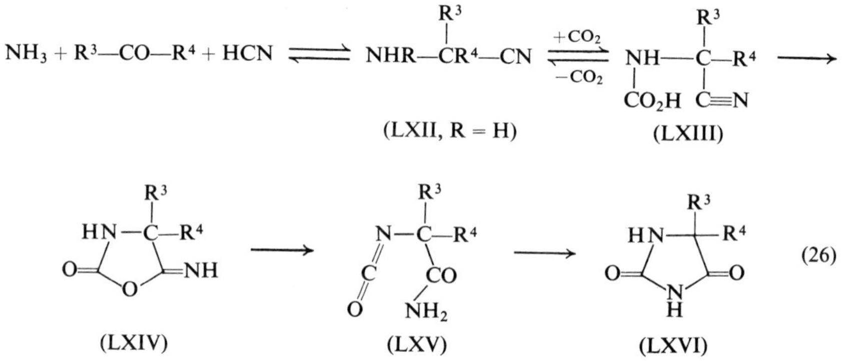

$$NH_3 + R^3\text{—CO—}R^4 + HCN \rightleftharpoons NHR\text{—}CR^4\text{—}CN \underset{-CO_2}{\overset{+CO_2}{\rightleftharpoons}} \underset{\underset{C\equiv N}{CO_2H}}{NH\text{—}\overset{R^3}{\underset{|}{C}}\text{—}R^4} \longrightarrow$$

(LXII, R = H)   (LXIII)

(LXIV)   (LXV)   (LXVI)   (26)

## VIII. N-ALKYL-3-ACYL-4-HYDROQUINOLINE-4-CARBONAMIDES

The quinolinium ions (LXVII) can be considered as analogs of the immonium ions both with respect to structure and electrophilic reactivity.

Just like the immonium ions, quinolinium ions in combination with carboxylate anions are capable of $\alpha$-addition to isonitriles. The $\alpha$-adducts (LXVIII) are further subject to O → C acyl transfer. In the $\alpha$-adduct, LXVIII, the 3-position of the 1,4-dihydroquinoline system corresponds to the nucleophilic $\beta$-position of an enamine to which the acyl group of the $O$-acylimide system is transferred by a cyclic mechanism (27).[65]

The products (LXIX) are only obtained from $N$-alkylquinolinium bromides and iodides (Table XI). The chlorides do not react in a corresponding manner.

The structure of LXIX follows from the independent synthesis (28) of LXIXe and the IR, UV, and NMR spectra of LXIXa–p.

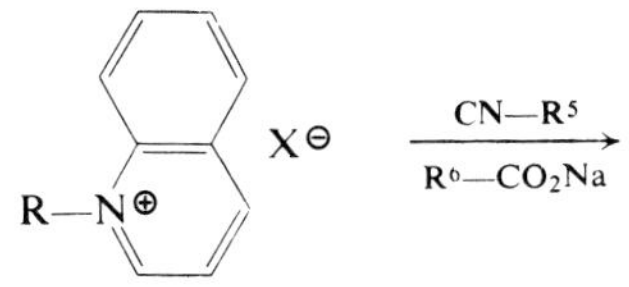

8. *Four-Component Condensations* 187

$$R-N^{\oplus}\diagdown X^{\ominus} \xrightarrow{\text{CN}-R^5}_{R^6-CO_2Na}$$

(LXVII)

$X^{\ominus} = Br^{\ominus}, I^{\ominus}$

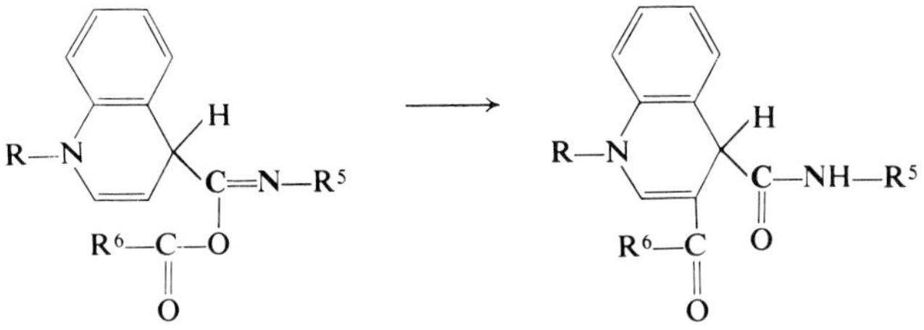

$$\text{(LXVIII)} \longrightarrow \text{(LXIX)} \quad (27)$$

TABLE XI

THE PREPARATION OF $N$-ALKYL-3-ACYL-4-HYDROQUINOLINE-4-CARBONAMIDES (LXIX) FROM $N$-ALKYL-QUINOLINIUM HALIDES (LXVII), SODIUM CARBOXYLATES, AND ISONITRILES IN METHANOL/WATER (1:1) AT 20°C[65]

| Designation no. (LXIX) | R (LXVII) | $R^5$ (III) | $R^6$ (XXXIII) | $X^{\ominus}$ | Yield (%) |
|---|---|---|---|---|---|
| a | $CH_3-$ | $i\text{-}C_3H_7-$ | $CH_3-$ | $I^{\ominus}$ | 13 |
| b | $C_2H_5-$ | $i\text{-}C_3H_7-$ | $H-$ | $Br^{\ominus}$ | 11 |
| c | $C_2H_5-$ | $i\text{-}C_3H_7-$ | $CH_3-$ | $I^{\ominus}$ | 24 |
| d | $C_2H_5-$ | $t\text{-}C_4H_9-$ | $CH_3-$ | $Br^{\ominus}$ | 10 |
| e | $C_2H_5-$ | $c\text{-}C_6H_{11}-$ | $CH_3-$ | $Br^{\ominus}$ | 29 |
| f | $C_2H_5-$ | $c\text{-}C_6H_{11}-$ | $CH_3-$ | $I^{\ominus}$ | 27 |
| g | $C_2H_5-$ | $2,6\text{-}(CH_3)_2\text{-}C_6H_3-$ | $CH_3-$ | $Br^{\ominus}$ | 23 |
| h | $C_2H_5-$ | $C_2H_5-$ | $C_6H_5-$ | $Br^{\ominus}$ | 22 |
| i | $C_2H_5-$ | $c\text{-}C_6H_{11}-$ | $C_6H_5-$ | $Br^{\ominus}$ | 32 |
| j | $C_2H_5-$ | $C_6H_5-$ | $C_6H_5-$ | $Br^{\ominus}$ | 11 |
| k | $C_6H_5-CH_2-$ | $i\text{-}C_3H_7-$ | $CH_3-$ | $Br^{\ominus}$ | 40 |
| l | $C_2H_5-CH_2-$ | $t\text{-}C_4H_9-$ | $CH_3-$ | $Br^{\ominus}$ | 29 |
| m | $C_2H_5-CH_2-$ | $c\text{-}C_6H_{11}-$ | $CH_3-$ | $Br^{\ominus}$ | 57 |
| n | $C_2H_5-CH_2-$ | $C_2H_5-$ | $C_2H_5-$ | $Br^{\ominus}$ | 37 |
| o | $C_2H_5-CH_2-$ | $c\text{-}C_6H_{11}-$ | $C_6H_5-$ | $Br^{\ominus}$ | 32 |
| p | $C_2H_5-CH_2-$ | $C_6H_5-CH_2-$ | $C_6H_5-$ | $Br^{\ominus}$ | 28 |

(LXX)

(LXXI)

(LXXII)

(LXIXe)

(28)

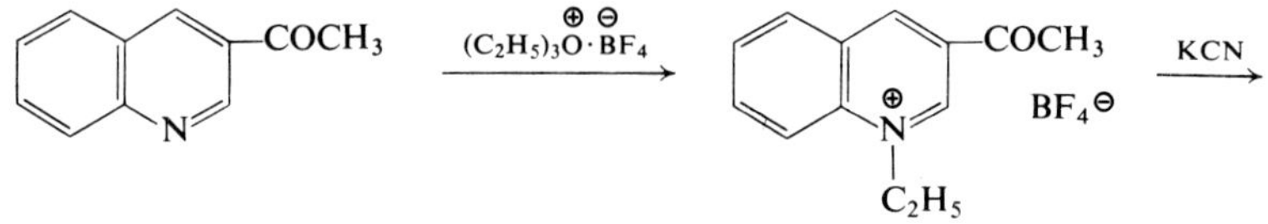

The addition of cyanide ion to the 4-position of LXX to form LXXI results from the fact that LXXII is not identical with LXXIV, which can be prepared from the methyl ester of the Einhorn-Woodward acid (LXXIII)[26,82] by ethylation (29).

(LXXIII) (LXXIV)

(29)

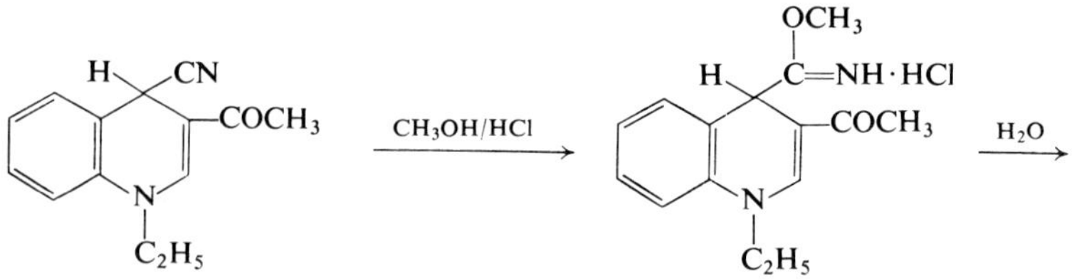

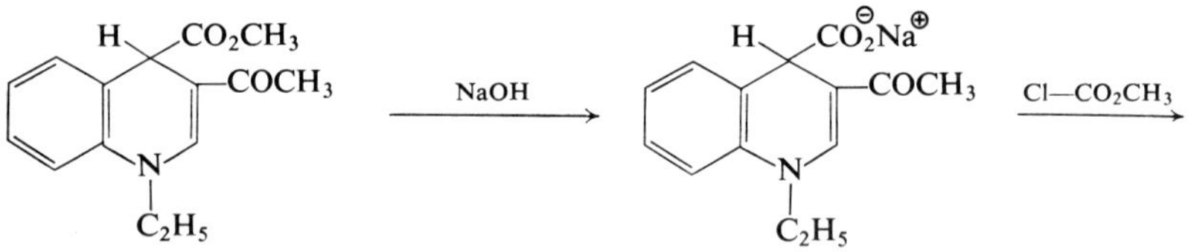

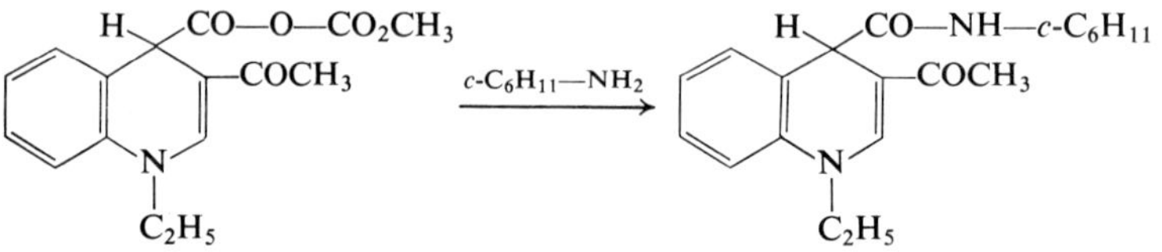

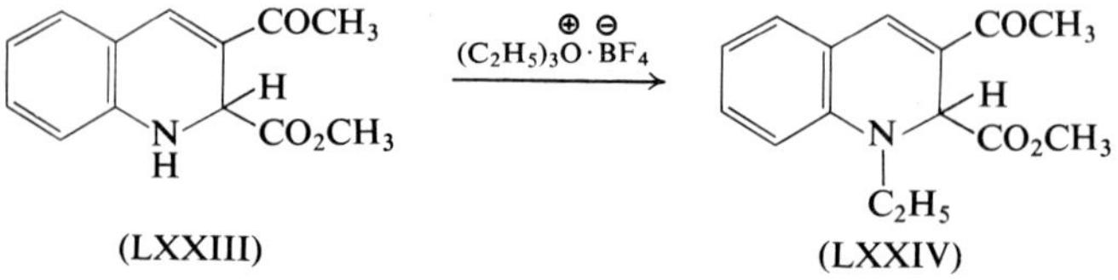

## IX. DIACYLIMIDES, AMIDES, THIOAMIDES, SELENOAMIDES, AND AMIDINES OF α-AMINO ACIDS

The formation of α-acylaminocarbonamides, XIII, from carboxylic acids, amines, carbonyl compounds, and isonitriles is only possible when ammonia or primary amines are used as amine components. With secondary amines, the reaction proceeds in a different way.

For instance, the enamine from piperidine and isobutyraldehyde reacts with cyclohexyl isocyanide and benzoic acid in methanol to yield α-piperidino-isovaleric $N$-cyclohexylamide (XXa) and methylbenzoate. The α-adduct (XIVa) transfers its benzoyl group to the solvent. In chloroform solution, the diacylimide (XIXa) results from a Mumm rearrangement[36] of XIVa. The zwitterionic species (LXXV) is a possible intermediate, because if it is an N → N' acyl transfer, it is possible by two cyclic steps.

$$\text{piperidinyl—CH}=\text{C(CH}_3)_2 + \text{C}_6\text{H}_5\text{—COOH} + \text{CN—}c\text{-C}_6\text{H}_{11} \longrightarrow$$

```
      i-C3H7                              i-C3H7
        |                                   |
 pip—CH—C=N—c-C6H11    CH3OH      pip—CH—C—NH—c-C6H11 + C6H5CO2CH3
        |             ———→                 ‖
 C6H5—CO—O                                 O
       (XIVa)                             (XXa)
```

↓

```
   ⊕  i-C3H7                              i-C3H7
   |   |                                   |
  pip—CH—C⋯N—c-C6H11           pip—N—CH—CO—N—c-C6H11      (30)
   |      ⋮ /⊖                                |
   CO     O                                   C=O
   |                                          |
   C6H5                                       C6H5
      (LXXV)                                 (XIXa)
```

It is remarkable that XIXa, in contrast to XIVa,* is stable for a short period of time even in boiling methanol; it is instantaneously cleaved by cold benzylamine to form XXa and $N$-benzylbenzamide.

1-Piperidino- and 1-morpholino-2-ethylhexadiene (LXXIX), carboxylic acids (acetic acid, benzoic acid, 3,5-dinitrobenzoic acid) and cyclohexylisocyanide react in ether solution in an analogous fashion (31) to form mixtures

---

* An active benzoylating species, probably XIVa is found in chloroform solutions at −20°C up to 1 hr after the components of (30) have been mixed.

of diacylamides (XIXb and c) which could easily be deacylated to the α-aminocarbonamides (XXb and c) with sodium hydroxide solution or secondary amines. If, in (28), formic acid is used ($R^6 = H$), then the α-aminocarbonamides (XXb and c) are obtained directly.[45]

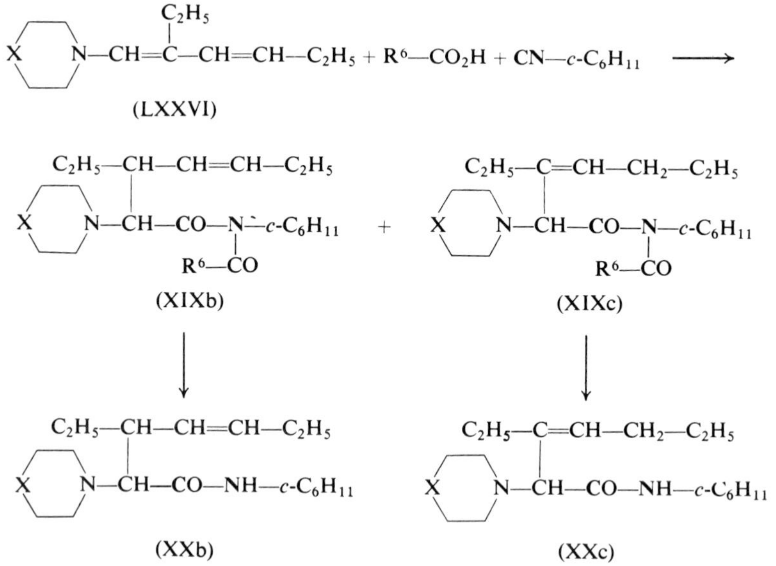

$$X\diagdown N-CH=\overset{\overset{\displaystyle C_2H_5}{|}}{C}-CH=CH-C_2H_5 + R^6-CO_2H + CN-c-C_6H_{11} \longrightarrow$$

(LXXVI)

$$C_2H_5-CH-CH=CH-C_2H_5$$
$$X\diagdown N-\overset{|}{C}H-CO-\overset{|}{N}-c-C_6H_{11}$$
$$R^6-CO$$

(XIXb)

$$C_2H_5-C=CH-CH_2-C_2H_5$$
$$X\diagdown N-\overset{|}{C}H-CO-\overset{|}{N}-c-C_6H_{11}$$
$$R^6-CO$$

(XIXc)

(31)

$$C_2H_5-CH-CH=CH-C_2H_5$$
$$X\diagdown N-\overset{|}{C}H-CO-NH-c-C_6H_{11}$$

(XXb)

$$C_2H_5-C=CH-CH_2-C_2H_5$$
$$X\diagdown N-\overset{|}{C}H-CO-NH-c-C_6H_{11}$$

(XXc)

$R^6 = CH_3-, C_6H_5-, 3,5-(NO_2)_2-C_6H_3-$; $X = O, CH_2$

TABLE XII

Condensation of Cyclohexyl Isocyanide with Isobutyraldehyde and Dimethylamine (both 0.40 moles/liter) in Methanol at 20°C[33]

| Exp. no. | Additive | Conc. [moles/liter] | Dimethylamine conc. [moles/liter] | Product |
|---|---|---|---|---|
| 1 | — | — | 0.80 | — |
| 2 | $(CH_3)_2NH \cdot HCl$ | 0.40 | — | XXd (32%), LXXIX (32%) |
| 3 | $(CH_3)_2NH \cdot HCl$ | 0.40 | 0.40 | LXXVIII (70%) |
| 4 | $CH_3CO_2H$ | 0.40 | 0.40 | XXd (35%), LXXVII (41%) |
| 5 | $CH_3CO_2H$ | 0.40 | 0.80 | XXd (94%), LXXVII (6%) |

## 8. Four-Component Condensations 191

McFarland[33] carried out some very instructive experiments on the condensation of cyclohexyl isocyanide with isobutyraldehyde and dimethylamine. He found that the reaction conditions (such as added acids) greatly influence the course of the reaction (see Table XII).

In the absence of acids none of the products, XXd, LXXVIII, or LXXIX, is formed. In the presence of acetic acid (exp. nos. 4 and 5), the product LXXVII of the Passerini reaction results, in addition to the α-aminocarbonamide (XXd). The latter is probably formed from an α-adduct (XIVd) which transfers its acyl group to available nucleophiles. The addition of dimethylamine hydrochloride leads to the formation of the α-hydroxycarbonamide (LXXIX) and of the α-aminoamidine (LXXVIII).

$$CH_3-CO-O-\underset{\underset{CH(CH_3)_2}{|}}{CH}-CO-NH-c\text{-}C_6H_{11}$$
(LXXVII)

$$(CH_3)_2N-\underset{\underset{CH(CH_3)_2}{|}}{CH}-CO-NH-c\text{-}C_6H_{11}$$
(XXd)

$$(CH_3)_2N-\underset{\underset{CH(CH_3)_2}{|}}{CH}-\underset{\underset{Z}{|}}{C}=N-c\text{-}C_6H_{11}$$

(LXXVIII) Z = (CH$_3$)$_2$N—
(XIVd) Z = CH$_3$—CO$_2$—

$$HO-\underset{\underset{CH(CH_3)_2}{|}}{CH}-CO-NH-c\text{-}C_6H_{11}$$
(LXXIX)

The syntheses of (XXe and XXf) from piperidine, formaldehyde, and cyclohexylisocyanide or isocyanodiphenylphosphine oxide,[27] respectively, are further examples of the preparation of α-aminocarbonamides by 4 C C. Similarly, products of this type can be obtained from the following combinations of reactants: XXg from piperidine, benzaldehyde, and vinyl isocyanide; XXh from diethylamine, formaldehyde, and 2,6-dimethylphenyl isocyanide; and XXi from N-methylaniline, formaldehyde, and cyclohexyl isocyanide.

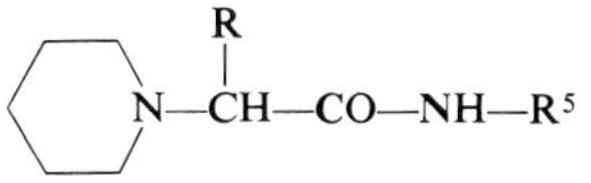

$$\underset{\displaystyle (XX)}{\left\langle\begin{array}{c}\\N\end{array}\right\rangle-\underset{\underset{R}{|}}{CH}-CO-NH-R^5}$$

e (78%): R = H—, R$^5$ = c-C$_6$H$_{11}$—
f (67%): R = H—, R$^5$ = (C$_6$H$_5$)$_2$P(O)—CH$_2$—[27]
g (50%): R = C$_6$H$_5$—, R$^5$ = CH$_2$=CH—[32]

(C$_2$H$_5$)$_2$N—CH$_2$—CO—NH—2,6-(CH$_3$)$_2$—C$_6$H$_3$
(XXh) (80%)

$$C_6H_5-\underset{\underset{CH_3}{|}}{N}-CH_2-CO-NH-c\text{-}C_6H_{11}$$
(XXi) (41%)

Mono-, di-, and trisubstituted hydrazines react with aldehydes and cyclohexyl isocyanide to form the hydrazine derivatives (XXi–p).[85]

$$CH_3-\underset{\underset{CH_2-CO-NH-c-C_6H_{11}}{|}}{N}-N(CH_2-CO-NH-c-C_6H_{11})_2$$

(XXi) (15%)

$$R_2^1N-N(CH_2-CO-NH-c-C_6H_{11})_2$$

(XX)

j (77%): $R^1 = CH_3-$

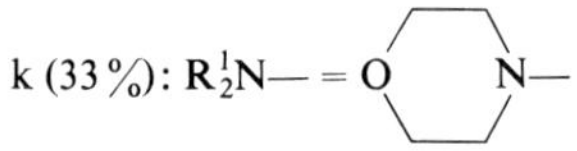

k (33%): $R_2^1N- = O\underset{\phantom{X}}{\diagup\!\!\!\diagdown} N-$

$$[c-C_6H_{11}-NH-CO-N(CH_3)-]_2$$

(XXl) (37%)

$$R_2^1N-NR^2-CHR^3-CO-NH-c-C_6H_{11}$$

(XX)

m (87%): $R^1 = R^2 = CH_3-$, $R^3 = H-$
n (70%): $R^1 = R^2 = CH_3-$, $R^3 = C_2H_5-$

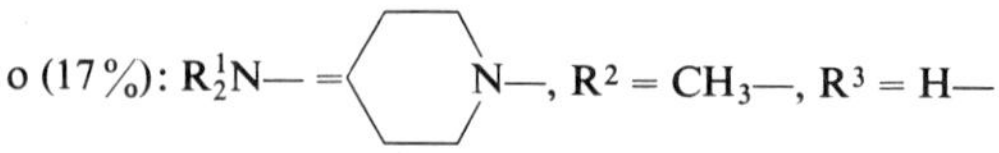

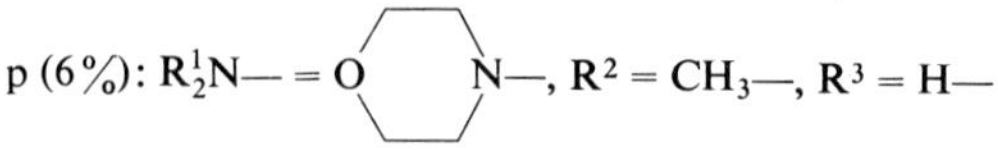

o (17%): $R_2^1N- = \diagup\!\!\!\diagdown N-$, $R^2 = CH_3-$, $R^3 = H-$

p (6%): $R_2^1N- = O\diagup\!\!\!\diagdown N-$, $R^2 = CH_3-$, $R^3 = H-$

The α-aminothio- and selenoamides, e.g., XXIa and b and XXIIa, are accessible by 4 C C of hydrogen thiosulfate or hydrogen selenide.[56,76]

$(CH_3)_2N-CH_2-CS-NH-R^5$           $\diagup\!\!\!\diagdown N-\underset{\underset{i-C_3H_7}{|}}{CH}-CSe-NH-c-C_6H_{11}$

(XXI)                                   (XXIIa)

a: $R^5 = c-C_6H_{11}-$
b: $R^5 = C_6H_5-CH_2-$

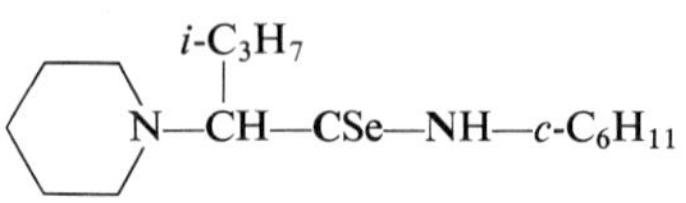

## X. TETRAZOLES

Reactive carbonyl components, such as the piperidine aminal of formaldehyde isobutyraldehyde or benzaldehyde react with amine hydrochlorides, cyclohexyl isocyanide, and sodium azide in aqueous acetone to form the tetrazoles (XXIIIa, c, d, g, h, k) (see Table XIII).[77] An analogous 4 C C of isocyanomethyldiphenylphosphine oxide produces 67% of XXIIIo.[27]

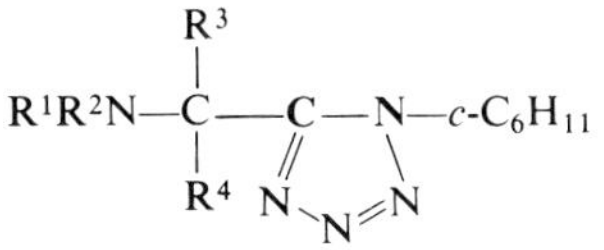

$$R^1R^2N-\underset{R^4}{\overset{R^3}{\underset{|}{C}}}-\underset{N\diagdown_N\diagup^N}{\overset{\diagup}{C}}-N-c\text{-}C_6H_{11}$$

(XXIIIa–n)

$$\text{piperidinyl}-N-CH_2-\underset{N\diagdown_N\diagup^N}{\overset{\diagup}{C}}-N-CH_2-P(O)(C_6H_5)_2$$

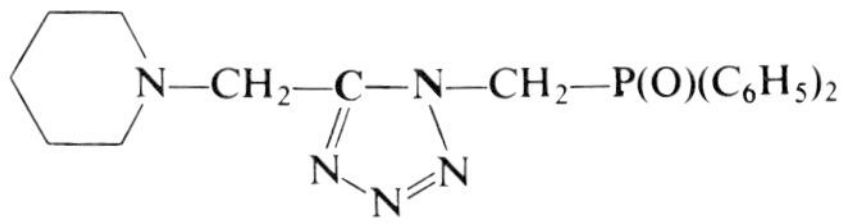

(XXIIIo)

Mostly, however, the synthesis of 1,5-disubstituted tetrazoles (see Table XIII) from isonitriles, amines, carbonyl compounds, and hydrazoic acid (method A) proceeds well only in anhydrous media.

The overall reaction rate of 4 C C is either dependent upon the rate of formation or their equilibrium concentration or equivalent species (see Section IV) of immonium ions; therefore, it is frequently found to be advantageous to use the condensation products of the carbonyl and amine components (method B). Immonium ions are formed faster and in a higher equilibrium concentration from condensation products* than from amines and carbonyl compounds.

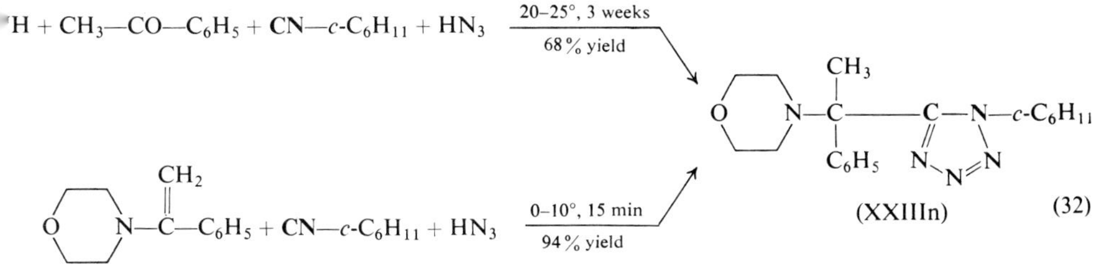

$$\text{morpholine-H} + CH_3-CO-C_6H_5 + CN-c\text{-}C_6H_{11} + HN_3 \xrightarrow[68\% \text{ yield}]{20-25°, 3 \text{ weeks}}$$

$$\text{morpholine-}N-\underset{C_6H_5}{\overset{CH_3}{\underset{|}{C}}}-\underset{N\diagdown_N\diagup^N}{\overset{\diagup}{C}}-N-c\text{-}C_6H_{11}$$

$$\text{morpholine-}N-\overset{CH_2}{\underset{\|}{C}}-C_6H_5 + CN-c\text{-}C_6H_{11} + HN_3 \xrightarrow[94\% \text{ yield}]{0-10°, 15 \text{ min}}$$

(XXIIIn)            (32)

For example, at 20–25°C in 3 weeks the 4 C C of morpholine, acetophenone, cyclohexyl isocyanide, and hydrazoic acid (each approx. 1 mole/liter) in benzene/methanol produces 68% of crude XXIIIn; 94% yield of the same product is obtained from α-morpholinostyrene, cyclohexyl isonitrile, and hydrazoic acid under analogous conditions in 15 min (see Table XIII).

* Schiff bases, enamines, aminals.

## TABLE XIII

THE PREPARATION OF 1-CYCLOHEXYL-5-AMINOALKYLTETRAZOLES (XXIIIa–n) BY 4 C C OF CARBONYL COMPOUNDS AND AMINES (METHOD A), OR THEIR CONDENSATION PRODUCTS (AMINALS, SCHIFF BASES, AND ENAMINES) (METHOD B) WITH CYCLOHEXYL ISOCYANIDE AND HYDRAZOIC ACID IN BENZENE/METHANOL AT 0–20°C[77]

| Designation no. (XXIII) | R³ (II) | R⁴ (II) | R¹ (I) | R² (I) | Yield (%) A | Yield (%) B |
|---|---|---|---|---|---|---|
| a | H— | H— | —(CH$_2$)$_5$— | —(CH$_2$)$_5$— | (92)[a] | 92 |
| b | H— | n-C$_3$H$_7$— | H— | 3-Cl—C$_6$H$_4$— | 90 | — |
| c | H— | i-C$_3$H$_7$— | H— | n-C$_3$H$_7$— | (53)[a] | 73 |
| d | H— | i-C$_3$H$_7$— | —(CH$_2$)$_5$— | —(CH$_2$)$_5$— | 90(71)[a] | 93 |
| e | H— | i-C$_3$H$_7$— | —(CH$_2$)$_2$—NH—(CH$_2$)$_2$— | —(CH$_2$)$_2$—NH—(CH$_2$)$_2$— | 87 | — |
| f | H— | i-C$_3$H$_7$— | H— | 2-C$_5$H$_4$N— | 64 | — |
| g | H— | 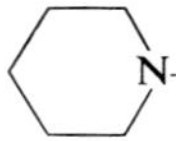 ⟨N—CO⟩ (piperidine) | —(CH$_2$)$_5$— | —(CH$_2$)$_5$— | — | 93[b] |
| h | H— | C$_6$H$_5$— | —(CH$_2$)$_5$— | —(CH$_2$)$_5$— | 90(34)[a] | — |
| i | CH$_3$— | CH$_3$— | H— | i-C$_3$H$_7$— | (84)[a] | 82 |
| j | CH$_3$— | CH$_3$— | —(CH$_2$)$_5$— | —(CH$_2$)$_5$— | 93 | — |
| k | CH$_3$— | CH$_3$— | H— | 4-CH$_3$O—C$_6$H$_4$— | 90 | — |
| l | —(CH$_2$)$_5$— | —(CH$_2$)$_5$— | H— | n-C$_4$H$_9$— | (68)[a] | 92 |
| m | —(CH$_2$)$_5$— | —(CH$_2$)$_5$— | —(CH$_2$)$_5$— | —(CH$_2$)$_5$— | — | 76 |
| n | CH$_3$— | C$_6$H$_5$—CH$_2$— | —(CH$_2$)$_2$—O—(CH$_2$)$_2$— | —(CH$_2$)$_2$—O—(CH$_2$)$_2$— | 69 | 87 |
| o | CH$_3$— | C$_6$H$_5$— | —(CH$_2$)$_2$—O—(CH$_2$)$_2$— | —(CH$_2$)$_2$—O—(CH$_2$)$_2$— | 68 | 94 |

[a] In aqueous acetone.
[b] From ref. 16.

## 8. Four-Component Condensations

The syntheses of XXIIIp and XXIIIg from formaldehyde-$t$-butylimide or isobutyraldehyde and piperidine with hydrazoic acid and 2,6-dimethylphenyl isocyanide, a rather unreactive isocyanide, proceed equally well (85 and 95%).[77]

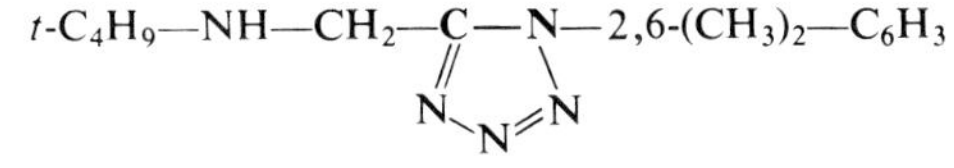

$$t\text{-}C_4H_9-NH-CH_2-C-N-2,6\text{-}(CH_3)_2-C_6H_3$$

(XXIIIp)

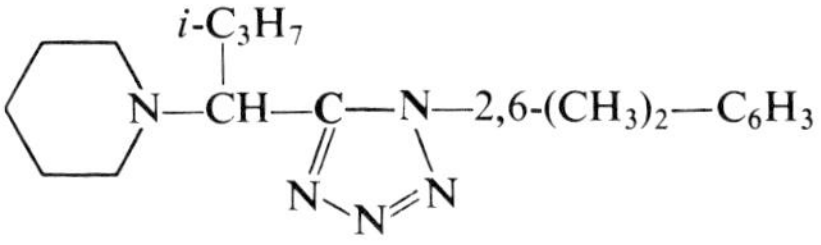

(XXIIIg)

Neidlein[37-39] obtained (XXIIIr–t) from diisonitriles and demonstrated the existence of the unstable diisocyanomethane by trapping it as the 4 C C product (XXIIIr) of $N\text{-}\Delta^1$-cyclohexenylpiperidine and hydrazoic acid (33).

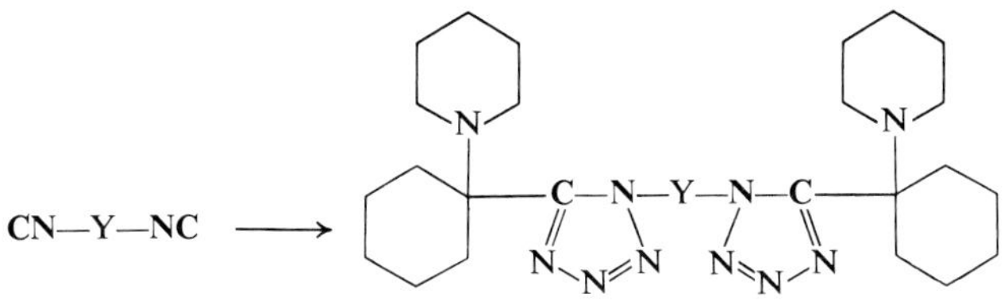

CN—Y—NC ⟶ (XXIIIr–t)  (33)

r (89%): Y = —CH$_2$—
s (98%): Y = —(CH$_2$)$_6$—
t (76%): Y = —1,4-$c$-C$_6$H$_{10}$—

According to Opitz and Merz,[44] the dieneamines (LXXVI), hydrazoic acid and cyclohexyl isocyanide form mixtures of the tetrazoles (XXIIIu–w) (see footnote, p. 145).

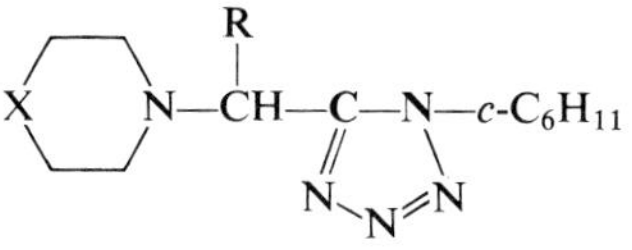

(XXIIIu–w)

u: R = C$_2$H$_5$—CH=CH—CH(C$_2$H$_5$)—
v: R = C$_2$H$_5$—CH$_2$—CH=C(C$_2$H$_5$)—
w: R = C$_2$H$_5$—CH$_2$—CH(N$_3$)—CH(C$_2$H$_5$)—
X = O, CH$_2$

Opitz et al.[43] also found that α-aminoalkylazides (LXXX) (see p. 195) are capable of "inserting" isonitriles to yield the tetrazoles (XXIII$x_1$–$x_7$, Table XIV) (34).

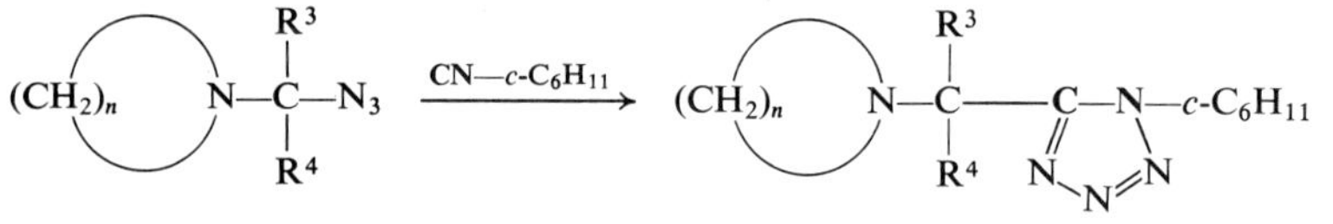

$$\text{(CH}_2)_n\underset{R^4}{\overset{R^3}{N-C-N_3}} \xrightarrow{CN-c-C_6H_{11}} \text{(CH}_2)_n\underset{R^4}{\overset{R^3}{N-C}}\underset{N\diagdown_{N}\diagup N}{\overset{}{-C-N-c-C_6H_{11}}} \quad (34)$$

(LXXX)  (XXIII$x_1$–$x_7$)

TABLE XIV

The Preparation of the Tetrazoles (XXIII$x_1$–$x_7$) from the α-Aminoalkylazides (LXXX) in Acetonitrile[43]

| Designation no. (XXIII) | $R^3$ (LXXX) | $R^4$ (LXXX) | $n$ | Yield (%) |
|---|---|---|---|---|
| $x_1$ | H— | $C_2H_5$— | 5 | 72 |
| $x_2$ | H— | $i$-$C_3H_7$— | 4 | 90 |
| $x_3$ | H— | $i$-$C_3H_7$— | 5 | 89 |
| $x_4$ | H— | $C_2H_5$—CH($C_2H_5$)— | 5 | 79 |
| $x_5$ | H— | $C_2H_5$—CH($n$-$C_4H_9$)— | 5 | 95 |
| $x_6$ | —$(CH_2)_5$— | —$(CH_2)_5$— | 4 | 23 |
| $x_7$ | —$(CH_2)_5$— | —$(CH_2)_5$— | 5 | 22 |

The 4 C C of $N_\beta$-acylhydrazones (XL), isonitriles, and hydrazoic acid leads to the tetrazoles (LXXXI) (35) (see Table XV).[64]

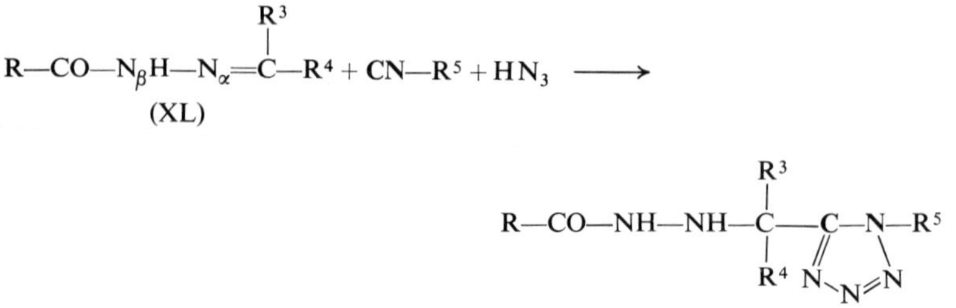

$$R-CO-N_\beta H-N_\alpha=\overset{R^3}{\underset{}{C}}-R^4 + CN-R^5 + HN_3 \longrightarrow$$
(XL)

$$R-CO-NH-NH-\underset{R^4}{\overset{R^3}{C}}\underset{N\diagdown_{N}\diagup N}{\overset{}{-C-N-R^5}} \quad (35)$$

(LXXXI)

Analogously, the azines (XLIII) yield the tetrazole derivatives (LXXXII).[64]

## 8. Four-Component Condensations

### TABLE XV

THE PREPARATION OF THE TETRAZOLES (LXXXIa–g$_5$) FROM $N_\beta$-ACYLHYDRAZONES (XL), ISONITRILES, AND HYDRAZOIC ACID[64]

| Designation no. (LXXXI) | R$^3$ (XL) | R$^4$ (XL) | R (III) | R$^5$ (III) | Yield (%) |
|---|---|---|---|---|---|
| a | H— | $n$-C$_3$H$_7$— | C$_6$H$_5$— | $c$-C$_6$H$_{11}$— | 80 |
| b$_1$ | H— | $i$-C$_3$H$_7$— | C$_6$H$_5$— | $c$-C$_6$H$_{11}$— | 67 |
| b$_2$ | H— | $i$-C$_3$H$_7$— | H— | $c$-C$_6$H$_{11}$— | 79 |
| c | H— | $c$-C$_6$H$_{11}$— | C$_6$H$_5$— | $c$-C$_6$H$_{11}$— | 66 |
| d$_1$ | CH$_3$— | CH$_3$— | H— | $c$-C$_6$H$_{11}$— | 76 |
| d$_2$ | CH$_3$— | CH$_3$— | C$_6$H$_5$— | $c$-C$_6$H$_{11}$— | 56 |
| d$_3$ | CH$_3$— | CH$_3$— | 4-C$_5$H$_4$N— | $c$-C$_6$H$_{11}$— | 70 |
| e | CH$_3$— | CH$_3$— | C$_6$H$_5$—SO$_2$—[a] | $c$-C$_6$H$_{11}$— | 75 |
| f$_1$ | CH$_3$— | C$_6$H$_5$—CH$_2$— | $t$-C$_4$H$_9$— | $t$-C$_4$H$_9$— | 50 |
| f$_2$ | CH$_3$— | C$_6$H$_5$—CH$_2$— | C$_6$H$_5$— | $c$-C$_6$H$_{11}$— | 68 |
| g$_1$ | —(CH$_2$)$_5$— | —(CH$_2$)$_5$— | C$_6$H$_5$— | $n$-C$_4$H$_9$— | 88 |
| g$_2$ | —(CH$_2$)$_5$— | —(CH$_2$)$_5$— | C$_6$H$_5$— | $t$-C$_4$H$_9$— | 86 |
| g$_3$ | —(CH$_2$)$_5$— | —(CH$_2$)$_5$— | C$_6$H$_5$— | $c$-C$_6$H$_{11}$— | 79 |
| g$_4$ | —(CH$_2$)$_5$— | —(CH$_2$)$_5$— | C$_6$H$_5$— | C$_6$H$_5$—CH$_2$— | 80 |
| g$_5$ | —(CH$_2$)$_5$— | —(CH$_2$)$_5$— | C$_6$H$_5$— | 2,6-(CH$_3$)$_2$—C$_6$H$_3$— | 85 |

[a] Acetone benzenesulfonylhydrazone was used.

$$R^3-\underset{\underset{R^4}{|}}{C}=N-N=\underset{\underset{R^3}{|}}{C}-R^4 + CN-c\text{-}C_6H_{11} + HN_3 \longrightarrow$$
(XLIII)

$$R^3-\underset{\underset{R^4}{|}}{C}=N-NH-\underset{\underset{R^4}{|}}{C}-\underset{N\diagdown_{N}\diagup^{N}}{C}-N-c\text{-}C_6H_{11} \quad (36)$$

(LXXXIIa–c)

a (83%): R$^3$ = H—; R$^4$ = $i$-C$_3$H$_7$—
b (67%): R$^3$ = R$^4$ = CH$_3$—
c (63%): R$^3$ = H; R$^4$ = $n$-C$_3$H$_7$—

### REFERENCES

1. Asinger, F., and Thiel, M., *Angew. Chem.* **70**, 667 (1958) (rev. art.).
2. Badin, E. J., and Pacsu, E., *J. Amer. Chem. Soc.* **67**, 1352 (1945).
3. Bergs, H., Ger. Pat. 566,094 (1929); *Chem. Abstr.* **27**, 1001 (1933).
4. Bodesheim, F., Ph.D. Thesis, University of Munich, 1962.
5. Boyd, D. R., and McKervey, M. H., *Quart. Rev.* (London) **22**, 95 (1968) (rev. art.).
6. Bucherer, H. T., and Barsch, H., *J. Prakt. Chem.* [N.S.] **140**, 151 (1934).

7. Bucherer, H. T., and Brandt, W., *J. Prakt. Chem.* [N.S.] **140**, 129 (1934).
8. Bucherer, H. T., and Fischbeck, H., *J. Prakt. Chem.* [N.S.] **140**, 69 (1934).
9. Bucherer, H. T., and Lieb, V. A., *J. Prakt. Chem.* [N.S.] **141**, 5 (1934).
10. Bucherer, H. T., and Steiner, W., *J. Prakt. Chem.* [N.S.] **140**, 291 (1934).
11. Cahn, R. S., Ingold, C. K., and Prelog, V., *Angew. Chem.* **78**, 413 (1966); *Angew. Chem., Int. Ed. Engl.* **5**, 385 (1966).
12. Cruse, R., *in* "Stereochemie der Kohlenstoffverbindungen" (E. L. Eliel, ed.), p. 537. Verlag Chemie, Weinheim, 1966.
13. Eliel, E. L., "Stereochemistry of Carbon Compounds." McGraw-Hill, New York, 1962.
14. Eyring, H., "Statistical Mechanics and Dynamics." Van Nostrand, Princeton, New Jersey, 1954.
15. Gamberjan, N. P., *Zh. Vses. Khim. Obshchestva im. D.I. Mendeleeva* **12**, 65 (1967) (rev. art.).
16. Gross, H., Gloede, J., Keitel, I., and Dunath, D., *J. Prakt. Chem.* [4] **37**, 192 (1968).
17. Herlinger, H., Kleimann, H., Offermann, K., Rücker, D., and Ugi, I., *Justus Liebigs Ann. Chem.* **692**, 94 (1966).
18. Herlinger, H., Rücker, D., and Kleimann, H., *Angew. Chem.* **76**, 757 (1964); *Angew. Chem., Int. Ed. Engl.* **3**, 808 (1964).
19. Hoffmann, P., Marquarding, D., and Ugi, I., *in* "The Chemistry of the Cyanogen Group" (S. Patai and Z. Rapoport, eds.), Wiley (Interscience), New York, 1969.
20. Hoffmann, P., Marquarding, D., and Ugi, I., unpublished results (1967).
21. Kelvin, Lord, "On the Molecular Tactics of Crystals, Baltimore Lectures," Appendix H, p. 619. Clay & Sons, London, 1904.
22. Kemp, D. S., *Tetrahedron* **23**, 2001 (1967).
23. Kemp, D. S., and Chien, S. W., *J. Amer. Chem. Soc.* **89**, 2743 (1967).
24. Kemp, D. S., and Woodward, R. B., *Tetrahedron* **21**, 3019 (1965).
25. Klabunowski, J. J., "Asymmetrische Synthese." Dtsch. Verlag Wiss., Berlin, *1963*.
26. Koller, G., Ruppersberg, H., and Strang, E., *Monatsh. Chem.* **52**, 59 (1929).
27. Kreutzkamp, N., and Lämmerhirt, K., *Angew. Chem.* **80**, 394 (1968); *Angew. Chem., Int. Ed. Engl.* **7**, 372 (1968).
28. Long, L. M., Miller, C. A., and Troutman, H. D., *J. Amer. Chem. Soc.* **70**, 900 (1948).
29. Marquarding, D., unpublished results (1967).
30. Marquarding, D., Hoffmann, P., and Ugi, I., unpublished results (1967).
31. Mathieu, J., and Weill-Raynal, J., *Bull. Soc. Chim. Fr.* p. 1211 (1968) (rev. art.).
32. Matteson, D. S., and Bailey, R. A., *Chem. Ind.* (*London*) p. 191 (1967).
33. McFarland, J. W., *J. Org. Chem.* **28**, 2179 (1963).
34. McKenzie, A., *Angew. Chem.* **45**, 59 (1932) (rev. art.).
35. Mislow, K., "Introduction to Stereochemistry." Benjamin, New York, 1966.
36. Mumm, O., Hesse, H., and Volquartz, H., *Chem. Ber.* **48**, 379 (1915).
37. Neidlein, R., *Angew. Chem.* **76**, 440 (1964); *Angew. Chem., Int. Ed. Engl.* **3**, 382 (1964).
38. Neidlein, R., *Arch. Pharm.* (*Weinheim*) **297**, 589 (1964).
39. Neidlein, R., *Arch. Pharm.* (*Weinheim*) **298**, 491 (1965).
40. Neidlein, R., *Arch. Pharm.* (*Weinheim*) **299**, 603 (1966).
41. Oda, R., and Shono, T., *J. Soc. Org. Syn. Chem., Tokyo* **22**, 695 (1964) (rev. art.).
42. Offermann, K., Ph.D. Thesis, University of Munich, 1963.
43. Opitz, G., Griesinger, A., and Schubert, H. W., *Justus Liebigs Ann. Chem.* **665**, 91 (1963).
44. Opitz, G., and Merz, W., *Justus Liebigs Ann. Chem.* **652**, 158 (1962).
45. Opitz, G., and Merz, W., *Justus Liebigs Ann. Chem.* **652**, 163 (1962).
46. Pracejus, H., *Fortschr. Forsch.* **8**, 493 (1967) (rev. art.).
47. Prelog, V., *Bull. Soc. Chim. Fr.* p. 987 (1956) (rev. art.).

## 8. Four-Component Condensations 199

48. Rosendahl, F. K., Ph.D. Thesis, University of Munich, 1962.
49. Ruch, E., and Ugi, I., *Theor. Chim. Acta* **4**, 287 (1966).
50. Ruch, E., and Ugi, I., *Top. Stereochem.* **4**, 99 (1969).
51. Rücker, D., M.A. Thesis, Tech. Hoch. T. H. Stuttgart, 1964.
52. Sheehan, J. C., and Corey, E. J., *Org. React.* **9**, 388 (1957).
53. Sjöberg, K., *Sv. Kem. Tidskr.* **75**, 493 (1963) (rev. art.).
54. Sjöberg, K., Ph.D. Thesis, Technical University, Stockholm, 1970.
55. Slotta, K., Behnisch, R., and Szyszka, G., *Chem. Ber.* **67**, 1529 (1934).
56. Steinbrückner, C., Ph.D. Thesis, University of Munich, 1961.
57. Steinbrückner, C., Betz, W., and Ugi, I., unpublished results (1961).
58. Turner, E. E., and Harris, M. M., *Quart. Rev. (London)* **1**, 299 (1947).
59. Ugi, I., *Angew. Chem.* **74**, 9 (1962); *Angew. Chem., Int. Ed. Engl.* **1**, 8 (1962) (rev. art.).
60. Ugi, I., *Jahrb. Akad. Wiss. Goettingen* p. 21 (1965) (rev. art.).
61. Ugi, I., *Z. Naturforsch.* B **20**, 405 (1965).
62. Ugi, I., in "Neuere Methoden der Präparativen Organischen Chemie" (W. Foerst, ed.), Vol. IV, p. 1 (rev. art.). Verlag Chemie, Weinheim, 1966.
63. Ugi, I., and Bodesheim, F., *Justus Liebigs Ann. Chem.* **666**, 61 (1963).
64. Ugi, I., and Bodesheim, F., *Chem. Ber.* **94**, 2797 (1961).
65. Ugi, I., and Böttner, E., *Justus Liebigs Ann. Chem.* **670**, 74 (1963).
66. Ugi, I., and Kaufhold, G., *Justus Liebigs Ann. Chem.* **709**, 11 (1967).
67. Ugi, I., Meyr, R., Fetzer, U., and Steinbrückner, C., *Angew. Chem.* **71**, 386 (1959).
68. Ugi, I., and Offermann, K., *Angew. Chem.* **75**, 917 (1963); *Angew. Chem., Int. Ed. Engl.* **2**, 624 (1963).
69. Ugi, I., and Offermann, K., *Chem. Ber.* **97**, 2276 (1964).
70. Ugi, I., and Offermann, K., *Chem. Ber.* **97**, 2996 (1964).
71. Ugi, I., Offermann, K., and Herlinger, H., *Angew. Chem.* **76**, 613 (1964); *Angew. Chem., Int. Ed. Engl.* **3**, 656 (1964).
72. Ugi, I., Offermann, K., and Herlinger, H., *Chimia* **18**, 278 (1964).
73. Ugi, I., Offermann, K., Herlinger, H., and Marquarding, D., *Justus Liebigs Ann. Chem.* **709**, 1 (1967).
74. Ugi, I., and Rosendahl, F. K., *Justus Liebigs Ann. Chem.* **666**, 65 (1963).
75. Ugi, I., Rosendahl, F. K., and Bodesheim, F., *Justus Liebigs Ann. Chem.* **666**, 54 (1963).
76. Ugi, I., and Steinbrückner, C., *Angew. Chem.* **72**, 267 (1960).
77. Ugi, I., and Steinbrückner, C., *Chem. Ber.* **94**, 734 (1961).
78. Ugi, I., and Steinbrückner, C., *Chem. Ber.* **94**, 2802 (1961).
79. Ugi, I., and Wischhöfer, E., *Chem. Ber.* **95**, 136 (1962).
80. Ware, E., *Chem. Rev.* **46**, 403 (1950) (rev. art.).
81. Wischhöfer, E., Ph.D. Thesis, University of Munich, 1962.
82. Woodward, R. B., and Kornfeld, E. C., *J. Amer. Chem. Soc.* **70**, 2508 (1948).
83. Woodward, R. B., and Olofson, R. A., *J. Amer. Chem. Soc.* **83**, 1007 (1961).
84. Woodward, R. B., Olofson, R. A., and Mayer, H., *J. Amer. Chem. Soc.* **83**, 1010 (1961).
85. Zinner, G., and Kleigel, W., *Arch. Pharm. (Weinheim)* **299**, 746 (1966).

# Chapter 9
# Peptide Syntheses

G. Gokel, P. Hoffmann, H. Kleimann,
H. Klusacek, G. Lüdke, D. Marquarding,
and I. Ugi

I. The General Concept . . . . . . . . . 201
II. The Present Status of Classical Methods . . . . . . . 202
III. Potential Advantages of the 4 C C Concept . . . . . . . 204
IV. Isonitrile and Amine Components—Model Reactions . . . . . 204
   A. Optically Active Isonitrile Components . . . . . . . 204
   B. Amine Components . . . . . . . . . . 206
V. Tactics of the 4 C C Peptide Synthesis . . . . . . . . 211
   References . . . . . . . . . . . . 213

## I. THE GENERAL CONCEPT

The stereoselective 4 C C of carboxylic acids (Chapter 8, Sections III and VI) provides a basis for a method of peptide synthesis. N-terminally protected amino acids or peptides (I), suitable optically active amine components (II), aldehydes (III) and α-isocyano esters or α-isocyanoacyl peptide esters (IV) undergo 4 C C (1) to produce peptide derivatives (V). The desired peptide (VI) can be obtained only from the condensation product (V) if the group $R^{1*}$ of the amine component (II) can be replaced by hydrogen under conditions which do not destroy or racemize peptides.

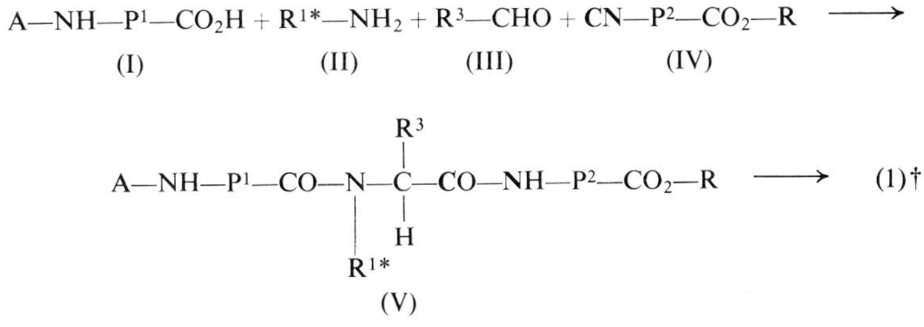

$$A\text{—NH—}P^1\text{—CO}_2H + R^{1*}\text{—NH}_2 + R^3\text{—CHO} + CN\text{—}P^2\text{—CO}_2\text{—R} \longrightarrow$$
$$\text{(I)} \qquad \text{(II)} \qquad \text{(III)} \qquad \text{(IV)}$$

$$A\text{—NH—}P^1\text{—CO—N}\underset{\underset{R^{1*}}{|}}{\overset{\overset{R^3}{|}}{C}}\text{—CO—NH—}P^2\text{—CO}_2\text{—R} \longrightarrow \quad (1)\dagger$$
$$\text{(V)}$$

† A and R are N— and C— terminal protecting groups,[6,43] $P^1$ and $P^2$ are residues of α-amino acid or peptides.

202    Gokel et al.

$$NH_2-P^1-CO-NH-\underset{\underset{H}{|}}{\overset{\overset{R^3}{|}}{C}}-CO-NH-P^2-CO_2H$$

(VI)

## II. THE PRESENT STATUS OF CLASSICAL METHODS

In order to evaluate the potential usefulness of the above synthetic concept, a brief discussion of the present status of peptide syntheses, as achieved by classical methods, is included.

"When at the beginning of this century Emil Fischer gave, in a summarizing report, an account of his work on the synthesis of peptides, the popular press, with its usual penchant for exaggeration, hailed his accomplishments as a prelude to the creation of living matter."[6]

It might appear, from the great successes recently experienced in the field of synthetic peptide chemistry, that now some of the vital ingredients of living matter have really become accessible and that some sort of pinnacle has been reached.

This viewpoint seems to be bolstered by the recent syntheses of a large number of naturally occurring peptides and their analogs. These peptides include oxytocin,[6] the vasopressins,[6] the angiotensins,[6] gramicidin A[41] and S,[6] bradykinin,[6,10] eledoisin,[6] gastrin,[1,32] α-MSH,[6] the corticotropins,[2,3,45] the melanotropins,[44] insulin,[6, 19, 20, 23, 33, 57, 58] glucagon,[55, 56] secretin,[7] thyreocalcitonin,[14] and as the most spectacular achievement yet, even a peptide which exhibits ribonuclease activity, by two independent groups, one by Hirschmann and Denkewalter et al.[15] with Leuch's anhydride, and the other by Merrifield's solid phase method.[13, 26]

All of the present-day successes in peptide synthesis are due to the classical synthetic concept[6,43] which was pioneered by Fischer[9] and by Bergmann and Zervas.[5]

The synthesis of peptides by classical methods (2) includes the formation of an amide bond between the carboxyl group of one amino acid and the amino group of another.

$$NH_2-\underset{\underset{R^1}{|}}{CH}-CO_2H + NH_2-\underset{\underset{R^2}{|}}{CH}CO_2H + \ldots NH_2-\underset{\underset{R^n}{|}}{CH}CO_2H \rightarrow$$

$$NH_2-\underset{\underset{R^1}{|}}{CH}-CO-NH-\underset{\underset{R^2}{|}}{CH}-CO-\ldots -NH-\underset{\underset{R^n}{|}}{CH}-CO_2H \quad (2)$$

This reaction requires "activation of the carboxyl group" by the introduction of an activating group Z and the protection of all functional groups not expected to participate in the desired reaction (3), e.g.,

$$\underset{(VII)}{A-NH-\overset{\overset{R}{|}}{C}H-COZ} + \underset{(VIII)}{H_2N-\overset{\overset{R'}{|}}{C}H-CO_2R} \xrightarrow{-ZH}$$

$$A-NH-\overset{\overset{R}{|}}{C}H-CO-NH-\overset{\overset{R'}{|}}{C}H-CO_2R \quad (3)$$

Today, the state of the art of synthetic peptide chemistry[6,8,43] is the result of almost forty years of discovery and development, based on this classical concept. New protecting and activating groups as well as new experimental approaches are the result of this work.

Although a large number of more or less different methods for the protection and the activation of amino acids have been developed during this time, relatively few of them are now generally used.[6,8,12,43] These few, however, generally suffice to form the peptide bond in satisfactory yield, with a tolerable amount of racemization and with a minimum of undesirable side reactions, but are particularly suitable for preparation of small amounts of short-chain peptides.

Each vital step in a classical peptide synthesis involves activated amino acid derivatives which are subject to racemization. The yields of such steps are therefore not only less than quantitative because of incompleteness of the main reaction and because of side reactions, but the products contain impurities ("epimers") which are extremely difficult to remove. Furthermore, conventional coupling steps are generally second-order reactions and some of the competing side reactions, like racemization, are either first order or pseudo-first order. As the size of the peptide increases, one is forced to work in increasingly dilute solutions, in which case the side reactions begin to become favored. The sum of these problems is that the longer the chain, the worse the yield, the greater the degree of racemization, the greater the number and amount of impurities, and the greater the difficulty of removing them.

The very brilliance of the Merrifield synthesis[13,31] also points out quite clearly the greatest problems in synthetic peptide chemistry, namely, the immense number of single steps and operations required to synthesize a moderately long peptide chain, and the near impossibility of ascertaining intermediate and product purities.

Recently methods have become available by which fragments can be coupled with a very low degree of racemization.[21,54] The syntheses of large peptides will probably be accomplished in the future by fragment strategies,[6] no matter what tactics[6] of protection and activation will be used, and whether the syntheses will be carried out in homogeneous or heterogeneous reaction media.

The elegant synthesis of glucagon by Wünsch et al.[55,56] is an instructive example of the advantages of the fragment strategy.

## III. POTENTIAL ADVANTAGES OF THE 4 C C CONCEPT

Stereoselective 4 C C present a promising supplement or even an alternative to the classical concept of peptide syntheses. Stereoselective 4 C C seem to be particularly well suited to the synthesis of peptides when combined with the fragment strategy, and can be used for the synthesis of fragments as well as the formation of larger peptides from fragments.

The potential advantages of 4 C C over the classical concept are the following:

1. A considerably lower number of synthetic steps is needed.
2. Some of the optically active amino acid units can be built up during the synthesis from simple aldehydes, a property which is particularly useful for producing amino acids that are not readily available, e.g., the nonnaturally occurring amino acids, like D-amino acids, or isotopically labeled amino acid units.
3. The activated forms of amino acids (see Chapter 8) are only intermediates which rearrange to the peptide derivatives by extremely fast first-order reactions, whereas the activated amino acid or peptide derivatives (VII) of classical peptide syntheses react slowly with the C-terminally protected amino acids and peptides (VIII) by second-order reactions (see Section II).
4. The 4 C C concept provides a basis for the synthesis of multiply S—S-bridged proteins and peptides, such as insulin, from starting materials which already contain the desired S—S bridges.

## IV. ISONITRILE AND AMINE COMPONENTS—MODEL REACTIONS

### A. Optically Active Isonitrile Components

For the general usefulness of the synthetic concept (1), all the starting materials (I–IV) must be readily available. This is, of course, the case for N-protected amino acids and peptides, which may be products of previous 4 C C peptide syntheses. The optically active α-isocyano esters and "isocyano peptide esters" (IV) can be prepared from the corresponding formylamino compounds by the phosgene method[49] (Chapter 2); no racemization occurs if the "phosgenation" is carried out at temperatures below −20°C and in the presence of pyridine or N-methyl morpholine[16] (MM). Although the direct synthesis of the isocyano derivatives of larger peptides from the corresponding formylamino precursors, like Gramicidin A,[40] is possible,[28] it is advisable to synthesize the isocyano derivatives of larger peptides from α-amino esters (VIII) or C-protected peptides and activated α-isocyano acids (IX) or N-α-isocyanoacylamino acid derivatives (X), respectively.

CN—CHR—CO—Z          CN—CHR—CO—NH—CHR'—CO—Z
     (IX)                              (X)

A variety of activated derivatives of N-formyl L-valine was phosgenated under conditions which afford a minimum of racemization. The only activated derivative of L-valine which did not suffer total racemization on phosgenation[16] (4) was the 8-oxyquinoline ester[17,18,46] (XI) (S).

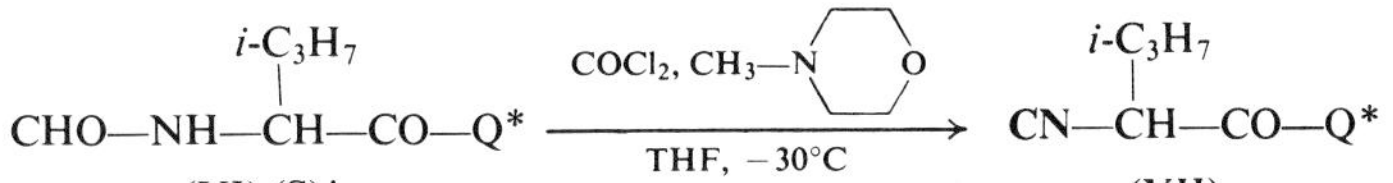

$$\underset{\underset{(XI)\ (S)\dagger}{CHO-NH-CH-CO-Q^*}}{\overset{i\text{-}C_3H_7}{|}} \xrightarrow[\text{THF, }-30°C]{COCl_2,\ CH_3-N\diagup O} \underset{(XII)}{\overset{i\text{-}C_3H_7}{\underset{|}{CN-CH-CO-Q^*}}} \quad (4)$$

$$(90\%\ S + 10\%\ R)$$

The 8-hydroxyquinoline derivatives (**X**, Z = Q) are useful for the synthesis of "isocyano peptide esters" (**IV**).

The methyl esters of optically active α-isocyano acids (**IX**, $Z = OCH_3$) and N-[α-isocyano-acyl]α-amino acids (**X**, $Z = OCH_3$) are obtained without racemization by the phosgene method.[16] Yet, the alkaline saponification racemizes the α-isocyano esters, and destroys (**X**, $Z = OCH_3$) completely. Therefore, reaction sequence (5) is presently the best way of preparing optically active activated α-isocyano acids; an analogous procedure is useful for the preparation of **X**.

$$CHO-NH-CHR-CO_2H \xrightarrow[\text{MM, THF, 0-30°C}]{(CH_3)_3SiCl}$$

$$CHO-NH-CHR-CO_2Si(CH_3)_3 \xrightarrow[\text{MM, }-40\text{ to }-20°C]{COCl_2} \quad (5)$$

$$CN-CHR-CO_2Si(CH_3)_3 \xrightarrow[H_2O]{(c\text{-}C_6H_{11})_2NH}$$

$$CN-CHR-CO_2^\ominus \cdot {}^\oplus NH_2(c\text{-}C_6H_{11})_2 \xrightarrow{KR\dagger} \underset{(IXa)}{CN-CHR-CO-K}$$

\* Q = 8-oxyquinolyl.
† KR: Kemp's reagent.[21] The abbreviation K is used here for the group

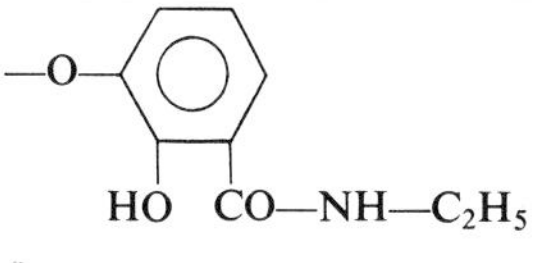

—O—⟨phenyl with HO and CO—NH—$C_2H_5$ substituents⟩

## B. Amine Components

In a tripeptide synthesis by 4 C C an N-terminally protected α-amino acid I, a suitable amine II, and aldehyde III, and an α-isocyano ester IV (see Section I) are combined to form a tripeptide derivative (VI).[53] With ammonia as the amine component (II, $R^3 = H$) we obtain the N- and C-terminally protected tripeptide directly; but the use of ammonia has two serious drawbacks. Ammonia not only undergoes 4 C C in generally lower yields than primary amines,[53] because with ammonia as the amine component there are not only irreversible side reactions, e.g., the Passerini reaction (Chapter 7), but with ammonia it is also not possible to influence the steric course of the 4 C C.

4 C C which involve unsymmetrically substituted carbonyl compounds ($R^3$—CO—$R^4$, where $R^3 \neq R^4$) lead to the formation of products with new centers of chirality. Because of this, the 4 C C in which chiral components take part lead to mixtures of diastereoisomers.

In the absence of asymmetric induction (Chapter 8, Section IV), the diastereomers are formed at equal rates, i.e., in equal amounts. The stereoselectivity of asymmetrically induced syntheses depends on the *chemical chirality*[38,39,47,48] of the chiral reference system; the asymmetric inducing power of an element of chirality is furthermore determined by its distance from the newly formed element of chirality, the chemical properties of the reactants, and the reaction conditions.

The dependence of the ratio $Q_{pn}$ of the *p*- and *n*-products (Chapter 8, Section IV) of stereoselective 4 C C upon all these factors has to be investigated in order to obtain the information which is needed for synthesizing the desired stereoisomers of 4 C C products in optimum yields.

In order to use the 4 C C approach to best advantage an optically active amine (II) must be employed whose asymmetric inducing power is high under conditions where the overall yield of the 4 C C is good. Furthermore the group $R^{1*}$ of the inducing optically active primary amine component (II) must, after the condensation, be replaceable by hydrogen under mild conditions. So the main precondition for successful stereoselective 4 C C peptide syntheses is finding the optimum asymmetrically inducing and cleavable amine component (II) and the reaction conditions under which this amine reacts with optimum stereoselectivity and overall yield. For peptide syntheses a stereoselectivity of $\geqslant 99\%$ should be achieved. In order to find the best amine and suitable reaction conditions, model reactions were investigated (Chapter 8, Section IV)[11,16,22,24,27] with both aspects in mind.

The results provide a basis for finding reaction conditions under which either one of the diastereomeric products of stereoselective 4 C C is formed in optimum yield,[50] and also criteria for selecting reaction conditions under which

9. Peptide Syntheses    207

the asymmetric inducing power of optically active amine components can be evaluated and compared with stereoselectivity data from other stereoselective reactions (e.g., by the *stereochemical linear free energy relationship*[39,48]).
This is desirable because it allows testing of new amine components by stereoselective reactions which are simple to carry out and evaluate, like the acylation of primary optically active amines (II) by phenylmethylketene (XIII) (6).[34-36] This reaction is, in chloroform at $-60°C$ to $0°C$, according to our criteria,[39,48] a simple pair of corresponding reactions.[25,36]

$$C_6H_5(CH_3)C\!=\!C\!=\!O + H_2N\!-\!R^* \rightarrow C_6H_5\!-\!CH(CH_3)\!-\!CO\!-\!NH\!-\!R^* \quad (6)$$
$$\text{(XIII)} \qquad \text{(II)} \qquad \qquad \text{(XIV)} \quad p+n$$

An optically active primary amine (II) which shows by this procedure a high relative power of asymmetric induction (i.e., a high chemical chirality $\equiv \chi$-value)[39,48] has, under suitable reaction conditions, also high potential as an amine component (II) of stereoselective 4 C C peptide syntheses (1).

The amine components of 4 C C peptide syntheses must have a high chemical chirality and must as well be easy to cleave according to V → VI under mild conditions.

There are three classes of primary amines which fulfill the second of these conditions.

1. RESONANCE STABILIZED VINYL AMINES

Resonance stabilized vinyl amines (XV) (with, e.g., $R = CH_3$, $R^1 = H$, or R, $R^1 = -(CH_2)_{3,4}-$ and $X = -CN$ or $-CO_2C_2H_5$) yield, according to (1) and (7), amino acid derivatives, e.g., V, $R^{1*} = R\!-\!C\!=\!CXR^1$, from

$$R\!-\!\underset{\underset{\text{(XV)}}{\uparrow\!\!\!_____}}{\overset{NH_2}{C}}\!\!=\!CXR^1 \xrightarrow{4\,C\,C} V \xrightarrow{H_3O^{\oplus}} R\!-\!CO\!-\!CHXR^1 + VI \qquad (7)$$
$$\qquad \qquad \qquad \qquad \qquad NH_3 \qquad \qquad \text{(XVI)}$$

which the residue $R^{1*}$ may then be cleaved by acids under mild conditions.[52] These vinyl amines (XV) have two disadvantages: the 4 C C yields are low and optically active vinyl amines like XVa generally have no strong asymmetric inducing power, because the distance between the asymmetrically inducing

(XVa)

elements of chirality of XV and the newly formed center of chirality in the selectivity determining transition states is generally large. The vinyl amines (XV) are, however, readily regenerated from XVI by treatment with ammonia; this would be a definite advantage.

2. $\beta$-ALANINE DERIVATIVES AND STEREOSELECTIVE CLEAVAGE

The 4 C C products, V, $R^{1*} = R-\overset{|}{C}H-CH_2-CO_2C_2H_5$, of XVII are

$$R-\overset{NH_2}{\underset{|}{C}H}-CH_2-CO_2C_2H_5 \xrightarrow{4CC} V \xrightarrow{Base} R-CH=CH-CO_2C_2H_5 + VI \quad (8)$$
$$(XVII) \hspace{6cm} (XVIII)$$

$$R = C_6H_5-, \quad 2,4,6-(i-C_3H_7)_3C_6H_2-$$

cleaved by alkali hydroxides or alkoxides under conditions by which the remainder of the peptide is not destroyed to any appreciable extent.

Model 4 C C show that the $\beta$-alanine derivatives, like XVII, have great potential with regard to their asymmetric inducing power.[22,52]

$$C_6H_5-CO-N-\overset{i-C_3H_7}{\underset{|}{C}H}-CO-NH-t-C_4H_9 \xrightarrow{NaOC_2H_5}$$
$$\underset{|}{R-CH-CH_2-CO_2C_2H_5}$$

$$C_6H_5-CO-NH-\overset{i-C_3H_7}{\underset{|}{C}H}-CO-NH-t-C_4H_9 + XVIII \quad (9)$$
$$(XXI)$$

(XIX) R = $C_6H_5-$
(XX) R = $2,4,6-(i-C_3H_7)_3C_6H_2-$

On reacting benzoic acid and $t$-butyl isonitrile with the Schiff base from isobutyraldehyde and the (R)— isomer of XVII (R = $C_6H_5$) in concentrated methanolic solution at $-80°C$ in the presence of tetraethylammonium benzoate 75% of the (R)(S)— isomer of XIX is obtained; this is subsequently cleaved by strong base to form the L-valine derivative (S)— (XXI).[52]

The Schiff base (R)— (XXII) reacts with benzoic acid and $t$-butyl isonitrile

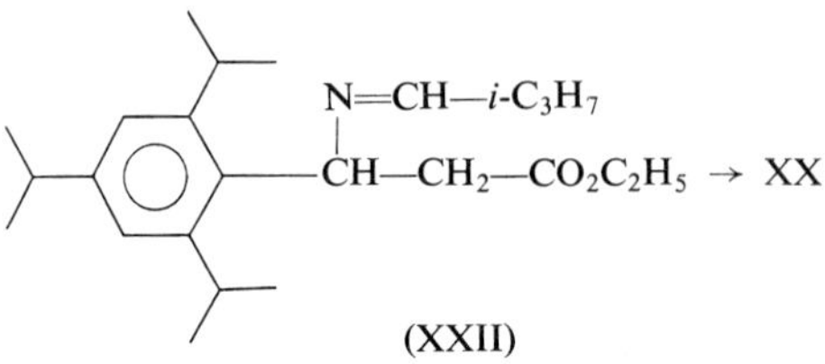

$$\text{(aryl)}-\overset{N=CH-i-C_3H_7}{\underset{|}{C}H-CH_2-CO_2C_2H_5} \rightarrow XX \quad (10)$$

(XXII)

in 0.7 $M$ methanol solution at $-60°C$ to form a 4 C C product (XXa) with $[\alpha]_{546}^{20} = +85.5°$, whereas the product XXb with $[\alpha]_{546}^{20} = +16.2°$ is formed in 0.03 $M$ methanol solution at $0°C$.[22] On treatment with potassium isopropoxide and isopropanol at $-20°C$, XXa yields 81% of XXIa with $[\alpha]_D^{20} = -54°$ (90% opt. purity). The 4 C C product XXb yields 6% XXIb with $[\alpha]_D^{20} = +55°$ on treatment with sodium methoxide in methanol for 15 min at $-20°C$, whereas 99% of XXIb with $[\alpha]_D^{20} = +28.1°$ results from prolonged treatment. These results can be interpreted as follows: The 4 C C products XXa and XXb are 95:5 and 27:73 mixtures, respectively, of the (R)(S)— (= $n$) and (R)(R)— (= $p$) isomers of XX. As XXa and XXb result from a complete conversion of XX into XXI the optical rotation of XXIa and XXIb can be used for computing the diastereoisomer ratios of XXa and XXb. This is possible because under the reaction conditions chosen, XXI is not racemized to any detectable extent. The high optical purity of XXIb has to be accounted for by a difference in rate of the formation of XXI from $p$-XX and $n$-XX, by which the "desired" product is obtained in higher purity than is expected from the diastereoisomer ratio of XX.

The stereoselective removal of the chiral auxiliary group $R^{1*}$ from a peptide derivative (V) has great potential with regard to obtaining very pure peptides from 4 C C products.

If, for instance, the desired stereoisomer of the 4 C C products, e.g., the $p$ isomer, is formed in a 95:5 preference, and if, furthermore, it is "cleaved" 10 times faster than the $n$ isomer, a product of a 99:1 antipode ratio results if the cleavage reaction is stopped at 90% conversion.

3. α-FERROCENYLALKYL AMINES

The most promising class of amines appear to be the α-ferrocenylalkyl amines (XXIII).

(XXIII)

These undergo 4 C C with high yields and good stereoselectivity and are readily cleaved V → VI[29] because of the easy formation of α-ferrocenylalkyl carbonium ions.[37] For α-ferrocenylalkyl amines with planar chirality (XXIII, $R^1 \neq H$) the optimum combination of R and $R^1$ has to be found empirically by model reactions, e.g., by (6).

210     Gokel et al.

The products (XXIV) of the stereoselective 4 C C (11) are cleaved on treatment with formic or trifluoroacetic acid: (S)(S)— (XXIV) yields (S)— (XXI) and (S)(R)— (XXIV) the corresponding (R)— antipode.[27]

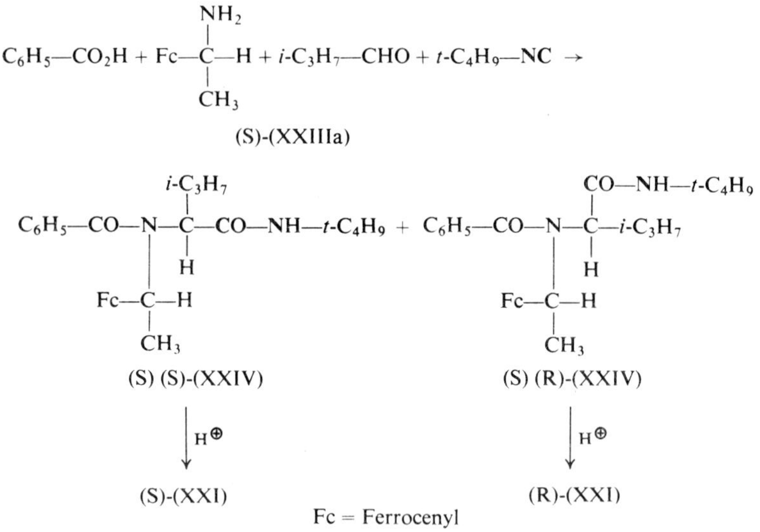

$$C_6H_5-CO_2H + Fc-\underset{\underset{CH_3}{|}}{\overset{\overset{NH_2}{|}}{C}}-H + i\text{-}C_3H_7-CHO + t\text{-}C_4H_9-NC \rightarrow$$

(S)-(XXIIIa)

$$C_6H_5-CO-N-\underset{\underset{\underset{CH_3}{|}}{\overset{\overset{i\text{-}C_3H_7}{|}}{|}}}{\overset{}{C}}-CO-NH-t\text{-}C_4H_9 \quad + \quad C_6H_5-CO-N-\underset{\underset{\underset{CH_3}{|}}{\overset{\overset{CO-NH-t\text{-}C_4H_9}{|}}{|}}}{\overset{}{C}}-i\text{-}C_3H_7 \qquad (11)$$

(S) (S)-(XXIV)                                  (S) (R)-(XXIV)

              ↓ H⊕                                         ↓ H⊕

(S)-(XXI)                                       (R)-(XXI)
              Fc = Ferrocenyl

These "cleavability tests" not only demonstrate the potential advantages of α-ferrocenylalkyl amines (XXIII) as asymmetrical inducing and cleavable amine components of stereoselective 4 C C peptide syntheses, but also allow one to reassign the absolute configuration of (+)-α-ferrocenylethyl amine as (S)— (XXIIIa)[27]; the (R)— configuration had been assigned to this amine before.[42] The absolute configuration of XXIIIa is of particular interest in this context as the α-ferrocenylalkyl amines with planar chirality (XXIII) ($R^1 \neq H$) are prepared from XXIIIa by stereospecific transformations, e.g. (12). The configurations are related to the configuration of XXIIIa in a conclusive manner.[30]

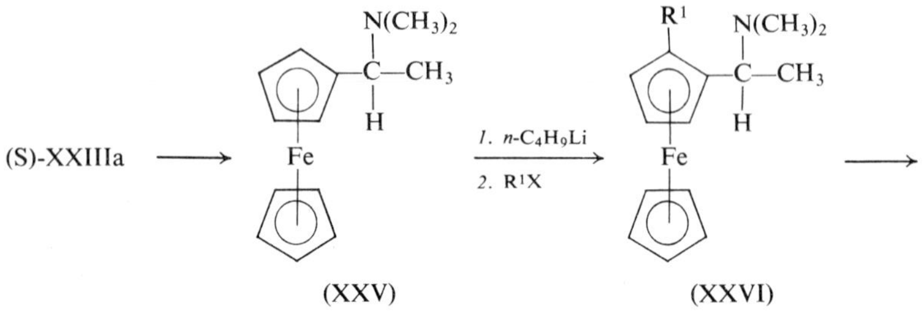

(S)-XXIIIa  ⟶  (XXV)  $\xrightarrow{\substack{1.\ n\text{-}C_4H_9Li \\ 2.\ R^1X}}$  (XXVI)  ⟶

XXIII    $R^1 = CH_3-, C_2H_5-, (CH_3)_3Si-$, etc.                    (12)

9. Peptide Syntheses    211

The optically active amines (XXIII) can be recovered from the cleavage product with full retention of the configuration, because either central chirality with an α-ferrocenyl ligand is involved or planar chirality; under favorable conditions nucleophilic substitutions via α-ferrocenylalkylcarbonium ions occur with retention.[11] Therefore a cyclic process as outlined by (13) can be anticipated for the synthesis of the peptides (VIa); this allows one to save the valuable optically active amine (XXIII).

## V. TACTICS OF THE 4 C C PEPTIDE SYNTHESIS

A wide variety of tactics[6] is possible for the 4 C C synthesis of large peptides by a fragment strategy.[6] There are not only many ways of choosing the fragments, but the fragments themselves can often be built up in different ways. This is due to the fact that amino acids can be introduced into fragments by 4 C C as N-protected amino acids ($N_2$) or dipeptides ($N_1, N_2$) functioning as acid components (I), as aldehyde components (III) (A), or as α-isocyanoesters ($I_1$) or α-isocyanoacyl amino acid esters ($I_1, I_2$) (IV). Furthermore amino acids (C) can be added to the N-terminal end of fragments by conventional acylation methods, prior to combining the fragments, according to (1), (3), or (14) [see Section III (3)].

I + III + R'—NC + NH$_2$—P$^2$—CO$_2$R →

A—NH—P$^1$—CO—N—P$^2$—CO$_2$R
                      |                              → A—NH—P$^1$—CO—NH—P$^2$—CO$_2$R    (14)
R$^3$—CH—CO—NH—R'

On this basis 139,000,000 different tactics of different efficiencies are possible for the synthesis of the A chain of insulin (XXVII). For planning optimum 4 C C syntheses of large peptides it is necessary to find the most effective choice of a small number of the best tactics by computer methods. The selection of optimum tactics can be achieved by calculating the total amount, $M$, of starting material that is needed for the synthesis of 1 mole of a large peptide; $M$ is represented as a function $F(q_{a_1, r_1} \ldots q_{a_n, r_n})$ of the chosen tactic. With the aid of a computer it is possible (most effectively by dynamic programming techniques[4]) to find tactics with the lowest $M$ values. This can be done by using inverse (average) yield parameters, $q_{a,r}$ for each amino acid, $a$, and any role ($r = N_1, N_2, I_1, I_2$, or C) that it can play in a 4 C C fragment synthesis. The $q_{a,r}$ values are obtained from yield data of model experiments, and correspond to the reciprocal yields of model reactions. By this procedure it is found that the optimum synthesis of the A-chain of insulin is represented by XXVII.[51]

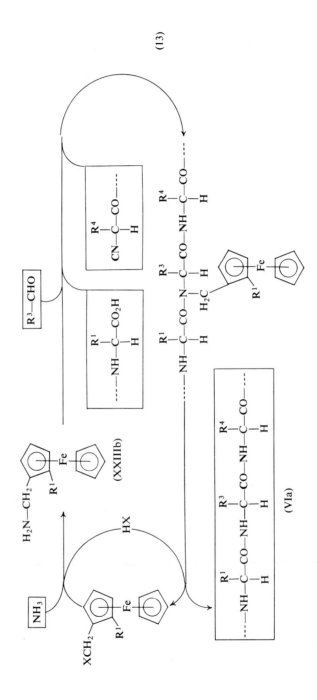

(13)

9. *Peptide Syntheses* 213

| Gly-Ileu-Val-Glu | Gln-Cys-Cys-Ala-Ser-Val | Cys-Ser-Leu-Tyr-Gln-Leu-Glu | Asn-Tyr-Cys-Asn |
|---|---|---|---|
| $N_1$ $N_2$ A I | C $N_1$ $N_2$ A $I_1$ $I_2$ | C C $N_1$ $N_2$ A $I_1$ $I_2$ | $N_1$ $N_2$ A I |
| 1 2 3 4 | 5 6 7 8 9 10 | 11 12 13 14 15 16 17 | 18 19 20 21 |

(XXVII)

The lines define the ends of the fragments which are synthesized by 4 C C and the capital letters below indicate how the corresponding amino acid was introduced into the peptide fragment.

For example, the synthesis (15) of the fragment 1–4 (XXVIII) is represented by Eq. (16).

$$\text{A-Gly-Ileu-OH} + i\text{-C}_3\text{H}_7\text{—CHO} + \text{C}{\equiv}\overset{\overset{\displaystyle OR^1}{|}}{\text{Glu}}\text{—OR} \xrightarrow{\text{II}}$$

$$\text{A-Gly-Ileu-}\underset{\underset{\displaystyle R^{1*}}{|}}{\overset{\overset{\displaystyle OR^1}{|}}{\text{Val}}}\text{-Glu-OR} \rightarrow \text{A-Gly-Ileu-}\overset{\overset{\displaystyle OR^1}{|}}{\text{Val}}\text{-Glu-OR} \quad (15)$$

(XXVIII)

$$\bar{q}_{1-4} = \tfrac{1}{2} \cdot q_{\text{Val,A}}(q_{\text{Gly},N_1} \cdot q_{\text{Ileu},N_2} + q_{\text{Glu},I_1}) \quad (16)$$

The combination of fragments is represented by $\bar{q}_{\text{comb}}$, and the overall inverse yield function $M$ of insulin A according to scheme XXVII is (17).

$$\bar{Q}_{iA} = \tfrac{1}{4}\bar{q}^2_{\text{comb}} \cdot (\bar{q}_{1-4} + \bar{q}_{5-10} + \bar{q}_{11-17} + \bar{q}_{18-21}) \quad (17)$$

It is doubtful that it will be possible in the near future to find optimal multistep syntheses for complicated compounds by computer methods. This would necessitate storing in a computer more pertinent information than is presently available on synthetic procedures including stereochemical aspects, and programming the computer to draw analogy conclusions in a semi-quantitative manner. In contrast to a general use of computers for multistep syntheses, the case of planning a 4 C C peptide synthesis by computer has potential. Here, one has only to deal with a few types of synthetic reactions and for these, yield estimates are possible by close analogy to a limited number of suitable model reactions.

## REFERENCES

1. Anderson, J. C., Barton, M. A., Gregory, R. A., Hardy, P. M., Kenner, G. W., MacLeod, J. K., Preston, J., and Sheppard, R. C., *Nature* **204**, 933 (1964).
2. Bajusz, S., Medzihradszky, K., Paulay, Z., and Lang, Z., *Acta Chim. Acad. Sci. Hung.* **52**, 335 (1967).
3. Bajusz, S., Paulay, Z., Lang, Z., Medzihradszky, K., Misfaludy, L., and Löw, M., in "Peptides" (E. Bricas, ed.), Vol. 4, p. 237. Wiley, New York, 1968.

4. Bellman, R. E., "Dynamic Programming." Princeton Univ. Press, Princeton, New Jersey, 1957.
5. Bergmann, M., and Zervas, L., *Chem. Ber.* **65**, 1192 (1932).
6. Bodanszky, M., and Ondetti, M. A., "Peptide Synthesis," p. 127. Wiley (Interscience), New York, 1966.
7. Bodanszky, M., Ondetti, M. A., von Saltza, M. H., Narayan, V. K., and Levine, S. D., Squibb, U.S. Pat. 3,400,118 (1968).
8. Bricas, E., ed., "Peptides." Wiley, New York, 1968.
9. Fischer, E., *Chem. Ber.* **39**, 530 (1906).
10. Fridkin, M., Patchornik, A., and Katchalski, E., *J. Amer. Chem. Soc.* **90**, 2953 (1968).
11. Gokel, G., Hoffmann, P., Klusacek, H., Marquarding, D., Ruch, E., and Ugi, I., *Angew. Chem.* **82**, 77 (1970), *Angew. Chem. Int. Ed.* **9**, 64 (1970).
12. Greenstein, J. P., and Winitz, M., "Chemistry of the Amino Acids." Wiley, New York, 1960.
13. Gutte, B., and Merrifield, R. B., *J. Amer. Chem. Soc.* **91**, 501 (1969).
14. Guttmann, S., Pless, J., Sandria, E., Jaquenoud, P. A., Bossert, H., and Willems, H., *Helv. Chim. Acta* **51**, 1155 (1968).
15. Hirschmann, R., Nutt, R. F., Veber, D. F., Vitali, R. A., Varga, S. L., Jacob, T. A., Holly, F. W., and Denkewalter, R. G., *J. Amer. Chem. Soc.* **91**, 507 (1969).
16. Hoffmann, P., Marquarding, D., and Ugi, I., unpublished results (1968).
17. Jakubke, H. D., and Voigt, A., *Chem. Ber.* **99**, 2419 (1966).
18. Jakubke, H. D., Voigt, A., and Burkhardt, S., *Chem. Ber.* **100**, 2367 (1967).
19. Katsoyannis, P. G., *Science* **154**, 1509 (1966).
20. Katsoyannis, P. G., Tometsko, A. M., Zalut, C., and Fukuda, K., *J. Amer. Chem. Soc.* **88**, 164, 5625 (1966).
21. Kemp, D. S., and Chien, S. W., *J. Amer. Chem. Soc.* **89**, 2743 (1967).
22. Kleimann, H., and Ugi, I., unpublished results (1967).
23. Klostermeyer, H., and Humbel, R. E., *Angew. Chem.* **78**, 871 (1966); *Angew. Chem., Int. Ed. Engl.* **5**, 807 (1966).
24. Lüdke, G., unpublished results (1969).
25. Lüdke, G., and Ugi, I., unpublished results (1969).
26. Marglin, A., and Merrifield, R. B., *J. Amer. Chem. Soc.* **88**, 5051 (1966).
27. Marquarding, D., Hoffmann, P., Heitzer, H., and Ugi, I., *J. Amer. Chem. Soc.* **92**, 1969 (1970).
28. Marquarding, D., Hoffmann, P., and Ugi, I., unpublished results (1968).
29. Marquarding, D., Hoffmann, P., and Ugi, I., unpublished results (1970).
30. Marquarding, D., and Ugi, I., unpublished results (1969).
31. Merrifield, R. B., *Fed. Proc.* **21**, 412 (1962).
32. Morley, J. S., *J. Chem. Soc.*, C p. 2410 (1967).
33. Nin, C., Wang, Y., *et al.*, *Sci. Sinica* **13**, 1343 and 2030 (1964); **14**, 1887 (1965); **15**, 231 (1966).
34. Pracejus, H., *Justus Liebigs Ann. Chem.* **634**, 23 (1960).
35. Pracejus, H., *Fortschr. Chem. Forsch.* **8**, 493 (1967).
36. Pracejus, H., and Tille, A., *Chem. Ber.* **96**, 854 (1963).
37. Rosenblum, M., "Chemistry of the Iron Group Metallocenes," Part 1, p. 129. Wiley, New York, 1965.
38. Ruch, E., and Ugi, I., *Theor. Chim. Acta* **4**, 287 (1966).
39. Ruch, E., and Ugi, I., *Top. Stereochem.* **4**, 99 (1969).
40. Sakiyama, F., and Witkop, B., *J. Org. Chem.* **30**, 1905 (1965).
41. Sarges, R., and Witkop, B., *J. Amer. Chem. Soc.* **86**, 1861 and 1862 (1964); **87**, 2027 (1965).

42. Schlögl, K., *Top. Stereochem.* **1**, 39 (1967).
43. Schröder, E., and Lübke, K., "The Peptides," Vol. 2. Academic Press, New York, 1967.
44. Schwyzer, R., *Naturwissenschaften* **53**, 189 (1966).
45. Schwyzer, R., and Sieber, P., *Nature* **199**, 172 (1963); *Helv. Chim. Acta* **49**, 134 (1966).
46. Šibnev, V. A., Porošiu, K. T., Čuvaeva, T. P., and Martynova, G. A., *Izv. Akad. Nauk SSSR, Ser. Khim.* p. 1144 (1968).
47. Ugi, I., *Jahrb. Akad. Wiss. Goettingen* p. 21 (1965).
48. Ugi, I., *Z. Naturforsch.* B **20**, 405 (1965).
49. Ugi, I., Eholzer, U., Knupfer, H., and Offermann, K., *Angew. Chem.* **77**, 492 (1965); *Angew. Chem., Int. Ed. Engl.* **4**, 472 (1965).
50. Ugi, I., and Kaufhold, G., *Justus Liebigs Ann. Chem.* **709**, 11 (1967).
51. Ugi, I., Kaufhold, G., Hoffmann, P., Kleimann, H., and Marquarding, D., unpublished results (1968).
52. Ugi, I., and Offermann, K., *Chem. Ber.* **97**, 2996 (1964).
53. Ugi, I., and Steinbrückner, C., *Chem. Ber.* **94**, 2802 (1961).
54. Weygand, F., and Ragnarsson, U., *Z. Naturforsch.* B **21**, 1141 (1966).
55. Wünsch, E., *Z. Naturforsch.* B **22**, 1269 (1967).
56. Wünsch, E., Wendlberger, G., Jaeger, E., and Scharf, R., *in* "Peptides" (E. Bricas, ed.), p. 229. Wiley, New York, 1968.
57. Zahn, H., Bremer, H., Zobel, R., Meienhofer, J., and Schnabel, E., *Z. Naturforsch.* B **20**, 653 and 661 (1965).
58. Zahn, H., Danto, W., Schmidt, G., Dahlmans, J., Costopanagiotis, A., and Engels, E., *in* "Peptides" (E. Bricas, ed.), p. 217. Wiley, New York, 1968.

# Chapter 10
# Coordinated Isonitriles

*Arnd Vogler*

I. Introduction . . . . . . . . . . . . 217
II. Structure and Bonding in Isonitrile Complexes . . . . . . 218
III. Reactions of Coordinated Isonitriles . . . . . . . . 222
IV. The Synthesis of Metal Isonitrile Complexes . . . . . . . 231
References . . . . . . . . . . . . 232

## I. INTRODUCTION*

The versatility of the isonitriles is reflected not only by the variety of their reactions in organic chemistry, but by their ability to function as bonding partners for metals in complexes. In some respects, the chemistry of isonitrile metal complexes is simply inorganic coordination chemistry; by other criteria, the chemistry of coordinated isonitriles is organometallic or organic chemistry, although any formal distinction is purely artificial. A close examination of these isonitrile–metal complexes can often yield significant information on the mechanisms of organic isonitrile reactions.

Since the essential features of the ligands remain unchanged in the complexes, it is reasonable to consider the physical and chemical properties of the free isonitrile in an explanation of the chemistry of isonitrile ligands bonded to metals. In this connection, we consider the bonded isonitrile to be simply a free isonitrile which is perturbed by coordination. The structure and reactivity of complexed isonitriles vary not only with the isonitrile but also with the metal and its oxidation state in the complex. An attempt is made below to correlate these phenomena in terms of the metal and its oxidation state.

* When this chapter was in the planning stage, very up-to-date review articles on metal–isonitrile chemistry were unavailable. At that time, it was the purpose of this chapter to catalog most or all of the references in this field and explain some of the important reactions. In the inevitable interim between conception and writing, the monograph *Isocyanide Complexes of Metals*[27] has appeared. Since this volume has successfully catalogued the references in this field, we have shifted the emphasis of this chapter to mechanism and explanation of some of the more important aspects of coordinated isonitrile chemistry.

## II. STRUCTURE AND BONDING IN ISONITRILE COMPLEXES

When an isonitrile molecule coordinates to a metal, the lone electron pair of the formally divalent carbon atom forms a $\sigma$-type donor bond to the metal. If the metal has filled $d$-orbitals which are capable of overlapping with low-lying, empty antibonding orbitals of the attached ligand atom, a second bond of the $\pi$-type may be formed. In the case of isonitrile ligands, formation of this "back donating" bond can take place only with considerable rearrangement of the bonding system within the ligand since no such orbitals are available in the free ligand at the C atom. Back donation occurs to reduce the negative charge on the metal which is accumulated by the formation of the donor $\sigma$-bond. In isonitrile complexes of the same coordination number but of different charge, back donation increases as the positive charge on the complex decreases. This means that isonitriles, like some other ligands (CO, $PF_3$, etc.), are able to stabilize the low oxidation states of metals.

Some of these conclusions and interpretations are consistent with all of the experimental data; others are still ambiguous because different approaches seem to lead to different answers which will be discussed later.

IR spectroscopy is a most helpful tool for elucidating the structure and charge distribution of isonitriles as well as their complexes. The bond order of the isonitrile **CN** bond lies between 2 and 3. In an oversimplified but useful model the donor $\sigma$-bond provides a bond order that is essentially 3 (I) for the

$$\overset{\ominus}{M}\leftarrow C\equiv\overset{\oplus}{N}-R \qquad M\overset{\frown}{\leftarrow C}=\bar{N}-R$$

(I)                      (II)

**CN** group which also prevails in the free ligand. A slight increase in the **CN** vibrational frequency seems to be due to a kinematic coupling of the M—C and **CN** bond.[10] Back donation and formation of a second M—C bond of the $\pi$-type is accompanied by a decrease of the **CN** stretching frequency corresponding to a contribution from structure II. While metals of the groups IB and IIB (Cu, Ag, Au, Zn, Cd, Hg) and rare earth metals[12] essentially form only $\sigma$-bonds with isonitriles, most of the other transition metals can also engage in $\pi$-bonding.

In a simple valence bond (VB) treatment,[10] structure II would require some deviation from linearity of the **CN**—R axis at the N atom as a result of the change in hybridization at the N atom, whereas molecular orbital (MO) considerations are consistent with a linear isonitrile molecule.[8]

In the case of trigonal bipyramidal $Co(CN-CH_3)_5^{+1}$ III[8] the **CNC** angle was found, by X-ray analysis, to be quite close to 180° (**CNC** < 177° ± 1.5 for axial ligands, **CNC** < 173° ± 2 for equatorial ligands).

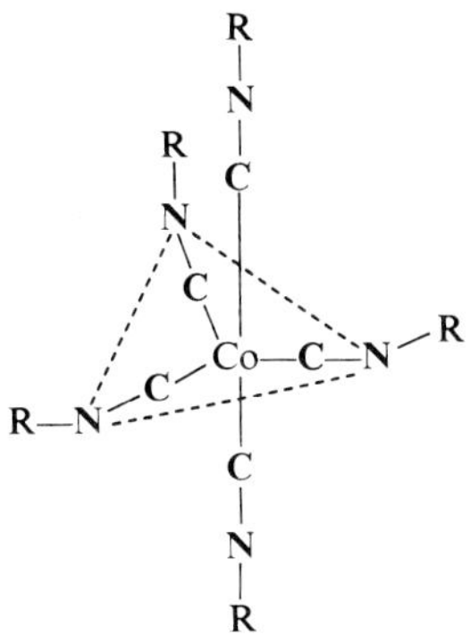

## 10. Coordinated Isonitriles 219

$$\begin{array}{c}
R\\
|\\
N\\
R\ \ |\\
|\ \ C\\
N\\
\diagup\diagdown\ \ |\\
C\ \ |\\
\diagdown\ \ \ \ \diagup\ \ \ \ \ R\\
C\!\!-\!\!Co\!\!-\!\!C\!\!\equiv\!\!N\!\!-\!\!\\
R\!\!-\!\!N\!\!\equiv\!\!\diagup\ \ C\ \ \diagdown\\
|\\
C\\
|\\
N\\
|\\
R
\end{array}$$

(III)

In a simple VB calculation this angle should decrease considerably, taking back donation into account, since the Co—C bond order was estimated to be 1.5 or greater. This estimate was made on the assumption that the Co—C bond was shortened compared to the sum of the covalent radii of cobalt and carbon.

Although this example is not in agreement with a simple VB calculation, there is at least one case known where a lowering of the CN bond order is accompanied by a bending of the CN—R axis. In complex IV[25] two iron atoms

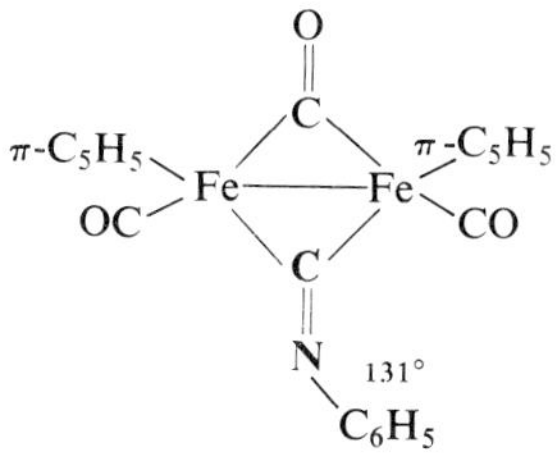

$$\begin{array}{c}
O\\
\|\\
\pi\text{-}C_5H_5\diagdown\ \ C\ \ \diagup\pi\text{-}C_5H_5\\
Fe\!\!-\!\!-\!\!-\!\!Fe\\
OC\diagup\ \ \diagdown\ \ \diagup\ \ \diagdown CO\\
C\\
\|\\
N\ \ \ 131°\\
\diagdown\\
C_6H_5
\end{array}$$

(IV)

are bridged by phenyl isocyanide. Since the C atom already forms two bonds to the two iron atoms, the CN bond order should not exceed 2. This point is substantiated by the IR spectrum which shows the NC vibration at the very low frequency of 1704 cm$^{-1}$ vs 2117 cm$^{-1}$ for the free ligand. The X-ray analysis shows indeed that the isonitrile molecule is no longer linear but has a CNC angle of 131°. For a pure double bond we would expect approximately 120°, and for a pure triple bond, 180°. The IR data indicate a double, rather than a triple bond and the question becomes why is the CNC angle 11° greater than it should be, and not why is it 49° smaller than expected. The deriviation from linearity might be smaller than expected due to contact angles or crystal forces; but one must also bear in mind that older X-ray data are subject to much greater uncertainty than that to which we are accustomed today,[8] and more precise data are needed before drawing any final conclusions.

The IR spectra of $Cr^0(CN-R)_6$ and $Ni^0(CN-R)_4$ could provide further criteria to test the validity of the VB and MO models. Since both complexes are highly symmetric (octahedral and tetrahedral configurations) they should exhibit only one NC stretching frequency unless back donation lowers the symmetry by bending the CNC axis. Back donation is also expected to reduce the negative charge on the zero valent metals. The fact that the IR spectra of both $Cr(CN-C_6H_5)_6$ (in $CHCl_3$, $\nu_{NC}$: 2070, 2012, 1965 cm$^{-1}$) and $Ni(CN-C_6H_5)_4$ (in $CHCl_3$, $\nu_{NC}$: 2050, 1990 cm$^{-1}$) show more than one CN—band, is taken as evidence that, in this case at least, the VB approach can afford an adequate explanation of the observations.[10] The IR spectra of Cr(O) and Ni(O) complexes are reported only for bulky isonitriles like phenyl isocyanide. In these complexes a deviation from the octahedral or tetrahedral symmetry might be caused simply by steric hindrance and subsequent deformation of the large ligand. Even for $Fe(CN-CH_3)_6^{2\oplus}$, where back donation is less likely to occur due to the positive charge on the central ion, two NC stretching frequencies have been reported ($\nu_{NC}$: 2234 and 2197 cm$^{-1}$).[11]

One must use caution in evaluating the data obtained from IR measurements because these spectra are frequently obtained on solutions of the compounds in chlorinated solvents. These particular solvents are known occasionally to cause decomposition of the complex or to give rise to unexplained and spurious bands.[3,25] Further experimental data are needed for a consistent interpretation of the IR spectra of these complexes.

The geometry of coordinated isonitrile ligands can be qualitatively evaluated by considering the irreducible representations of those orbitals involved in bonding. If one views the complex along the axis formed by a CN bond, and this axis has a symmetry of $C_n$ ($n \geqslant 3$) then the R—N—C bond angle should be ca. 180°, even if back donation occurs.[33] If the symmetry is lower than threefold, i.e., twofold, $\sigma$ plane, etc., then some deviation from linearity can be expected. In this case the degeneracy of the two empty antibonding $\pi$-orbitals of the ligand which are responsible for back-donation splits, and this may cause the preference of a mirror plane symmetry for the ligand. The structures of the complexes $[Co(CN-CH_3)_5]ClO_4$ (III) and $(C_5H_5)_2Fe_2(CO_3)(CN-C_6H_5)$ IV as resolved by X-ray analysis (see pages 218 and 219) support this explanation.*

The occurrence of back donation is not only influenced by the metal and its oxidation state. In addition, in the complex M—C≡N—R the residue R may also affect the charge distribution. In a complex like $(CH_3)_3SiNCMo(CO)_5$

---

\* The author gratefully acknowledges helpful discussions of the symmetry aspects of coordinated isonitriles with Professor Ernst Ruch.

the nitrogen is bound to a Si atom which has empty $d\pi$-orbitals. MO considerations show that back donation lowers the energy of the entire set of $\pi$-electrons which belong to the M—C—N—Si group.[9] As a consequence the difference in the CN stretching frequencies of [(CH$_3$)$_3$SiNC]Fe(CO)$_4$ and (CH$_3$)$_3$Si—NC is considerably bigger than the corresponding difference for [$t$-C$_4$H$_9$—NC]—Fe(CO)$_4$ and $t$-C$_4$H$_9$—NC.

For complexes with phenyl isocyanide as ligands the question arises as to how strong the interaction between the benzene ring and the M—CN moiety might be. The acceptor strength of C$_6$H$_5$—NC seems to be much greater than that of CH$_3$—NC as is indicated by a larger drop in the NC stretching frequency of phenyl isocyanide on complexation. This observation was interpreted as a contribution from formula VI[4,10]:

$$\overset{\ominus}{M} \leftarrow C \equiv \overset{\oplus}{N} - \!\!\bigcirc \qquad \longleftrightarrow \qquad M \leftarrow \overset{\oplus}{\overset{\displaystyle |}{C}} = \overset{\oplus}{N} = \!\!\bigcirc \!\!\ominus$$

(V)                    (VI)

Electron withdrawing substituents in the *para* position of the benzene ring should favor formula VI and thereby decrease the NC stretching frequency, whereas electron-donating groups should have the opposite effect. The IR spectrum, however, does not indicate such an effect. Therefore, it can be concluded that interaction of the $\pi$-electrons of the benzene and the NC group is weak in the complex.[10]

On the other hand, NMR experiments indicate that there is at least some interaction between the $d$-electrons of the metal and the $\pi$-electrons of the benzene ring. In solutions of Co(II)— or Ni(II)— acetylacetonate and phenyl isocyanide, isocyanide–metal complex formation occurs.[24] Unpaired spin density of the paramagnetic metal ions is transferred to the benzene ring via the C and N atoms, resulting in changes in the chemical shifts of the ring protons. The direction and magnitude of those effects are only consistent with the occurrence of back donation by $\pi$-bonding with participation of the benzene ring.

No attempt was made to measure the NMR shifts and NC stretching frequencies for this same system; such studies would eventually lead to a correlation of the extent to which back donation and coupling with the benzene ring takes place.

In addition to the particular metal, its oxidation state and coordination number, and the residue R in the isonitrile R—NC, there is one further factor which influences the reorganization of the isonitrile bonding system when the free ligand coordinates. In a mixed complex M(CN—R)$_m$L$_n$ the extent to which back donation to the isonitrile takes place also depends on the π-acceptor strength of the L ligands.[4,7,10,23,35] In the case of CO, for example, the π-acceptor strength of CO is so much greater than that of isonitriles, that the CO groups acquire almost all of the negative charge from the metal, leaving the CN—R group to form a σ-donor bond. This means that VIII (below) is preferred over VII.

$$|\overset{\oplus}{O}{\equiv}C \rightarrow \overset{2\ominus}{M} \leftarrow \overset{\oplus}{C}{\equiv}N-R \qquad \overset{-}{O}{=}C{=}\overset{\ominus}{M} \leftarrow \overset{\oplus}{C}{\equiv}N-R$$

$$\text{(VII)} \qquad\qquad\qquad \text{(VIII)}$$

Substitution of one CO by phenyl isocyanide ($\nu_{NC} = 2117$ cm$^{-1}$) in Cr(CO)$_6$ slightly increases the NC stretching frequency for Cr(CO)$_5$CN—C$_6$H$_5$; ($\nu_{NC} = 2125$ cm$^{-1}$). The dipole moment of Fe(CO)$_4$CN—CH$_3$ (5.07 D)[22] is exceptionally large as compared to 3.5 D for the free isonitrile, and this is further evidence that in monosubstituted metal carbonyls the isonitrile does not participate in π-bonding. It also means that, considering the high symmetry of the unsubstituted Fe(CO)$_5$ (trigonal bipyramid), the distance between the charges in $\overset{\ominus}{C}{\equiv}\overset{\oplus}{N}$—CH$_3$ increases on forming a σ-donor bond $\overset{\ominus}{M} \leftarrow \overset{\oplus}{C}{\equiv}N$—CH$_3$ whereas back donation

$$M \leftarrow C{=}N-R$$

might even decrease the dipole moment. Substitution of more than one CO in metal carbonyls by isonitrile ligands also permits the isonitriles to participate in back donation.

## III. REACTIONS OF COORDINATED ISONITRILES

As a consequence of the structural changes which occur when the isonitrile coordinates, the reactivity and chemical properties are different for the coordinated isonitrile than for the free ligand. In many cases the metal–isonitrile bond can be activated thermally or photochemically; this process generally leads to dissociation of one or more isonitrile ligands and substitution on the complex:

$$\text{Cr(CN—C}_6\text{H}_5)_6 + {}^{14}\text{CN—C}_6\text{H}_5 \rightarrow (\text{C}_6\text{H}_5\text{—NC}^{14})\text{Cr(CN—C}_6\text{H}_5)_5 + \text{CN—C}_6\text{H}_5 \qquad (1)^6$$

$$\text{Fe(CN—CH}_3)_6{}^{2\oplus} \xrightarrow[\text{H}_2\text{O}]{h\nu} \text{Fe(CN—CH}_3)_4(\text{H}_2\text{O})_2{}^{2\oplus} + 2\text{CN—CH}_3 \qquad (2)^5$$

It should be possible to form the metal complex of an isonitrile in order to protect the free ligand from reaction conditions which would otherwise destroy it.

10. *Coordinated Isonitriles* 223

By choosing a suitable complex one can change the charge distribution within the isonitrile to suit a particular reaction. Examples will be given to demonstrate this point.

Saegusa[34] (see Chapter 4) discovered that the α-addition of H—Y—R$^1$ compounds (Y = heteroatoms like O, N, P, Si, etc.) to isonitriles CN—R$^2$ to yield R$^1$—Y—CH=N—R$^2$ are catalyzed by group IB and IIB metals (Cu, Ag, Zn, Cd, and Hg in various oxidation states). The first step in the reaction (3) is presumed to be coordination of the isonitrile to the metal by a σ-donor bond, as indicated by an increased stretching frequency in IX. Since negative charge is drained from the C atom to the metal the carbon is now accessible to nucleophilic attack by a base, such as an alcohol R$^1$OH, which leads to the intermediate X. A subsequent proton transfer yields XI.

$$M + |\overset{\ominus}{C}\equiv\overset{\oplus}{N}-R^2 \rightarrow M-\overset{\ominus}{C}\equiv\overset{\oplus}{N}-R^2 \rightarrow$$
$$(IX)$$

$$\underset{(X)}{\overset{\ominus}{M}-C\underset{O^{\oplus}\diagdown R^1}{\overset{N\diagup R^2}{\diagdown H}}} \rightarrow \underset{(XI)}{M=C\underset{O\diagdown R^1}{\overset{N\diagup R^2}{\diagdown H}}} \rightarrow \underset{(XII)}{|C\underset{O\diagdown R^1}{\overset{N\diagup R^2}{\diagdown H}}} \rightarrow \underset{(XIII)}{R^1-O-CH=NR^2} \quad (3)$$

This, actually, is a carbene complex. For IB and IIB metals such complexes (XI) are quite unstable and tend to dissociate, leaving a high electron density at the carbon atom of XII. The free carbene finally stabilizes by proton transfer and forms XIII, the product of the catalysis.

The assumption of a carbene complex as an intermediate in this catalytic reaction (3) is supported by the fact that complexes with similar structures were recently synthesized by reactions like (4) of platinum isonitrile complexes with nucleophiles like alcohols and primary amines.[2]

$$Cl_2Pt[P(C_2H_5)_3][CN-C_6H_5] + C_2H_5OH \rightarrow Cl_2[P(C_2H_5)_3]Pt=C\underset{OC_2H_5}{\overset{NH-C_6H_5}{\diagup}} \quad (4)$$
$$(XIV) \qquad\qquad\qquad (XV)$$

This structure for XV was confirmed by X-ray analysis. A small contribution from an ylide form XVa is indicated by a shortening of the NC bond to 1.32 Å. The existence of stable Pt carbene complexes may also explain why metals like platinum do not catalyze Saegusa's α-additions (see Chapter 4, Section I), because when the metal carbene complex is formed Pt binds the carbene ligand much more tightly than either Cu or Zn.

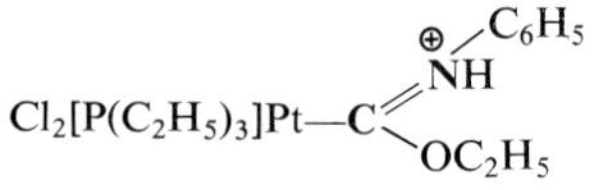

$$Cl_2[P(C_2H_5)_3]Pt-C\underset{OC_2H_5}{\overset{\overset{\oplus}{N}H-C_6H_5}{\diagup}}$$

(XVa)

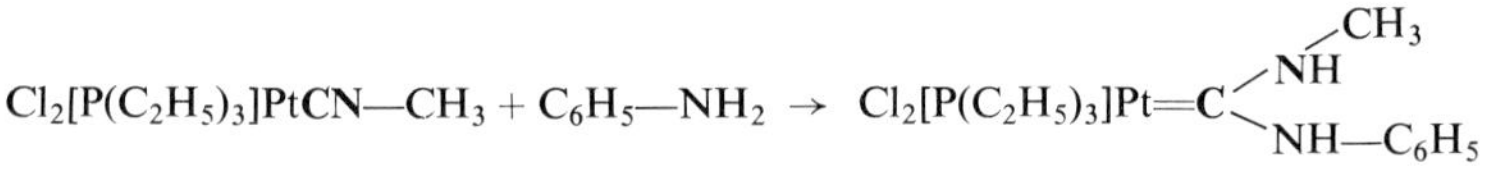

$$Cl_2[P(C_2H_5)_3]PtCN-CH_3 + C_6H_5-NH_2 \rightarrow Cl_2[P(C_2H_5)_3]Pt=C\underset{NH-C_6H_5}{\overset{NH-CH_3}{\diagup}} \quad (5)$$

(XIVa) (XVI)

Strong bases like phenyllithium react with $Cr(CO)_6$ according to reaction (6). Methylation of the complex XVII leads to stable compound XVIII, the first carbene complex to be reported.[13]

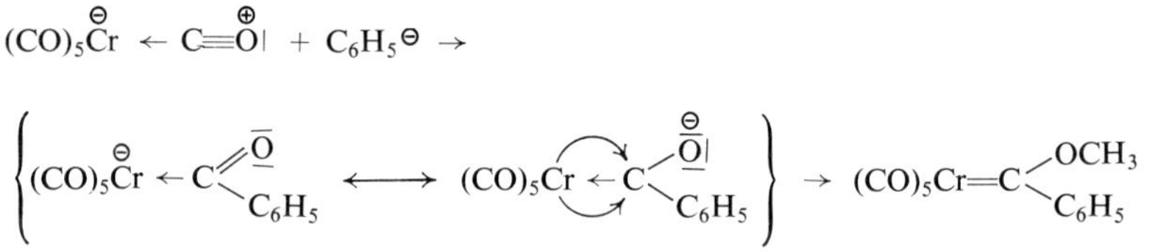

$$(CO)_5\overset{\ominus}{Cr} \leftarrow C\overset{\oplus}{\equiv}O| + C_6H_5^{\ominus} \rightarrow$$

$$\left\{ (CO)_5\overset{\ominus}{Cr} \leftarrow C\underset{C_6H_5}{\overset{\overline{O}}{\diagup}} \longleftrightarrow (CO)_5\overset{\ominus}{Cr} \leftarrow C\underset{C_6H_5}{\overset{\overset{\ominus}{O}|}{\diagup}} \right\} \rightarrow (CO)_5Cr=C\underset{C_6H_5}{\overset{OCH_3}{\diagup}} \quad (6)$$

(XVII) (XVIII)

Reaction (7) is related to the formation of carbene complexes. The nucleophilic $C_6F_5^{\ominus}$ anion attacks the isonitrile carbon of XIX forming XX, which is a stable compound and can be considered an iminoacyl derivative of the iron complex.[36] The complex XX exists in two isomers, in which the methyl group

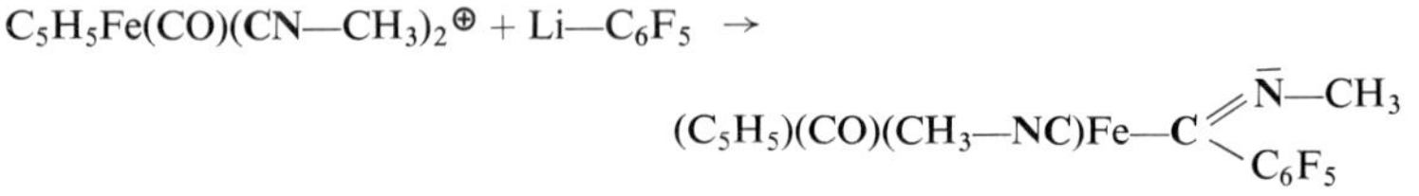

$$C_5H_5Fe(CO)(CN-CH_3)_2^{\oplus} + Li-C_6F_5 \rightarrow$$

$$(C_5H_5)(CO)(CH_3-NC)Fe-C\underset{C_6F_5}{\overset{\overline{N}-CH_3}{\diagup}} \quad (7)$$

(XIX) (XX)

is either *trans* or *cis* with regard to the pentafluorophenyl group. The NC stretching frequency of XX (1580 cm$^{-1}$) is in accordance with the assumed structure; back donation from the metal to iminoacyl group is not likely to occur because it would put a negative charge on the nitrogen (XXa).

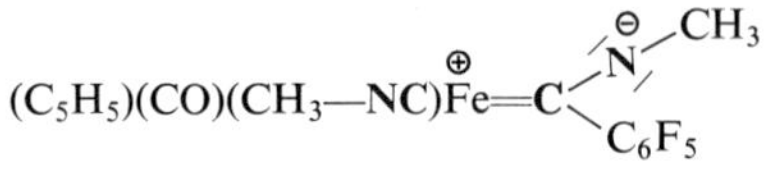

$$(C_5H_5)(CO)(CH_3-NC)\overset{\oplus}{Fe}=C\underset{C_6F_5}{\overset{\overset{\ominus}{N}-CH_3}{\diagup}}$$

(XXa)

Protonation could perhaps yield the stable carbene complex cation XXb.

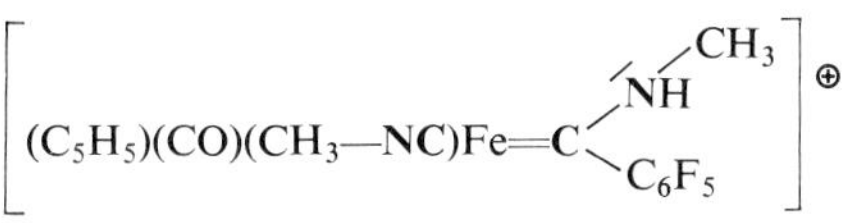

10. Coordinated Isonitriles 225

$$\left[(C_5H_5)(CO)(CH_3-NC)Fe=C\underset{C_6F_5}{\overset{NH-CH_3}{\diagdown}}\right]^{\oplus}$$

(XXb)

A similar mechanism was proposed to explain the insertion of cyclohexyl isocyanide into the nickel alkyl bond[40] of XXI [reaction (8)].

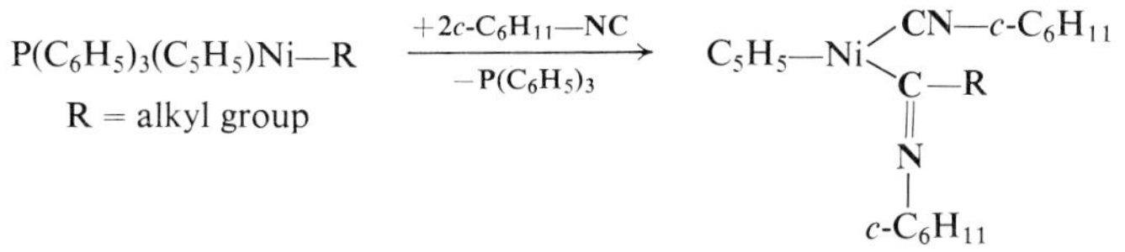

$$P(C_6H_5)_3(C_5H_5)Ni-R \quad \xrightarrow[-P(C_6H_5)_3]{+2c\text{-}C_6H_{11}-NC} \quad C_5H_5-Ni\underset{C-R}{\overset{CN-c\text{-}C_6H_{11}}{\diagdown}} \quad (8)$$
$$R = \text{alkyl group} \qquad\qquad\qquad\qquad\qquad \underset{c\text{-}C_6H_{11}}{\overset{\|}{N}}$$

(XXI) (XXII)

The alkyl anion is substituted by the isonitrile, and the subsequent nucleophilic addition of the alkyl group to the coordinated isonitrile yields the acyl imino complex XXII.

Reaction (9) was carried out to justify this mechanism.

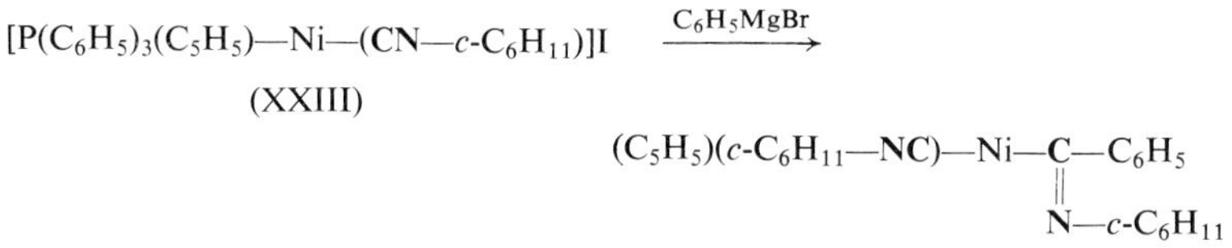

$$[P(C_6H_5)_3(C_5H_5)-Ni-(CN-c\text{-}C_6H_{11})]I \quad \xrightarrow{C_6H_5MgBr}$$

(XXIII)

$$(C_5H_5)(c\text{-}C_6H_{11}-NC)-Ni-\underset{N-c\text{-}C_6H_{11}}{\overset{C-C_6H_5}{\|}} \quad (9)$$

(XXIV)

The palladium complex XXV reacts in analogy to the above iron and nickel complexes XIX–XXIII [reaction (10)]. Two alternative mechanisms are presented below. The methyl anion of XXV is mobile. As a strong nucleophile, it attacks one of the coordinated isonitriles. The resulting iminoacyl palladium complex is coordinately unsaturated, and this gap may be filled by a suitable donor, like an isonitrile, to stabilize the complex and yield XXVI.

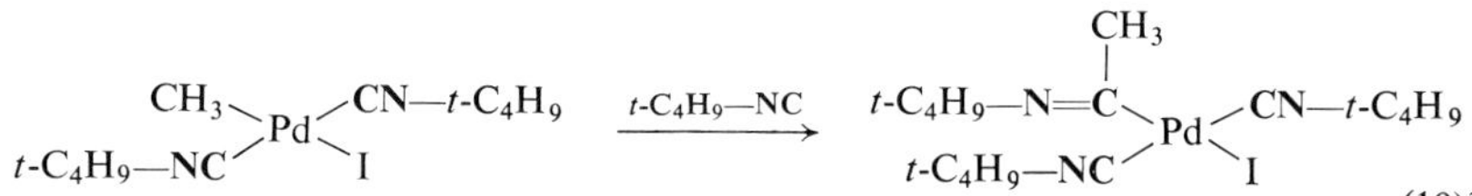

$$\underset{t\text{-}C_4H_9-NC}{\overset{CH_3}{\diagdown}}Pd\underset{I}{\overset{CN-t\text{-}C_4H_9}{\diagup}} \quad \xrightarrow{t\text{-}C_4H_9-NC} \quad \underset{t\text{-}C_4H_9-NC}{\overset{t\text{-}C_4H_9-N=\overset{CH_3}{\overset{|}{C}}}{\diagdown}}Pd\underset{I}{\overset{CN-t\text{-}C_4H_9}{\diagup}} \quad (10)^{32}$$

(XXV) (XXVI)

The confirmed *trans* configuration of XXVI seems to favor the following alternative mechanism. The isonitrile ligand migrates first, and substitutes the methyl group which subsequently adds to the isonitrile.

226  Arnd Vogler

The attempt to prepare the homologous nickel complex leads to surprising results. The reaction of XXVIII with methyl iodide in the presence of $t$-butyl isocyanide did not afford the expected product XXVII; a multiple successive insertion reaction of unknown mechanism leads to XXIX.

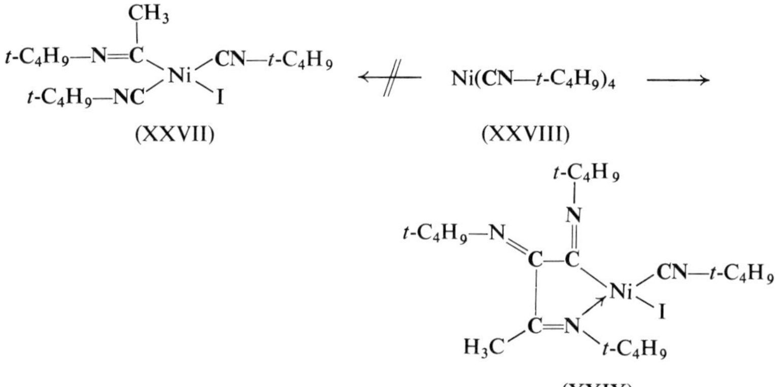

$$t\text{-}C_4H_9\text{—}N{=}C(CH_3)\text{—}Ni(CN\text{—}t\text{-}C_4H_9)(I) \quad \xleftarrow{\;\;/\!\!/\;\;} \quad Ni(CN\text{—}t\text{-}C_4H_9)_4 \longrightarrow$$

(XXVII)          (XXVIII)

$$\text{(XXIX)} \quad \quad (11)$$

This structure is suggested by NMR and IR data ($\nu_{N\equiv C} = 2172$ cm$^{-1}$; $\nu_{C=N} = 1667, 1634, 1610$ cm$^{-1}$).

The complex XXIX seems to serve as an intermediate in the catalytic polymerization of isonitriles[39]; upon heating with an excess of isonitrile, XXX is formed.

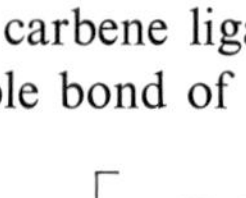

$$\left[\begin{array}{c} R\text{—}N \\ \| \\ \text{—}C\text{—} \end{array}\right]_n$$

(XXX)

Another unusual insertion reaction occurs when an isonitrile reacts with the metal carbene complex.[1] The resulting complex XXXIII suggests that the isonitrile first substitutes the carbene ligand which (because of its high reactivity) adds to the NC multiple bond of the isonitrile ligand of XXXII.

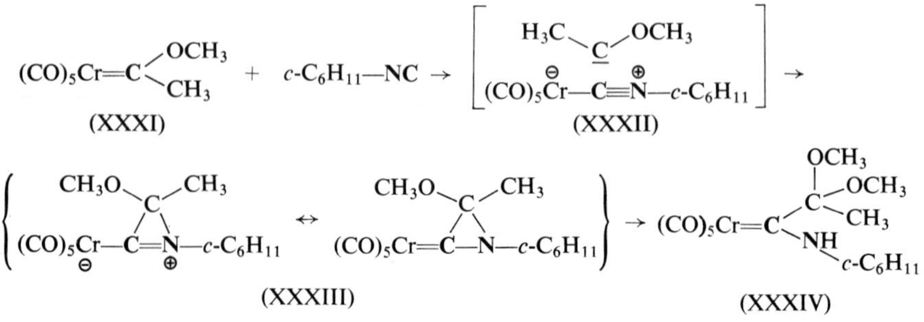

$$(CO)_5Cr{=}C(OCH_3)(CH_3) \;+\; c\text{-}C_6H_{11}\text{—}NC \;\rightarrow\; \left[(CO)_5Cr\text{—}C{\equiv}N\text{—}c\text{-}C_6H_{11}\right] \;\rightarrow$$

(XXXI)                                (XXXII)

$$\left\{ (CO)_5Cr\text{—}C{=}N\text{—}c\text{-}C_6H_{11} \;\leftrightarrow\; (CO)_5Cr{=}C\text{—}N\text{—}c\text{-}C_6H_{11} \right\} \;\rightarrow\; (CO)_5Cr{=}C(OCH_3)(NH\text{-}c\text{-}C_6H_{11})$$

(XXXIII)                                                       (XXXIV)

(12)

## 10. Coordinated Isonitriles 227

Nucleophiles like methanol attack the complex at the asymmetric carbon splitting the NC bond. As mentioned previously, a metal complex may serve as a protecting group for isonitriles. So it is even possible to stabilize isonitriles as ligands of complexes which do not exist in the free state. For example, the parent compound of all isonitriles "hydrogen isocyanide" XXXV cannot be isolated and exists only in equilibrium (13) with hydrogen cyanide XXXVa.

$$H-\overset{\oplus}{N}\equiv\overset{\ominus}{C} \rightleftharpoons H-C\equiv N \qquad (13)$$
$$(XXXV) \qquad (XXXVa)$$

But protonation of a complex XXXVI containing $CN^\ominus$ as a ligand can lead to a stable and well-characterized complex XXXVII with hydrogen isocyanide as a ligand[26]:

$$[(CO)_5Cr-NC]^\ominus + H^\oplus \rightarrow (CO)_5Cr-CN-H \qquad (14)$$
$$(XXXVI) \qquad (XXXVII)$$

In analogy to this protonation, alkylation and arylation of metal cyanides is a more general way to prepare metal isonitrile complexes and finally free isonitriles.[21]

Olefins which tend to form carbonium ions by protonation alkylate cyanide ion in the presence of Cu(I)Br to form isonitriles.[30] The reactive intermediate is certainly a copper cyanide complex (see Chapter 2). When the copper salt is absent the reaction yields only nitriles.

Heldt has demonstrated another important application of metal complexes as protecting groups for isonitriles. Benzyl isocyanide is readily hydrolyzed by acids and hence cannot take part in electrophilic substitution reactions which have to be carried out in acidic medium. Since the complex XXXVIII is very resistant even to strong acids, it is possible to subject the coordinated

$$[(C_6H_5-CH_2-NC)_5FeCN]^\oplus$$
$$(XXXVIII)$$

isonitrile to electrophilic substitution by $HNO_3$, $SO_3$, and $CH_2O$ in concentrated sulfuric acid, and with bromine in acetic acid, leading to substitution at the aromatic ring.[19,21] Surprisingly, the substitution takes place almost exclusively at the *para* position of the benzene ring and only to a small extent at the *meta* position when more than one aromatic nucleus of the complex is being substituted. Since the NC stretching frequency of benzyl isocyanide ($\nu_{NC} = 2146$ cm$^{-1}$) increases to 2200 cm$^{-1}$ in the complexed ligand, back donation appears not to take place and the complex XXXVIII is best represented

$$\overset{\ominus}{\text{fe}}-\text{C}\equiv\overset{\oplus}{\text{N}}-\text{CH}_2-\text{C}_6\text{H}_5 \longleftrightarrow \text{fe}=\text{C}=\text{N}-\text{CH}_2-\text{C}_6\text{H}_5$$
$$\text{(XXXVIIIa)}^* \qquad\qquad\qquad \text{(XXXVIIIb)}^*$$

by formula XXXVIIIa. As Heldt pointed out, the positive charge at the nitrogen should be strongly *meta* directing in analogy to trimethylbenzylammonium chloride which on electrophilic substitution yields 88% of the *meta* isomer and only 12% of the *ortho* and *para* isomers. The anchimeric effect which was invoked to explain the reactivity of this complex is not very likely because it would require a fairly high contribution from formula XXXVIIIb.

According to the observations described above, the interaction of the NC group and the benzene ring, separated by a methylene group, does not seem to be very strong. The *para* position could just as well be favored to undergo substitution reactions because the *ortho* and *meta* positions are blocked by the steric bulk of the other ligands of XXXVIII. The fact that the mononitration of XXXVIII proceeds about 300 times faster than the nitration of benzene was taken as evidence for the enhanced reactivity of the coordinated isonitrile. Even so, one has to consider that the benzene is suspended only in sulfuric acid, and has to compete with the dissolved complex. The difference in the physical states of the competing aromatic rings might account for this observation. As described for the Saegusa reaction, nucleophiles attack the carbon atom of a coordinated isonitrile.

Heldt investigated the reactions of various nucleophiles with XXXIX,

$$[(\text{C}_6\text{H}_5-\text{CH}_2-\text{NC})_5\text{Fe}-\text{CN}]\text{Br}$$
$$\text{(XXXIX)}$$

which take a different course, depending upon the nature of the nucleophile.[17] These experiments provide no evidence for back donation. As a consequence of the drain of electron density from the isonitrile carbon to the central metal of XXXIX, the complexed cyanide becomes a better leaving group, XL, by which the formation of a resonance stabilized benzyl cation is favored.

$$\text{fe}-\text{C}\equiv\text{N}-\text{CH}_2-\text{C}_6\text{H}_5 \rightleftharpoons [\text{fe}-\text{C}\equiv\text{N}|]^\ominus + \overset{\oplus}{\text{CH}_2}-\text{C}_6\text{H}_5 \qquad (15)$$
$$\text{(XXXIXa)} \qquad\qquad \text{(XL)} \qquad\quad \text{(XLI)}$$

Most of the reactions of XXXIX with nucleophiles seem to be typical of the reactions of XLI.

Weak nucleophiles like $\text{HOCH}_3$, $\text{HOCH}(\text{CH}_3)_2$, $\text{HNH}_2$, $\text{HNH}(\text{CH}_2)_2\text{CH}_3$, and $\text{HSCH}_2\text{CH}_3$, which undergo α-additions in the Saegusa reaction, are alkylated by $[(\text{C}_6\text{H}_5-\text{CH}_2-\text{NC})_5\text{Fe}-\text{CN}]\text{Br}$. Potassium thiocyanate yields benzyl thiocyanate which isomerizes to some extent to benzyl isothiocyanate.

* We use lower case letters to represent a metal that is coordinately saturated, but where some of the ligands are not shown.

10. Coordinated Isonitriles    229

Bases like cyanide, which tend to coordinate strongly, react with the complex predominantly in a substitution reaction replacing the benzyl isocyanide, which hydrolyzes to N-benzylformamide under the reaction conditions.

Tracer experiments [reaction (16)] with labeled C*N$^\ominus$ demonstrate that the benzyl, cyanide, and isonitrile groups of XXXIXb scramble. This is further evidence for the intermediacy of the benzylcarbonium ion.

$$[(C_6H_5\text{—}CH_2\text{—}NC)_5Fe\text{—}CN]C^*N \rightleftharpoons [(C_6H_5\text{—}CH_2\text{—}NC)_5Fe\text{—}C^*N]CN \rightleftharpoons$$
(XXXIXb)
$$\begin{bmatrix}(C_6H_5\text{—}CH_2\text{—}NC)_4 \\ (C_6H_5\text{—}CH_2\text{—}NC^*)\end{bmatrix}Fe\text{—}CN \Big| CN \quad (16)$$

The formation of the benzyl cation can be utilized in a transalkylation reaction to synthesize new isonitrile complexes and subsequently new isonitriles.[18] If one adds to XXXIX an alkyl halide RX, which boils higher than the corresponding benzyl halide $C_6H_5$—$CH_2$—X, the benzyl halide can be removed from an equilibrium mixture by continuous distillation during formation of XLII.

$$[(R\text{—}NC)_5Fe\text{—}CN]^\oplus$$
(XLII)

XLII can be hydrolyzed to yield the free isonitrile, R—NC. The tendency of XLIII to lose $t$-butyl carbonium ions is so great that the complex ion even

$$[Pt(CN\text{—}t\text{-}C_4H_9)_4{}^{2\oplus}]$$
(XLIII)

alkylates water to form $t$-butanol.[37]

$$[Pt(CN\text{—}t\text{-}C_4H_9)_4]^{2\oplus} + 2H_2O \rightarrow Pt(CN)_2(CN\text{—}t\text{-}C_4H_9)_2 + 2t\text{-}C_4H_9OH + 2H^\oplus \quad (17)$$
(XLIII)

Paraformaldehyde in concentrated sulfuric acid reacts with XXXIX to yield polymeric products.[20] The formaldehyde attacks the coordinated benzyl isocyanide at the *para* position of the aromatic ring to form the alcohol XLV which can be isolated as an intermediate.

$$\text{fe}\text{—}CN\text{—}CH_2\text{—}\langle\bigcirc\rangle + CH_2O \longrightarrow \text{fe}\text{—}CN\text{—}CH_2\text{—}\langle\bigcirc\rangle\text{—}CH_2OH \xrightarrow[-H_2O]{+H^\oplus}$$
(XXXIXa)    (XLV)

$$\text{fe}\text{—}CN\text{—}CH_2\text{—}\langle\bigcirc\rangle\text{—}CH_2{}^\oplus \quad (18)$$
(XLVI)

The next step is assumed to be the formation of a carbonium ion. An analysis of the polymer indicates that the carbonium ion XLVI may undergo three different reactions. It can react with benzyl isonitrile of the initial complex with formation of a methylene bridge which connects both aromatic rings at their *para* positions. It may attack the alcohol intermediate to yield an ether. Most surprising is the formation of an ethane bridge which requires the reduction of a carbonium ion to a radical and subsequent dimerization of two such radicals. The $Fe^{2+}$ is oxidized to $Fe^{3+}$, reducing the carbonium ion.

Since one ethane bridge was found per iron atom the complex ion has to undergo this internal redox process twice—but hardly simultaneously because it is impossible to form $Fe^{4+}$ under the reaction conditions. It is feasible that after the first reduction takes place the $Fe^{3+}$ is reduced to $Fe^{2+}$, possibly by an excess of formaldehyde.

The conversion of isocyanates to carbodiimides and $CO_2$ is catalyzed by metal carbonyls [$Fe_2(CO)_9$, $Fe(CO)_5$, $W(CO)_6$, $Mo(CO)_6$]. This process represents a further illustration of the reactivity of isonitrile complexes which are intermediates in this reaction.[38]

The first step may be a nucleophilic attack of an isocyanate on a metal carbonyl (19).

$$R-\bar{N}=C=\bar{O} + M=C=\bar{O} \longrightarrow \begin{matrix} R-\bar{N}=\overset{\oplus}{C}-\bar{O}| \\ \overset{\ominus}{M}-C=\bar{O} \end{matrix} \left( \text{or} \begin{matrix} R-\bar{N}=C-\bar{O}| \\ | \quad | \\ M-C=\bar{O} \end{matrix} \right) \longrightarrow$$

(XLVII) \quad\quad (XLVIII)

$$\overset{\ominus}{M}-\overset{\oplus}{C}\equiv N-R \longrightarrow \begin{matrix} |\bar{O}-\overset{\oplus}{C}=\bar{N}-R \\ | \\ \ominus M-C=\bar{N}-R \end{matrix} \longrightarrow \begin{matrix} M=C=\bar{O} \\ \\ R-\bar{N}=C=\bar{N}-R \end{matrix} \quad (19)$$

(XLIX) \quad\quad (L)

It is questionable whether the cyclic intermediate XLVIII is involved as Ulrich has suggested, because in the cases of $Mo(CO)_6$ or $W(CO)_6$ the formation of an unusual, heptacoordinated monosubstituted metal carbonyl is required. Also the simultaneous bond making and bond breaking of XLVII to form the isonitrile complex XLIX and release of $CO_2$ seems to be quite plausible.

A second isocyanate now attacks the isonitrile carbon of XLIX to form L which is finally stabilized, releasing the metal carbonyl and the carbodiimide.

When the metal carbonyl is employed in excess, the reaction yields metal carbonyls in which some of the carbonyl groups are substituted by isonitrile[28,38] which confirms the formation of an isonitrile complex as an intermediate in the catalytic conversion of isocyanate to carbodiimide.

10. *Coordinated Isonitriles*   231

In the absence of catalysis molecular oxygen does not oxidize isonitriles to isocyanates. This reaction is catalyzed by certain transition metals in low oxidation states ($Ni^0$, $Co^0$, $Rh^\circledast$).[31] The isolation of a Ni isonitrile peroxo complex LII which is formed by reaction (20) suggests the formation of LII

$$Ni(CN-t-C_4H_9)_4 \xrightarrow{O_2} \underset{t-C_4H_9-NC}{\overset{t-C_4H_9-NC}{>}}Ni\underset{O}{\overset{O}{<}}\bigg| \xrightarrow{\text{excess } t-C_4H_9-NC}$$

(LI)                (LII)

$$2t-C_4H_9-NCO + LI \quad (20)$$

as an intermediate in the catalytic oxidation of isonitriles.

## IV. THE SYNTHESIS OF METAL ISONITRILE COMPLEXES

The synthesis of isonitrile metal complexes is described in great detail by Malatesta and Bonati.[27] Therefore, only a brief general outline of the methods is given here.

There are three general ways to prepare isonitrile complexes.

1. The synthesis of an isonitrile complex by alkylation of silver ferrocyanide by Freund (21)[14] (see also Chapter 2) is representative for a general way to

$$Ag_4Fe(CN)_6 + 4C_2H_5-I \rightarrow (C_2H_5-NC)_4Fe(CN)_2 + 4AgI \quad (21)$$

obtain isonitrile complexes and furthermore free isonitriles, since the coordinated ligands can be released by substitution reactions.

The mechanism of the alkylation is not known. It may involve the formation of a carbonium ion ($C_2H_5^\oplus$) which reacts with the cyanide complex.

The reaction is very slow and often does not give well-defined products, possibly indicating an equilibrium reaction.[21]

Another alkylating agent is dimethylsulfate.[15] $H_4Fe(CN)_6$ reacts with diazomethane[29] and even with alcohol[16] to give isonitrile complexes.

The transalkylation reaction mentioned above extends the limited application of the simple alkylation reaction.

2. A general and very useful procedure for the preparation of isonitrile complexes simply requires mixing of a metal salt and the isonitrile.

$$MX_n + mR-NC \rightarrow [M(CN-R)_m]X_n \quad (22)$$

Sometimes the reaction is carried out in a solvent when mild conditions are required in order to avoid polymerization. Often an excess of isonitrile reduces

the metal to a lower oxidation state which is stable in the isonitrile complex.
3. Substitution of the ligand L in the complex by isonitriles also leads to

$$ML_x + yR\text{—}NC \longrightarrow ML_{x-y}(CN\text{—}R)_y + yL \qquad (23)$$

isonitrile complexes. The application of this method is limited by the fact that frequently—especially in the case of metal carbonyls—not all ligands L are replaced by isonitrile. The reaction of isocyanate with metal carbonyls can be considered to be a special case of this method.[28,38] The mechanism of this reaction has already been discussed.

## REFERENCES

1. Aumann, R., and Fischer, E. O., *Chem. Ber.* **101**, 954 (1968).
2. Badley, E. M., Chatt, J., Richards, R. L., and Sim, G. A., *Chem. Commun.* p. 1322 (1969).
3. Bamford, C. H., Eastmond, G. C., and Hargreaves, K., *Nature* **205**, 385 (1965).
4. Bigorgne, M., and Bouquet, A., *J. Organometal. Chem.* **1**, 101 (1963).
5. Carassiti, V., Condorelli, G., and Condorelli-Costanzo, L. L., *Ann. Chim. (Rome)* **55**, 329 (1965).
6. Cetini, G., and Gambino, O., *Ann. Chim. (Rome)* **53**, 236 (1961).
7. Cotton, F. A., *Inorg. Chem.* **3**, 703 (1964).
8. Cotton, F. A., Dunne, T. G., and Wood, J. S., *Inorg. Chem.* **4**, 318 (1965).
9. Cotton, F. A., and Parish, R. V., *J. Chem. Soc.* p. 1440 (1960).
10. Cotton, F. A., and Zingales, F., *J. Amer. Chem. Soc.* **83**, 351 (1961).
11. Fabbri, G., and Cappellina, F., *Ann. Chim. (Rome)* **48**, 909 (1958).
12. Fischer, E. O., and Fischer, H., *J. Organometal Chem.* **6**, 141 (1966).
13. Fischer, E. O., and Maasböl, A., *Angew. Chem.* **76**, 645 (1964).
14. Freund, M., *Chem. Ber.* **21**, 931 (1888).
15. Hartley, E. G. J., *J. Chem. Soc.* **97**, 1066 (1910).
16. Heldt, W. Z., *J. Org. Chem.* **26**, 3226 (1961).
17. Heldt, W. Z., *J. Inorg. Nucl. Chem.* **24**, 73 (1962).
18. Heldt, W. Z., *J. Inorg. Nucl. Chem.* **24**, 265 (1962).
19. Heldt, W. Z., *J. Org. Chem.* **27**, 2604 (1962).
20. Heldt, W. Z., *J. Org. Chem.* **27**, 2608 (1962).
21. Heldt, W. Z., *Advan. Chem. Ser.* **37**, 99 (1963).
22. Hieber, W., and Weiss, E., *Z. Anorg. Allg. Chem.* **287**, 223 (1956).
23. Horrocks, W. D., and Taylor, R. C., *Inorg. Chem.* **2**, 723 (1961).
24. Horrocks, W. D., Taylor, R. C., and LaMar, G. N., *J. Amer. Chem. Soc.* **86**, 3031 (1964).
25. Joshi, K. K., Mills, O. S., Pauson, P. L., Shaw, B. W., and Stubbs, W. H., *Chem. Commun.* p. 181 (1965).
26. King, R. B., *Inorg. Chem.* **6**, 25 (1967).
27. Malatesta, L., and Bonati, F., "Isocyanide Complexes of Metals." Wiley, New York, 1969.
28. Manuel, T. A., *Inorg. Chem.* **3**, 1703 (1964).
29. Meyer, J., Domann, H., and Mueller, W., *Z. Anorg. Allg. Chem.* **230**, 336 (1937).
30. Otsuka, S., Mori, K., and Yamagami, K., *J. Org. Chem.* **31**, 4170 (1966).
31. Otsuka, S., Nakamura, A., and Tatsuno, Y., *Chem. Commun.* p. 836 (1967).
32. Otsuka, S., Nakamura, A., and Yoshida, T., *J. Amer. Chem. Soc.* **91**, 7196 (1969).

33. Ruch, E., private communication (1969).
34. Saegusa, T., Ito, Y., Kobayashi, S., Hirota, K., and Yoshika, H., *Tetrahedron Lett.* p. 6121 (1966).
35. Strohmeier, W., and Hellmann, H., *Chem. Ber.* **97**, 1877 (1964).
36. Treichel, P. M., and Stenson, J. P., *Inorg. Chem.* **8**, 2563 (1969).
37. Tschugaeff, L., and Teearu, P., *Chem. Ber.* **47**, 2643 (1914).
38. Ulrich, H., Tucker, B., and Sayigh, A. A. R., *Tetrahedron Lett.* **18**, 1731 (1967).
39. Yamamoto, Y., and Hagihara, N., *Bull. Chem. Soc. Jap.* **39**, 1084 (1966).
40. Yamamoto, Y., Yamazaki, H., and Hagihara, N., *J. Organometal. Chem.* **18**, 189 (1969).

# Addendum
# Recent Developments in Isonitrile Chemistry

G. W. Gokel

In the inevitable time lag between the submission of the first chapters and the proofreading stage, a substantial number of papers has been published which are pertinent to this fast-growing field. This addendum has been included in an effort to provide the reader with references that hopefully cover the literature up to the end of 1969, and in a few instances, into 1970. Each reference is presented in abstract form for the reader's convenience and the abstracts appear, insofar as possible, in the order of the subdivisions of the book. Clearly, some of the references will apply to more than one of the chapters, but are referenced only under that section to which the greatest part of the information seems to apply.

## TO CHAPTER 1

*Vibration and rotation spectra of ethyl isocyanide.* K. Bolton, N. L. Owen, and J. Sheridan, *Spectrochim. Acta* **25A**, 1, (1969).

Vibrational and rotational spectra were measured for ethyl isocyanide in order to determine its molecular structure. A number of the spectral bands are assigned and other constants like the centrifugal distortion constants are given.

*The constitution, vibrational spectra and proton resonance spectra of trimethylsilyl cyanide and isocyanide.* M. R. Booth and S. G. Frankiss, *Spectrochim. Acta* **26A**, 859 (1970).

Trimethylchlorosilane and silver cyanide reacted to produce an equilibrium mixture of the cyanide and isocyanide. The mole fraction of the latter was determined to be $0.0015 \pm 0.0005$ for the liquid at 25°. Infrared and proton resonance spectra are obtained for the equilibrium mixture using samples

## Addendum

enriched in both $^{13}$C and $^{15}$N. The magnetic resonance data indicate a rapid exchange of CN groups between the cyanide and isocyanide forms.

*The rotation and vibration spectra of vinyl isocyanide.* K. Bolton, N. L. Owen, and J. Sheridan, *Spectrochimica Acta* **26A**, 909 (1970).

Microwave spectra are used to determine rotational constants for vinyl isocyanide in the ground vibrational state and for the first excited state. The dipole moment is determined to be $3.56 \pm 0.06$ D. The normal vibrations are also assigned on the basis of the infrared spectrum from 200 to 4400 cm$^{-1}$.

*Nitrogen-14 nuclear magnetic resonance determination of the quadrupole coupling constants of methyl isocyanide and ethyl isocyanide.* W. B. Moniz and C. F. Poranski, *J. Phys. Chem.* **73**, 4145 (1969).

$^{14}$N Quadrupole coupling constants were determined from NMR line widths for methyl and ethyl isocyanides; the values were 0.27 and 0.30, respectively. These values are in disagreement with earlier findings in the vapor phase. The differences are believed to be due to differences between the liquid and vapor phases and not fundamental differences in the molecules.

*Molecular g values, magnetic susceptibility anisotropies, diamagnetic and paramagnetic susceptibilities, second moment of the charge distribution and molecular quadrupole moments of $H_3CCN$ and $H_3CNC$.* J. Pochan, R. Shoemaker, R. Stone, and W. Flygare, *J. Chem. Phys.* **52**, 2478 (1970).

The physical data named in the title are measured both for methyl isocyanide and acetonitrile.

## TO CHAPTER 2

*Synthesis of sodium cyanotrihydroborate and sodium isocyanotrihydroborate.* R. C. Wade, E. A. Sullivan, J. R. Berscheid, and K. F. Purcell, personal communication (submitted for publication).

Sodium borohydride reacts with hydrogen cyanide to give an approximately 4:1 mixture of sodium cyanotrihydroborate and sodium isocyanotrihydroborate in 91 % yield, the latter being the only known example of a $C_1$ isonitrile (see table, Chapter 2). The rearrangement of the nitrile to the isonitrile is also discussed.

$$NaBH_4 + HCN \xrightarrow[25°]{THF} NaBH_3CN + NaBH_3NC$$

*Claisen condensation of ethyl isocyanoacetate.* D. Marquarding and I. Ugi, unpublished results (1969).

Ethyl isocyanoacetate in ether undergoes a Claisen condensation on treatment with *n*-butyllithium according to the following reaction scheme:

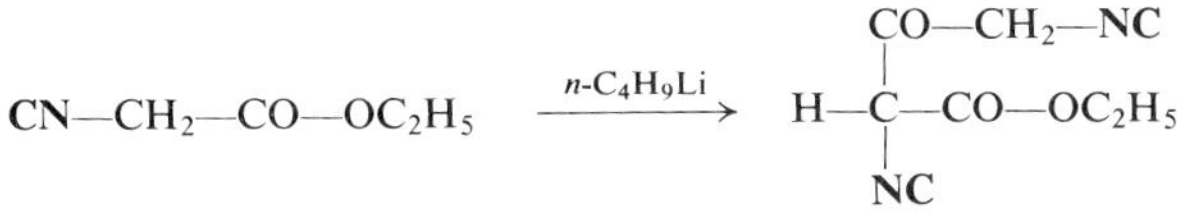

$$CN-CH_2-CO-OC_2H_5 \xrightarrow{n-C_4H_9Li} \underset{NC}{\overset{CO-CH_2-NC}{H-C-CO-OC_2H_5}}$$

*Isonitriles from alkyl halides and onium dicyanoargentates.* J. Songstad, L. Stangeland, and T. Austad, *Acta. Chem. Scand.* **24**, 355 (1970).

Refluxing $R(C_6H_5)_3AsCl$ ($R = CH_3$, $C_6H_5$) in a mixture of silver cyanide in ethanol gave compounds of the type $R(C_6H_5)_3AsAg(CH)_2$. Treatment of organic halides like benzhydryl bromide or methyl iodide with the complexes led to formation of the corresponding isocyanides (see Chapter 2, Section III, B).

$$R(C_5H_6)_3AsCl \xrightarrow[\text{ethanol}]{AgCN} R(C_5H_6)_3AsAg(CN)_2$$

$$R(C_5H_6)_3AsAg(CN)_2 + (C_5H_6)_2CHBr \longrightarrow (C_5H_6)_2CH-NC$$

## TO CHAPTER 3

*Existence of a novel mechanism in the thermal rearrangement of isocyanide to cyanide.* S. Yamada, K. Takashima, T. Sato, and S. Terashima, *Chem. Commun.* p. 811 (1969).

In rearrangements of optically active isocyanides to determine relative retention of configuration, it was noted that the thermal reaction in nonpolar media went well but with racemization. This led to the hypothesis that the rearrangement was a radical reaction. Conducting the reaction in the presence of *p*-benzoquinone gave apparent confirmation of this fact.

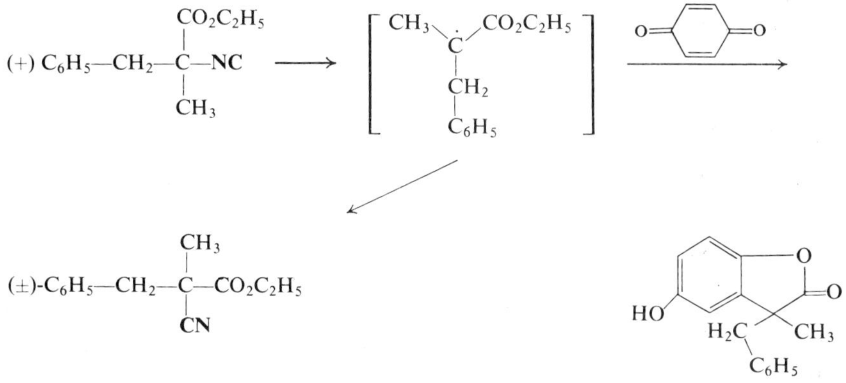

$$(+)\ C_6H_5-CH_2-\underset{CH_3}{\overset{CO_2C_2H_5}{C-NC}} \longrightarrow \left[ \underset{C_6H_5}{\overset{CH_3 \quad CO_2C_2H_5}{\underset{CH_2}{C}}} \right] \xrightarrow{O=\bigcirc=O}$$

$$(\pm)-C_6H_5-CH_2-\underset{CN}{\overset{CH_3}{C-CO_2C_2H_5}}$$

*Note:* This same result was observed by D. Marquarding and I. Ugi, using a similar compound in 1967. The optically active isocyanide racemized on heating to 220° C without solvent.

*Isomerization of methyl isocyanide sensitized by vibrationally excited ethane.* D. H. Shaw, B. K. Dunning, and H. D. Pritchard, *Can. J. Chem.* **47**, 669 (1969).

It is shown that methyl radicals can induce methyl isocyanide to isomerize by two different mechanisms. The mechanisms are direct free radical displacement (see Chapter 3, refs. 24, 25) and energy transfer from vibrationally excited ethane molecules. A pressure-quenching effect is observed on the latter.

*Energy transfer in thermal methyl isocyanide isomerization. Dependence of relative efficiency of helium on temperature.* S. C. Chan, J. T. Bryant, and B. S. Rabinovitch, *J. Phys. Chem.* **74**, 2055 (1970).

The thermal isomerization of methyl isocyanide is studied between 210° and 326°C. An apparent lowering in the activation energy is observed which is small (1.5 kcal/mole) and although this is in accord with theory, the study would have to be carried out over a wide range of temperature in order to observe appreciable effects.

*Energy transfer in thermal methyl isocyanide isomerization. Relative cross sections of fluoroalkanes and nitriles.* S. C. Chan, J. T. Bryant, L. D. Spicer, and B. S. Rabinovitch, *J. Phys. Chem.* **74**, 2058 (1970).

$n$-Perfluoroalkanes and $n$-nitriles are used as inert gases in the study of this effect on the thermal methyl isocyanide rearrangement. Homologs up to $C_6$ are studied in each case, at 280.5°C. Relative collision diameters for these molecules are obtained and a dipolar orientation effect on the collision diameter of the nitriles was observed. Some additions and corrections to earlier work are included.

## TO CHAPTER 4

*N-Cyclohexyldiphenylketenimine from cyclohexyl isocyanide and diphenyldiazomethane.* J. H. Boyer and W. Beverung, *Chem. Commun.* p. 1377 (1969).

Cyclohexyl isocyanide and diphenyldiazomethane were irradiated for 75 hours in cyclohexane; subsequent chromatography gave 50% of *N*-cyclohexyldiphenylacetamide **2**, the hydration product of the ketenimine **1**. The presence of **1** in the reaction mixture was demonstrated by the appropriate IR absorption.

$$c\text{-}C_6H_{11}\text{—NC} + (C_6H_5)_2CN_2 \xrightarrow{h\nu} [(C_6H_5)_2C\!=\!C\!=\!N\text{-}c\text{-}C_6H_{11}]$$
$$\mathbf{1}$$

$$\xrightarrow{H_2O} (C_6H_5)CH\text{—}CO\text{—}NH\text{-}c\text{-}C_6H_{11}$$
$$\mathbf{2}$$

*Ketenimines via the photolysis of diphenyldiazomethane in the presence of isonitriles.* J. A. Green and L. A. Singer, *Tetrahedron Lett.* p. 5093 (1969).

Diphenyldiazomethane and *t*-butyl or cyclohexyl isocyanide were photolyzed in petroleum ether solution to yield 40–50% or 25–35% of the corresponding ketenimines after isolation by crystallization or chromatography.

$$(C_6H_5)_2CN_2 \xrightarrow[R\text{—NC}]{h\nu} (C_6H_5)_2C\!=\!C\!=\!N\text{—}R + (C_6H_5)_2C\!=\!C(C_6H_5)_2$$
$$6\text{–}10\%$$

*A ketenimine from the addition of a carbene to an isocyanide.* E. Ciganek, *J. Org. Chem.* **35**, 862 (1970).

Methyl phenyldiazoacetate and *t*-butyl isocyanide were heated in a sealed Carius tube for 6 hours and 51% of phenyl methoxycarbonylketene-*N*-*t*-butylimine isolated by distillation.

$$C_6H_5\text{—}C(N_2)\text{—}CO\text{—}OCH_3 + t\text{-}C_4H_9\text{—}NC \xrightarrow[\Delta]{-N_2}$$

$$CH_3O\text{—}CO\text{—}C(C_6H_5)\!=\!C\!=\!N\text{-}t\text{-}C_4H_9$$

*Carbodiimides from nitrenes and isonitriles.* W. Lwowski, M. Grassmann, and T. Shingaki, unpublished results (1970).

In analogy to the generation of ketenimines from carbenes and isonitriles, nitrenes and isonitriles react to give moderate yields of carbodiimides. Thus, carbethoxynitrene reacts with *t*-butyl isocyanide to give *N-t*-butyl-*N'*-carbethoxycarbodiimide in about 65% yield.

$$C_2H_5O\text{—}CO\text{—}N_3 \xrightarrow[-N_2]{h\nu} C_2H_5O\text{—}CO\text{—}\overline{N}| \xrightarrow{t\text{-}C_4H_9\text{—}NC}$$

$$C_2H_5O\text{—}CO\text{—}N\!=\!C\!=\!N\text{-}t\text{-}C_4H_9$$

*Reactions between isocyanides and nitrones.* B. Zeeh, *Synthesis* **1**, 37 (1969).

Boron trifluoride catalyzes the reactions between isonitriles and nitrones which give amides as products. For example, $N$-phenyl-$p$-chlorophenyl nitrone (**1**) reacts with cyclohexyl isocyanide (**2**) to give the corresponding amide (**3**).

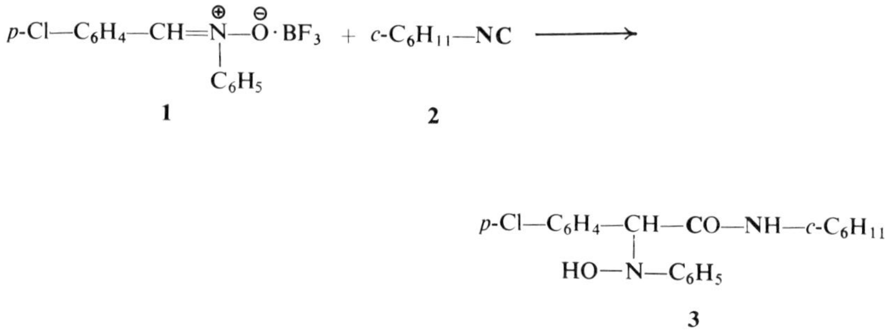

$$p\text{-Cl}-C_6H_4-CH=\overset{\oplus}{\underset{\underset{C_6H_5}{|}}{N}}-\overset{\ominus}{O}\cdot BF_3 + c\text{-}C_6H_{11}-NC \longrightarrow$$

$$\quad\quad\quad\quad\quad\quad\quad\quad\quad\text{1}\quad\quad\quad\quad\quad\quad\quad\text{2}$$

$$p\text{-Cl}-C_6H_4-\underset{\underset{HO-N-C_6H_5}{|}}{CH}-CO-NH-c\text{-}C_6H_{11}$$

$$\quad\quad\quad\quad\quad\quad\quad\quad\quad\quad\quad\text{3}$$

*Reaction of isocyanides with 1-halogeno-acetylenes: N-substituted 3-halogeno-acrylamides.* F. Johnson, A. H. Gulbenkian, and W. A. Nasutavicus, *Chem. Commun.* p. 608 (1970).

Isonitriles and 1-halogenoacetylenes react in boiling 10% aqueous methanol solution to give fairly low yields of $N$-substituted 3-halogenoacrylamides, possibly via nucleophilic attack of the isonitrile on the triple bond as illustrated below.

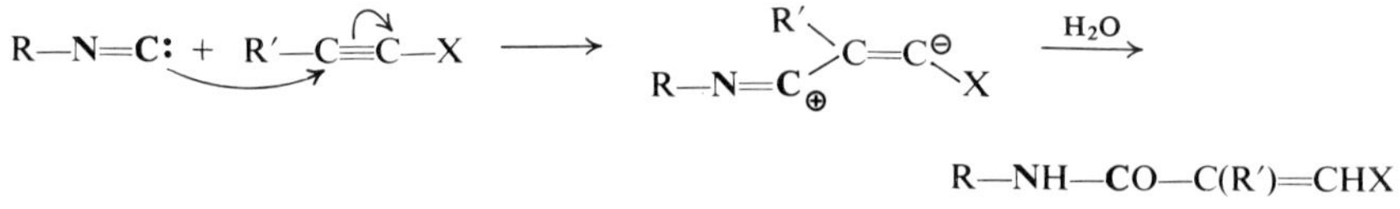

$$R-N\equiv C: + R'-C\equiv C-X \longrightarrow \underset{R-N=C_\oplus}{\overset{R'}{\diagdown}}C=C^\ominus_{\diagdown X} \xrightarrow{H_2O}$$

$$\quad\quad\quad\quad\quad\quad\quad\quad\quad\quad\quad\quad\quad\quad\quad\quad R-NH-CO-C(R')=CHX$$

*The adduct of triphenyl phosphine and hydrogen fluoborate with cyclohexyl isocyanide.* P. Hoffmann, unpublished results (1967).

The hydrogen fluoborate adduct of triphenyl phosphine adds to cyclohexyl isocyanide according to the reaction

$$[(C_6H_5)_3PH]^\oplus BF_4^\ominus + c\text{-}C_6H_{11}-NC = [(C_6H_5)_3P-CH=N\text{-}c\text{-}C_6H_{11}]^\oplus BF_4^\ominus$$

$$\quad\quad\quad\quad\text{1}\quad\quad\quad\quad\quad\quad\text{2}\quad\quad\quad\quad\quad\quad\quad\quad\text{3}$$

On treatment with base, **3** is decomposed to form **1** and **2**.

*Lithium aldimines. A new synthetic intermediate.* H. M. Walborsky and G. E. Niznik, *J. Amer. Chem. Soc.* **91**, 7778 (1969).

The appropriate lithium reagent adds to 1,1,3,3-tetramethylbutylisocyanide (no α-hydrogen) to give a lithium aldimine which can be hydrolyzed (water or deuterium oxide) to give the corresponding aldehyde or α-deuterioaldehyde in high yield **1**. Carbonation of the intermediate aldimine with $CO_2$ gives the corresponding glyoxylic acid derivative, **2**.

$$R-Li + R'-NC \longrightarrow R'-N=C(Li)-R \xrightarrow[2.\ H_3O^{\oplus}]{1.\ D_2O} R-CO-D \quad (1)$$

$$R'-N=C(Li)-R \xrightarrow[2.\ H_3O^{\oplus}]{1.\ CO_2} R-CO-CO-OH \quad (2)$$

*Polarographic reduction of isonitriles.* H. M. Walborsky, unpublished results (1970).

| Compound | $E_{1/2}$ vs Calomel | n | v. |
|---|---|---|---|
| 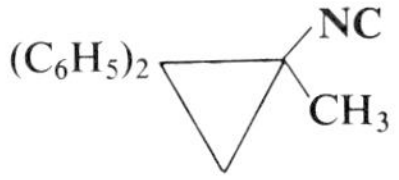 $(C_6H_5)_2$-cyclopropyl-NC, CH$_3$ | 2.91 | 1 | 0.135 |
| $C_2H_5-\underset{CH_3}{\overset{C_6H_5}{C}}-NC$ | 2.83 | 1 | 0.110 |
| $CH_3-C(CH_3)_2-C(CH_3)_2-NC$ | 3.00 | 1 | 0.090 |
| 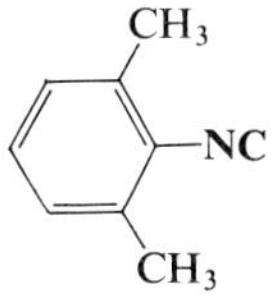 2,6-dimethylphenyl-NC | 2.72 | 2 | n(0.110) |

*Organic synthesis by electrolysis. II. Anodic methoxylation of isocyanide.* T. Shono and Y. Matsumura, *J. Amer. Chem. Soc.* **90**, 5937 (1968).

The anodic methoxylation of cyclohexyl isocyanide in a mixture of sodium methoxide in methanol led to a 20% yield of five products which were separated by preparative glpc.

*Reactions of nitrogen oxides with isocyanide.* T. Saegusa, S. Kobayashi, and Y. Ito., *Bull. Chem. Soc. Jap.* **43**, 275 (1970).

Isonitriles and nitrogen oxides undergo a redox reaction in which the isonitrile is oxidized to the corresponding isocyanate and the nitrogen oxide is reduced to $N_2$. Yields are moderate and $CO_2$ and $N_2O$ are formed as by-products.

$$c\text{-}C_6H_{11}\text{—NC} + NO \xrightarrow[\text{toluene}]{95°C/10hr} c\text{-}C_6H_{11}\text{—N}\!=\!C\!=\!O$$
$$56\%$$

## TO CHAPTER 5

*The rapid synthesis of 1-substituted tetrazoles.* D. M. Zimmerman and R. A. Olafson, *Tetrahedron Lett.* p. 5081 (1969).

The classical tetrazole synthesis by addition of an isonitrile to hydrazoic acid is improved by acid catalysis. A trace of sulfuric acid in the ethereal solution of $HN_3$ causes the tetrazole to be formed from the isonitrile rapidly and in high yield without formation of any formamide as in the case of $BF_3$ catalysis.

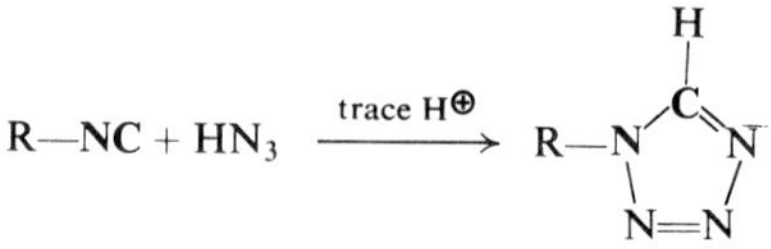

$$R\text{—NC} + HN_3 \xrightarrow{\text{trace } H^\oplus} R\text{—N}\underset{N=N}{\overset{\overset{H}{\underset{|}{C}}}{\diagup\diagdown}}N$$

*Diaziridones. IV. Formation by condensation of alkyl isocyanide with nitrosoalkane. Evidence for a carbodiimide N-oxide.* F. D. Greene and J. F. Pazos, *J. Org. Chem.* **34**, 2269 (1969).

2-Methyl-2-nitrosopropane (**1**) reacts with aliphatic isocyanides (**2**) at moderate temperatures affording good yields of disubstituted diaziridones **3**. [Cf. Chapter 5, Eq. (2a).] The reaction proceeds best when a tertiary alkyl isocyanide is used. The side products in the reaction are carbodiimide **4** and nitroalkane **5**. A carbodiimide-*N*-oxide **6** is postulated as the intermediate.

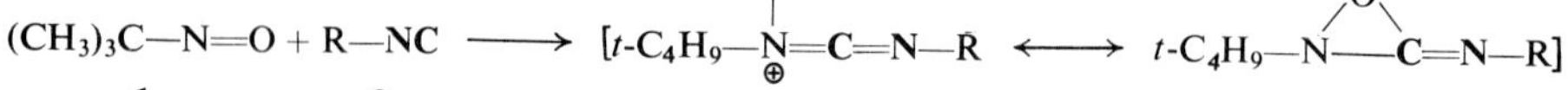

$$(CH_3)_3C\text{—N}\!=\!O + R\text{—NC} \longrightarrow [t\text{-}C_4H_9\overset{O^\ominus}{\underset{\oplus}{\text{—N}}}\!=\!C\!=\!N\text{—R} \longleftrightarrow t\text{-}C_4H_9\text{—N}\overset{O}{\diagup\diagdown}C\!=\!N\text{—R}]$$
$$\quad\;\, \textbf{1} \qquad\qquad \textbf{2} \qquad\qquad\qquad\qquad\qquad\qquad\qquad\qquad \textbf{6}$$

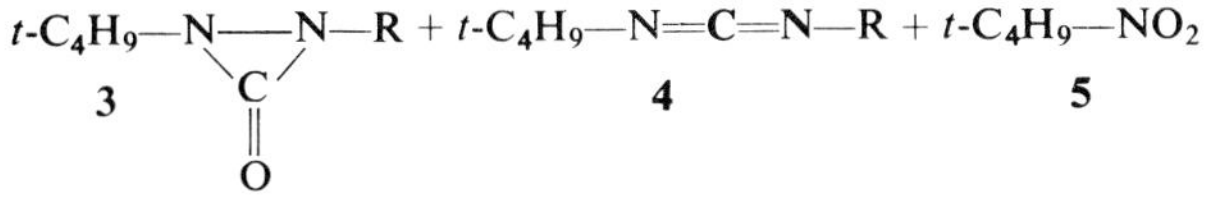

$t$-C$_4$H$_9$—N——N—R + $t$-C$_4$H$_9$—N=C=N—R + $t$-C$_4$H$_9$—NO$_2$
       \C/
3     ‖               4                   5
        O

*Reactions of imines with t-butyl isocyanide.* J. A. Deyrup, M. M. Vestling, W. V. Hagan, and H. Y. Yun, *Tetrahedron* **25**, 1467 (1969).

$t$-Butyl isocyanide and $N$-arylimines react at 120°C (sealed tube, CCl$_4$) to afford either 1:1 (3-$t$-butylamino-2-phenylindoles) or 2:1 (2,3-bis($t$-butylimino)azetidines) adducts in low to fair yield. The products were identified by combustion analysis, IR, NMR, and mass spectral data and confirmed by degradation and alternate synthesis.

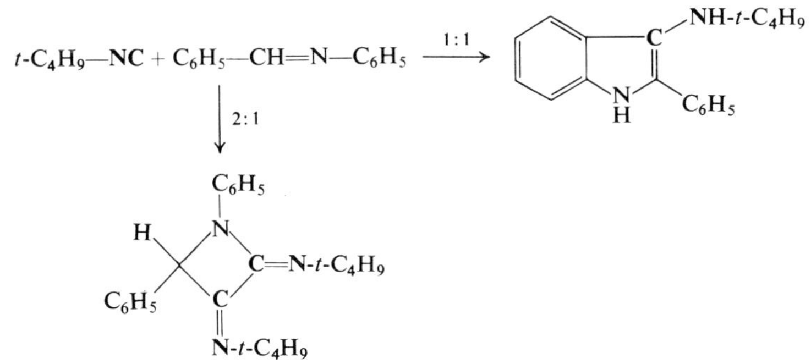

*Structure and reactions of 2:1-Adducts from acetylenedicarboxylate diesters and isonitriles.* E. Winterfeldt, D. Schumann, and H.-J. Dillinger, *Chem. Ber.* **102**, 1656 (1969).

Dicarbomethoxyacetylene reacts with isonitriles to give nonpolar adducts which react further with ester to give products such as compound **1** ($R = t$-C$_4$H$_9$ or $c$-C$_6$H$_{11}$, $R' = $ C≡C—CO—OCH$_3$). **1** undergoes a number of reactions of the C=N group; for example, chromatography on wet silica gel leads to partial hydrolysis.

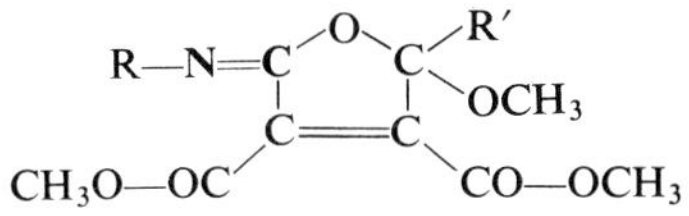

*The adduct from dicarbomethoxyacetylene and cyclohexyl isocyanide.* M. V. George and J. Z. Gougoutas, unpublished results (1970).

Dicarbomethoxyacetylene reacts with cyclohexyl isocyanide to give a yellow adduct, m.p., 157°C, which analyzes for $C_{32}H_{40}O_{12}N_2$ (molecular weight by mass spectrometry, 644). This adduct can be converted on heating to a red isomer which gives a monoperchlorate. The latter adduct undergoes thermal isomerization to a new yellow adduct of the same molecular formula. X-ray work on this problem is underway.

*A novel cycloaddition reaction of 4-bromo-2,6-dimethylphenyl isonitrile with acetylene derivatives.* T. Takizawa, N. Obata, Y. Suzuki, and T. Yanagido, *Tetrahedron Lett.* p. 3407 (1969).

Carbomethoxyphenylacetylene (**1**) reacts with 2,6-dimethyl-4-bromophenyl isocyanide (**2**) to give 41% of 4-carbomethoxy-5-phenyl-cyclopentene-1,2,3-tri(*N*-2,6-dimethyl-4-bromophenyl)imine (**3**). Dimethylacetylene dicarboxylate reacts with **2** in the cold to give bisketenimine **4** in addition to **5**, the hydrate of **3**. Therefore, **4** is thought to be the reactive intermediate which incorporates another molecule of isocyanide to give the triiminocyclopentene.

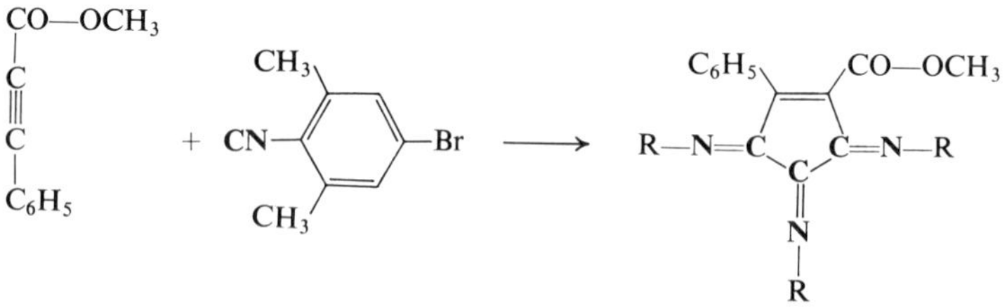

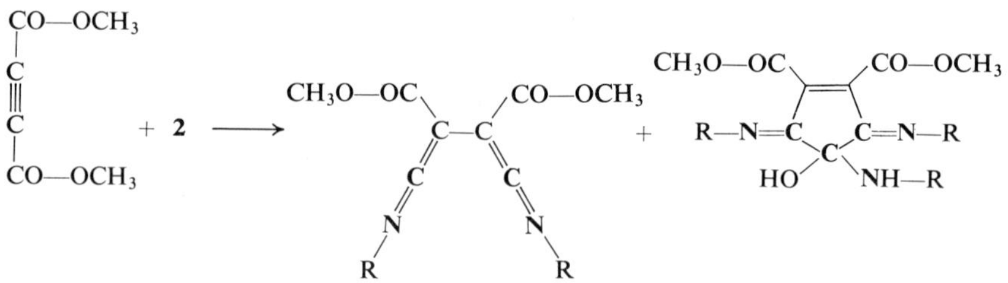

*Addition reactions between isonitriles and double bond systems.* B. Zeeh, *Synthesis* p. 65 (1969).

Review article dealing with additions of isonitriles to double bond systems, primarily the work of Kabbe (Cf. Chapter 5, refs. 11–13) and Zeeh (Cf. Chapter 5, refs. 21, 41 and 42) on boron trifluoride catalyzed additions of isonitriles to ketones. Additions to C=N, N=O, etc., are also mentioned. The reactions considered are of types 2b and 2c (see Chapter 5).

*The cycloaddition reaction of 2,6-dimethylphenylisonitrile with dimethyl acetylenedicarboxylate.* Y. Suzuki, N. Obata, and T. Takizawa, *Tetrahedron Lett.* p. 2667 (1970).

2,6-Dimethylphenyl isocyanide (**1**) undergoes a cycloaddition reaction with dicarbomethoxyacetylene (**3**) to give a substituted pyridine (**4**) which rearranges on heating above 150° to give compound **5**, whose structure (using 4-bromo-2,6-dimethylphenyl isocyanide) was determined by X-ray crystallography.

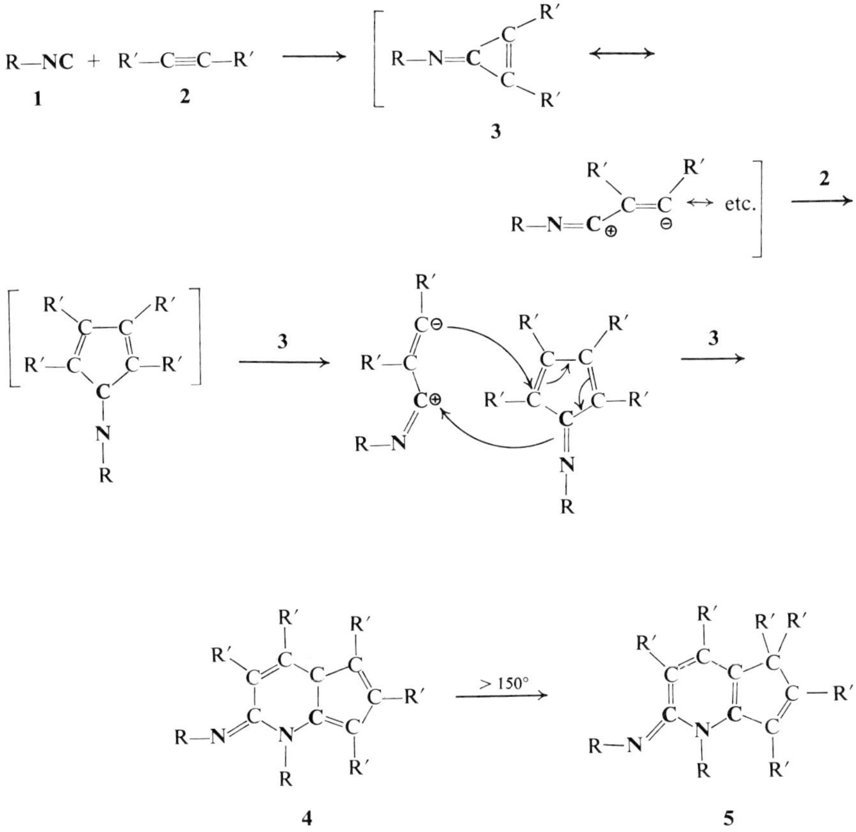

*Reactions of α,β-unsaturated steroidal ketones with isonitriles to give oxetanes.* B. Zeeh, *Tetrahedron Lett.* p. 113 (1969).

Cholest-4-en-3-one reacts with two equivalents of *t*-butyl isocyanide and one equivalent of $BF_3$ to give about 20% of the corresponding diiminooxetane (cf. Chapter 5) and about 5% of the 3-*t*-butoxy-5α-cyano derivative. Formation of the latter product is rationalized as a 1,4 addition of isocyanide to the α,β-unsaturated ketone system followed by an N to O alkyl transfer.

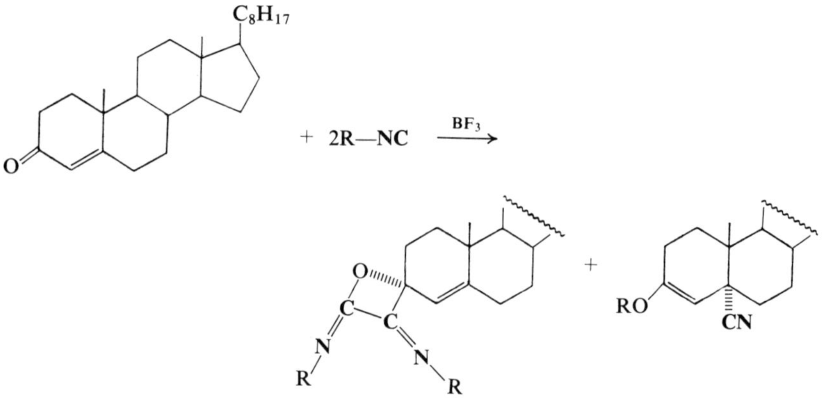

*Note on the dimerization of tert-butylisocyanide.* H.-J. Kabbe, *Chem. Ber.* **102**, 1447 (1969).

*t*-Butyl isocyanide dimerizes in the presence of a catalytic amount of boron trifluoride etherate to give cyanoimide **1**. **1** was reduced to **2** with lithium aluminum hydride and reacted with *o*-diaminobenzene to give **3**.

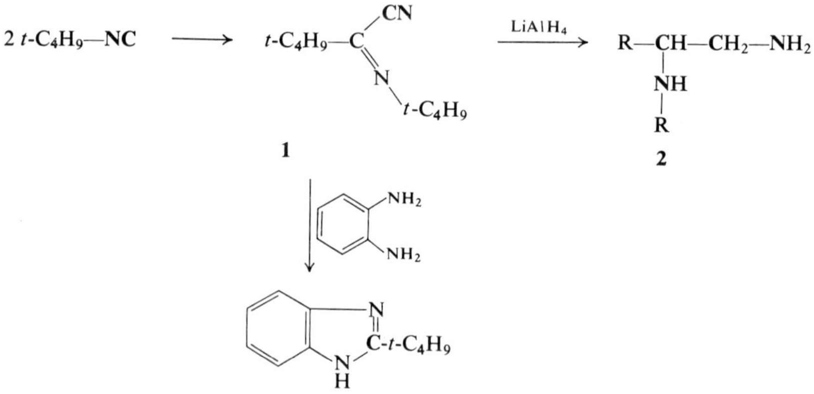

*Reactions of 2,6-dimethylphenylisonitrile with some small ring ketones.* N. Obata and T. Takizawa, *Tetrahedron Lett.* p. 3403 (1969).

Three equivalents of 2,6-dimethylphenylisocyanide (**1**), react with diphenyl-cyclopropenone (**2**) to give 76% of 4,5-bis(2,6-dimethylphenylimino)-2,3-diphenylcyclopenten-1-one (**3**). The structure of **3** was confirmed by spectroscopic means and degradation. **1** also reacted with 3,4-diphenylcyclobutenedione apparently incorporating two molecules of isocyanide and decarbonylating to give 66% of **3**.

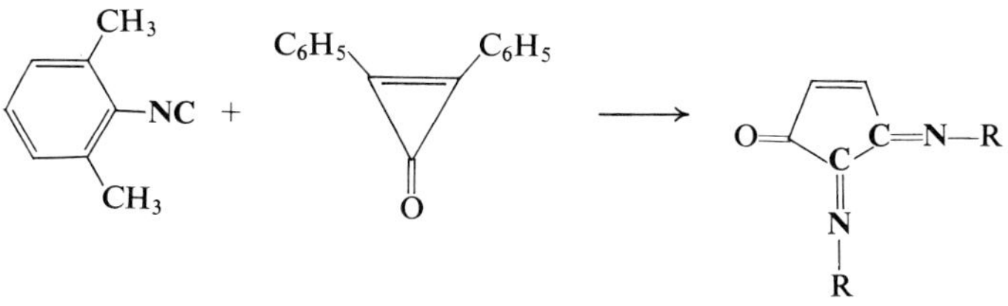

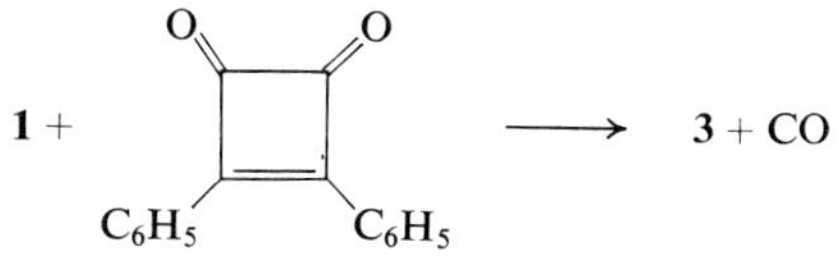

*Novel ring enlargement reaction of diphenylcyclopropenone with 2,6-dimethylphenylisonitrile in the presence of triphenyl phosphine.* N. Obata and T. Takizawa, *Tetrahedron Lett.* p. 2231 (1970).

The reaction of 2,6-dimethylphenyl isocyanide with diphenylcyclopropenone (**1**) in the presence of triphenyl phosphine (**2**) gave *N*-(2,6-dimethylphenyl)diphenyliminocyclobutenone (**4**). The intermediate is presumed to be a ketene-ylide (**3**).

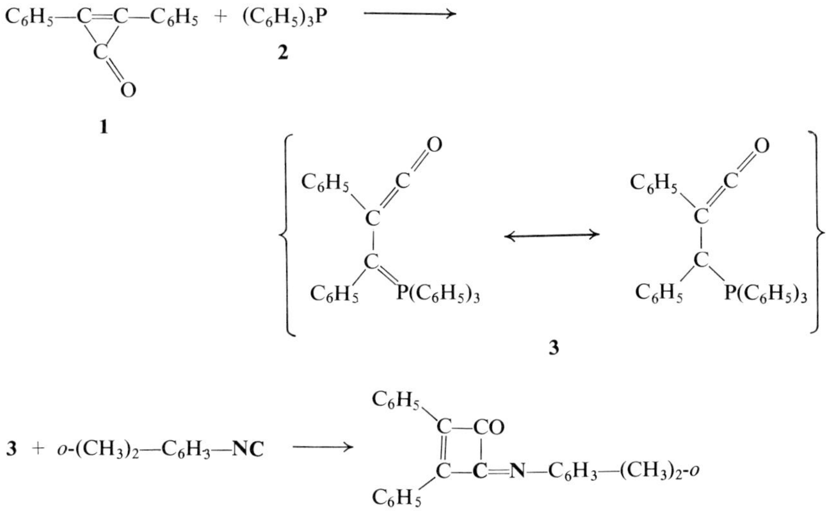

*Isomerization of o-biphenylyl isocyanide into 1-azabenz[b]azulene and the formation of both from o-biphenylyl isothiocyanate.* J. H. Boyer and J. DeJong, *J. Amer. Chem. Soc*, **91**, 5929 (1969).

UV irradiation of *o*-biphenylylisothiocyanate (**1**) gave *o*-biphenylylisocyanide (**2**), 1-azabenz[b]azulene (**3**) and sulfur. That **2** was also the intermediate in the reaction **1** → **3**, was confirmed by irradiation of **2** which gave 62% conversion to **3**.

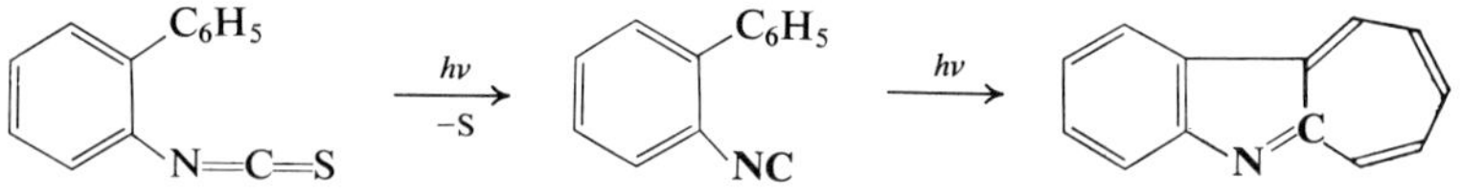

*Acid-catalyzed reaction of isocyanide with a Schiff base. New and facile synthesis of imidazolidines.* T. Saegusa, N. Taka-ishi, I. Tamura, and H. Fujii, *J. Org. Chem.* **34**, 1145 (1969).

Alkyl isocyanides react with benzaldehyde-*N*-alkyl imines in the presence of $AlCl_3$ to produce an imidazolidine derivative. (Cf. Chapter 5, reaction 2b.) Yields are generally high, being in the range 80–98%. $SnCl_4$ and $BF_3 \cdot O(C_2H_5)_2$ also catalyze the reaction although in decreased yield.

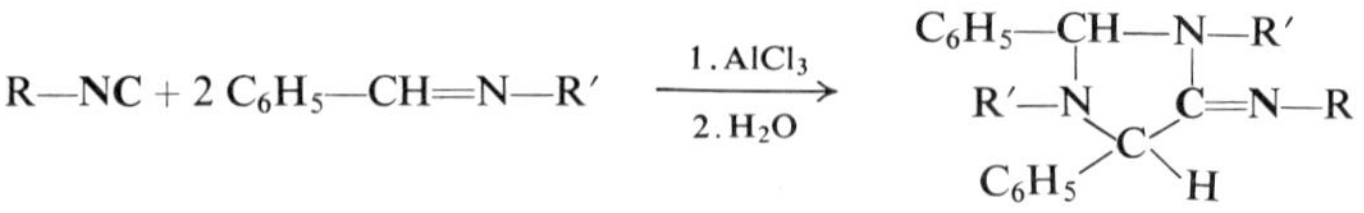

*Δ²-Oxazolines from α-metallated isocyanides and carbonyl compounds.* F. Gerhart and U. Schöllkopf, *Tetrahedron Lett.* p. 6231 (1968).

Isonitriles with an α-hydrogen are metallated with butyllithium and can react with carbonyl compounds to form $\Delta^2$-oxazolines **1**. The reaction is carried out at low temperature and yields range from 13–72% (average about

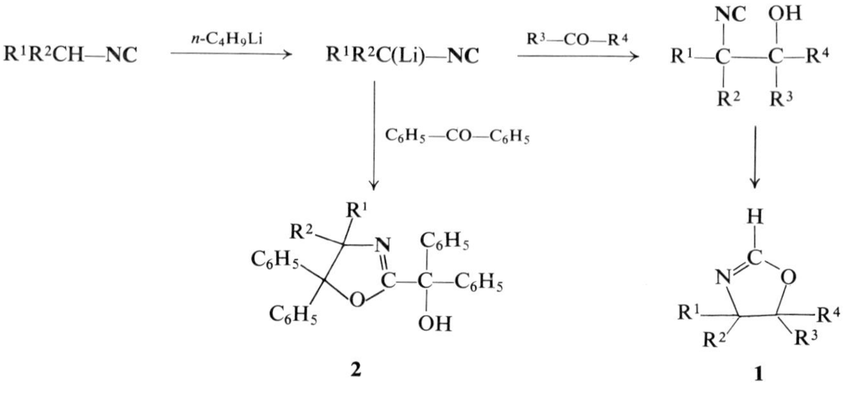

60%). The α-metallated isonitriles are also shown to react with two equivalents of benzophenone to yield 2-(diphenylhydroxymethyl)-$\Delta^2$-oxazolines (2).

*α-Hydroxy isocyanides from carbonyl compounds and α-metallated isocyanides.* W. A. Böll, F. Gerhart, A. Nürrenbach, and U. Schöllkopf, *Angew. Chem.* **82**, 482 (1970).

α-Lithiated isocyanides (1) attack carbonyl compounds (2) as nucleophiles to give α-hydroxy disubstituted isocyanides (4). The yields for the three cases reported are 75-80% and include cases with substitution on the isocyanide as well as on the carbonyl compound. Schöllkopf and co-workers have previously reported the cyclization of α-hydroxyisocyanides to $\Delta^2$-oxazolines (see preceding communication).

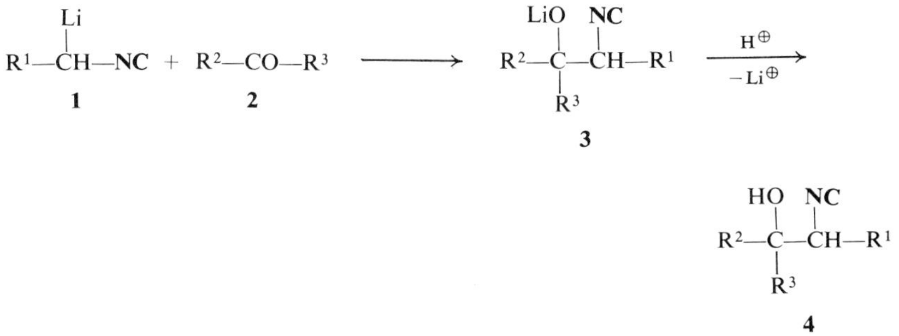

$$R^1-\underset{1}{\overset{\text{Li}}{\text{CH}}}-\text{NC} + R^2-\underset{2}{\text{CO}}-R^3 \longrightarrow R^2-\underset{\underset{3}{R^3}}{\overset{\overset{\text{LiO}}{|}\overset{\text{NC}}{|}}{C}}-\text{CH}-R^1 \xrightarrow[-\text{Li}^\oplus]{H^\oplus}$$

$$R^2-\underset{\underset{4}{R^3}}{\overset{\overset{\text{HO}}{|}\overset{\text{NC}}{|}}{C}}-\text{CH}-R^1$$

*Some novel properties of dihydro-1,3-oxazines and their use in formylation or organometallics.* A. I. Meyers and H. W. Adickes, *Tetrahedron Lett.* p. 5151 (1969).

Treatment of dihydro-1,3-oxazines 1 and 2 with butyllithium gives the open chain γ-hydroxyisocyanide (3) in 80% yield. This can be cyclized to 1 on heating.

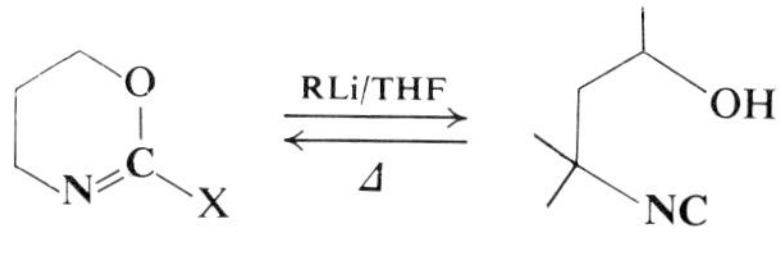

1, X = H
2, X = Br

*Small ring systems from isocyanides. I. The reaction of isocyanides with hexafluorobutynes-2.* T. R. Oakes, H. G. David, and F. J. Nagel, *J. Amer. Chem. Soc.* **91**, 4761 (1969).

Isonitriles react with two equivalents of hexafluorobutyne-2 in dichloromethane to give ketenimine 1. In ethereal solution, two 1:1:1 adducts are formed, iminoester **2** and ketenimine **3**. No isolation of simple 1:1 adducts was possible.

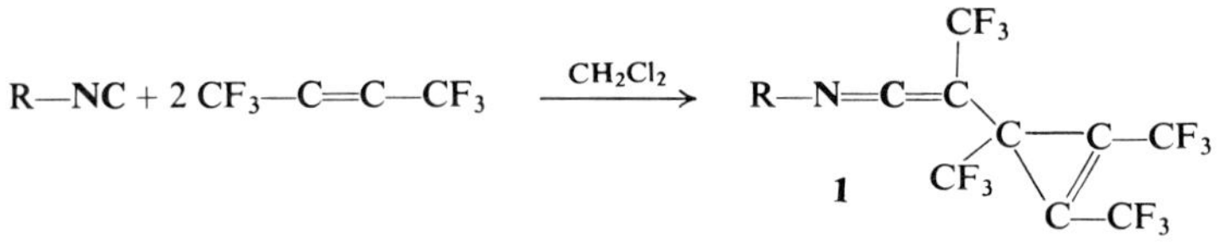

$$R-NC + 2\,CF_3-C\equiv C-CF_3 \xrightarrow{CH_2Cl_2} \underset{\mathbf{1}}{R-N=C=C(CF_3)\text{-cyclopropene}(CF_3)_3}$$

$$R-NC + CF_3-C\equiv C-CF_3 \xrightarrow{ethanol} \underset{\mathbf{2}}{R-N=C(OC_2H_5)-C(CF_3)=CH-CF_3}$$
$$+ \underset{\mathbf{3}}{R-N=C=C(CF_3)-CH(OC_2H_5)-CF_3}$$

*4-Carbethoxy-2-oxazolines from ethyl isocyanoacetate and carbonyl compounds.* D. Hoppe and U. Schöllkopf, *Angew. Chem.* **82**, 290 (1970).

Ethyl isocyanoacetate reacts with carbonyl compounds in the presence of ethanolic sodium cyanide to give 4-carbethoxy-2-oxazolines in yields ranging from 15.86% depending on the carbonyl compound, but generally greater than 50%. The reaction is presumed to proceed as shown below.

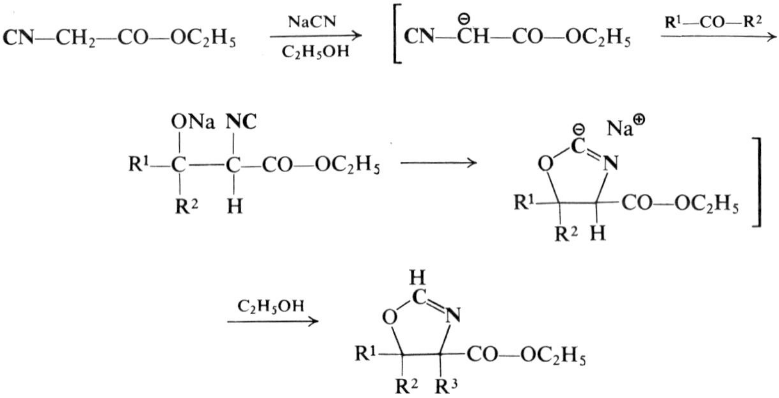

$$CN-CH_2-CO-OC_2H_5 \xrightarrow[C_2H_5OH]{NaCN} \left[ CN-\overset{\ominus}{C}H-CO-OC_2H_5 \xrightarrow{R^1-CO-R^2} \right.$$

$$\underset{R^2\;\;H}{\overset{ONa\;\;NC}{R^1-C-C-CO-OC_2H_5}} \longrightarrow \left. \text{oxazoline-Na}^{\oplus}\text{-CO-OC}_2H_5 \right]$$

$$\xrightarrow{C_2H_5OH} \text{oxazoline with } R^1, R^2, R^3, CO-OC_2H_5$$

Schöllkopf and co-workers have also found that phenyl isocyanate and isocyanomethyllithium react to give (after hydrolysis) 50% of imidazole derivatives.

$C_6H_5-N=C=O + LiCH_2-NC \longrightarrow$

$\underset{\substack{\text{crystal:}\\\text{only keto}}}{\begin{array}{c}C_6H_5-N-C\overset{O}{\diagdown}\\HC\diagdown_N\diagup CH_2\end{array}} \rightleftharpoons \underset{\substack{\text{DMSO: only}\\\text{enol}}}{\begin{array}{c}C_6H_5\diagdown_N-C\diagup^{OH}\\HC\diagdown_N\diagup CH\end{array}}$

α-(Formylamino)acrylic esters from α-metallated isocyanoacetic esters and carbonyl compounds. U. Schöllkopf, F. Gerhart, and R. Schröder, Angew. Chem. **81**, 701 (1969); Int. Ed. Engl. **8**, 672 (1969).

α-Metallated ethyl isocyanoacetate (**1**) (M = Na or Li) reacts with ketones and aldehydes (**2**) in yields ranging from 13–87% (generally greater than 70%) presumably via the oxazoline anion **3** to yield α-formylamino acrylic esters (**4**).

$$C_2H_5O-CO-\underset{\underset{M^\oplus}{\ominus}}{CH}-NC + R^1-CO-R^2 \longrightarrow \left[\begin{array}{c}R^1\\R^2-C-O\\H-C\diagdown_N\diagup C^\ominus \ M^\oplus\\C_2H_5O-OC\end{array}\right]$$

$\qquad\qquad\qquad$ **1** $\qquad\qquad$ **2** $\qquad\qquad\qquad\qquad\qquad$ **3**

$$\longrightarrow \underset{\mathbf{4}}{\begin{array}{c}R^1\diagdown_{\phantom{C}}\diagup CO-OC_2H_5\\R^2\diagup C=C\diagdown NH-CHO\end{array}}$$

β-Substituted N-Formyl-S-Benzylcysteines. U. Schollköpf and D. Hoppe, Angew. Chem. **82**, 253 (1970), Int. Ed. Engl. **9**, 236 (1970).

α-Metallated ethyl isocyanoacetate reacts with a variety of carbonyl compounds to give an intermediate 1-carbethoxy-1-formylaminoethylene derivative which undergoes Michael addition of benzyl mercaptan and subsequent hydrolysis affords N-formyl-S-benzylcysteine derivatives.

$$CN-\underset{\underset{\ominus}{\overset{M^\oplus}{C}H}}-CO-OC_2H_5 \xrightarrow[\text{2. H}_2\text{O}]{\text{1. R}-CO-R'} \left[\begin{array}{c}R\diagdown_{\phantom{C}}\diagup CO-OC_2H_5\\R\diagup C=C\diagdown NH-CHO\end{array}\right]$$

$$\xrightarrow[\text{2. H}_2\text{O/OH}^\ominus]{\text{1. C}_6\text{H}_5-\text{CH}_2-\text{SH}} \begin{array}{c}\phantom{R-}CH_2-C_6H_5\\\phantom{R-}|\\R-\overset{\phantom{|}}{\underset{|}{C}}-\overset{S}{\underset{|}{C}}\overset{NH-CHO}{\phantom{|}}\\\phantom{R-}R' \ H \phantom{-CO-OH}\end{array} \begin{array}{c}\\\\-CO-OH\\\\\end{array}$$

## Addendum

*The formation of imidazole derivatives by base catalyzed cyclization of isonitriles and related reactions.* D. Marquarding and I. Ugi, unpublished results (1969).

On treatment with sodamide, the carbanion of benzyl isocyanide (1) is formed, which undergoes addition and subsequent ring closure with benzyl isocyanide and also with other isonitriles, like *t*-butyl isocyanide. The ratio of **2** and **3** depends upon the reaction conditions and the relative amounts of the reactants. The benzyl isocyanide carbanion reacts with diethyl oxalate to form 4-phenyl-5-ethoxycarbonyl-oxazole (**4**).

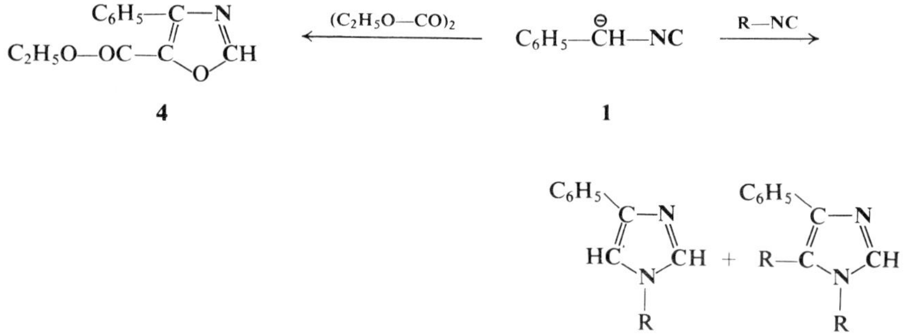

$$C_6H_5-\overset{C-N}{\underset{C_2H_5O-OC-C\diagdown_O\diagup CH}{}} \xleftarrow{(C_2H_5O-CO)_2} C_6H_5-\overset{\ominus}{C}H-NC \xrightarrow{R-NC}$$

     **4**           **1**

       $C_6H_5$      $C_6H_5$
       imidazole    imidazole
       **2**        **3**

## TO CHAPTER 6

*Isonitrile-trihaloboranes and hexahalo-2,5-diboradihydropyrazines.* A. Meller and H. Batka, *Monatsh. Chem.* **100**, 1823 (1969).

Hexahalo-2,5-diboradihydropyrazines resulting from the reaction of $BCl_3$ or $BBr_3$ with aceto- or propionitrile are believed to be formed from intermediate isocyanide-trihaloborane adducts. (see Chapter 6, Section I, A).

## TO CHAPTER 8

*Notice on the Ugi reaction with hydrazines.* I. G. Zinner and W. Kliegel, *Arch. Pharm.* (*Weinheim*) **299**, 746 (1966).

Hydrazine **1** and its alkyl derivatives **2**, react with cyclohexyl isocyanide and aldehydes in acidic solution to give cyclohexyl carbamoylalkyl hydrazines (**3**).

$R_2N-NRH + c\text{-}C_6H_{11}-NC + R'-CHO \longrightarrow$

           $R_2N-NR-CHR'-CO-NH\text{-}c\text{-}C_6H_{11}$
                 **3**

     **1**: R = H
     **2**: R = Alkyl

Addendum 253

In the case of hydrazine, the product may rearrange to give derivatives of 1,3,6,8-tetraazobicyclo[4.4.0] decane (4). 1,1,2-trimethyl-2-cyclohexylcarbamoyl methyl hydrazine formed in 87% yield from trimethyl hydrazine, formaldehyde, and cyclohexyl isocyanide in aqueous methanolic HCl.

*Hydroxylamine derivatives. XXXVII. Hydroxylamines in the Ugi four-component condensation.* G. Zinner, D. Moderhack, and W. Kliegel, *Chem. Ber.* **102**, 2536 (1969).

Methoxymethylamine reacts with aldehydes and cyclohexyl isocyanate to give derivatives like 1. In a similar fashion, monosubstituted hydroxylamines give products of type 2 and 3. Hydroxylamine itself gave products resulting from a double Ugi four-component condensation reaction 4.

$$CH_3-N(CH_3)-CHR^1-CO-NH-R^2$$
1

$$HO-NR^3-CHR^4-CO-NH-R^5$$
2

$$R^6O-NH-CHR^7-CO-NHR^8$$
3

$$HO-N[CH_2-CO-NH-N(R^8)_2]_2$$
4

*Reactions of diphenylphosphinylmethylisocyanide.* N. Kreutzkamp, unpublished results (1969).

(a) Diphenylphosphinyl methyl isocyanide (1) undergoes a 4 C C with $N_\alpha$-cyclohexyl-$N_\beta$-benzylhydrazone (2) and hydrazoic acid to give a tetrazole derivative (3) (see Chapter 8, Section X).

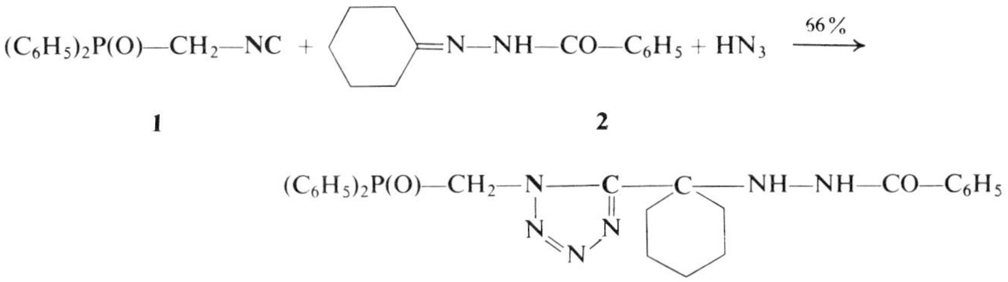

$$(C_6H_5)_2P(O)-CH_2-NC + \underset{2}{\text{C}_6H_{10}}=N-NH-CO-C_6H_5 + HN_3 \xrightarrow{66\%}$$
1

$$(C_6H_5)_2P(O)-CH_2-N\underset{N\diagdown N}{\overset{C}{\diagup}}C-C_6H_{10}-NH-NH-CO-C_6H_5$$
3

(b) Isocyanide 1 condenses with isobutyraldehyde and ammonium acetate to afford 66% of valine derivative 4.

1 + $i$-C$_3$H$_7$—CHO + CH$_3$COONH$_4$ ⟶

$$(C_6H_5)_2P(O)-CH_2-NH-CO-\underset{i\text{-}C_3H_7}{CH}-NH-CO-CH_3$$
4

(c) Isobutyraldehyde, acetic acid and **1** react to give 42% of **5**.

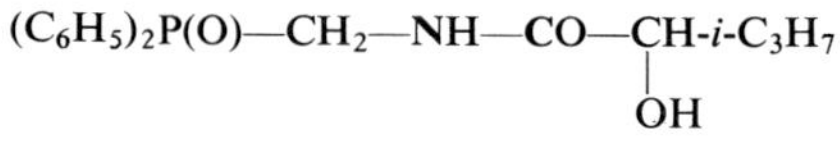

$$(C_6H_5)_2P(O)-CH_2-NH-CO-\underset{\underset{OH}{|}}{CH}\text{-}i\text{-}C_3H_7$$

**5**

## TO CHAPTER 10

*Revised structure for Chugaev's salt* $(PtC_8H_{15}N_6)_xCl_x$. G. Rouschias and B. Shaw, *Chem. Commun.* p. 183 (1970).

Chemical reactivity and NMR data lead to a revised structural formulation for Chugaev's salt as an isocyanide complex.

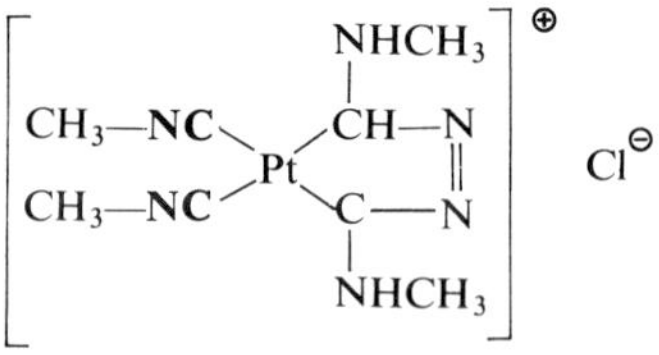

$$\left[\begin{array}{c} \phantom{XX}NHCH_3 \\ CH_3-NC \diagdown \phantom{X} \diagup CH-N \\ \phantom{XXX}Pt \phantom{XXX} \| \\ CH_3-NC \diagup \phantom{X} \diagdown C-N \\ \phantom{XX}NHCH_3 \end{array}\right]^{\oplus} Cl^{\ominus}$$

*Vibrational spectra of some isonitrile complexes of cobalt* (I) *and* (II). P. Boorman, P. Craig, and T. Swaddle, *Can. J. Chem.* **48**, 838 (1970).

Low-frequency infrared and Raman spectra have been measured for a number of cobalt(1) and (II) complexes like $Co(RNC)_5ClO_4$. A number of assignments and structural formulations are made.

*Copper(II) alkyl isocyanide complexes.* R. W. Stephany and W. Drenth, *Rec. Trav. Chim. Pays-Bas* **89**, 305 (1970).

Addition of a tertiary alkyl isocyanide to a solution of copper(II) tetrafluoroborate or perchlorate in polar solvents like water, ethanol, or ethanol/ether led to the separation of a purple solid apparently having the composition $Cu(CN-R)_4(H_2O)_2X_2$. IR, ESR, and some X-ray data are provided.

$$t\text{-}C_4H_9-NC + Cu(II) \longrightarrow Cu(CN-R)_4(H_2O)_2X_2$$

## New publications in the field of isonitrile–metal complexes

*Isocyanide complexes of iron (II).* F. Bonati, S. Cenini, and R. Ugo, New Aspects in the Chemistry of Metal Carbonyl Derivatives, International Symposium Proceedings, see *Chem. Abstr.* **72**, 12855g.

*Sign of the electric field gradient at the iron-57 nucleus in trans- and cis-$FeCl_2(p\text{-}CH_3O\text{—}C_6H_4\text{—}NC)_4$.* G. Bancroft, R. Garrod, A. Maddock, M. Mays, and B. Pratter, *Chem. Commun.* p. 200 (1970).

*Pentacoordination: electronic structure of perchloropentakis-(methyl isocyanide)cobalt(I).* Y. Dartiguenave, M. Dartiguenave, and Harry B. Gray, *Bull. Soc. Chim. Fr.* p. 4225 (1969).

*Reactions of tetrakis(alkyl isocyanide) nickel(0) complexes with triphenylphosphine and triphenylarsine.* R. Nast and H. Schulz, *Chem. Ber.* **103**, 785 (1970).

*New isocyanide complexes of iron(II).* F. Bonati and G. Minghetti, *J. Organomet. Chem.* **22**, 195 (1970).

*Nuclear magnetic resonance studies of rates of electron exchange between isonitrile complexes of manganese(I) and (II).* D. S. Matteson and R. Bailey, *J. Amer. Chem. Soc.* **91**, 1975 (1969).

*Preparation and reactions of some new isocyanide complexes of iron(II).* M. Mays and B. Prater, *J. Chem. Soc. (A)* p. 2525 (1969).

*Crystal structure of $K_2[ReN(NC)_4] \cdot H_2O$.* W. O. Davies, N. P. Johnson, P. M. Johnson, and A. J. Graham, *Chem. Commun.* p. 736 (1969).

*Stable adducts of copper(I) acetylacetonate.* R. Nast and W. Lepel, *Chem. Ber.* **102**, 3224 (1969).

*Reactions of isocyanides with carbon tetrahalides in the presence of dicobalt octacarbonyl.* Y. Yamamoto and N. Hagihara, *Bull. Chem. Soc. Jap.* **42**, 2077 (1969).

*Palladium and platinum complexes resulting from the addition of hydrazine to coordinated isocyanide.* A. Burke, A. L. Balch, and J. H. Enemark, *J. Amer. Chem. Soc.* **92**, 2555 (1970).

*Insertion reactions involving isocyanide ligands in platinum alkyl and aryl complexes.* P. M. Treichel and R. W. Hess, *J. Amer. Chem. Soc.* **92**, 4731 (1970).

*3,7,11-Cyclotridecatrien-1-one and 11-vinyl-3,7-cycloundecadien-1-one. (Isocyanide nickel organic compounds).* H. Breil and G. Wilke, *Angew. Chem. Int. Ed.* **9**, 367 (1970).

*Organic syntheses by means of noble metal compounds. XLI. Reactions of isocyanide with π-allylpalladium chloride.* T. Kajimoto, H. Takahashi, and J. Tsuji, *J. Organomet. Chem.* **23**, 275 (1970).

## Recent patents in the isonitrile field

Japan, 69 01,223 to N. Hagiwara, I. Yamamoto, and S. Otsuka (Japan Synthetic Rubber Company). Polyisonitriles. Group VIII metal catalysts used to polymerize isonitriles; *Chem. Abstr.* **70**, 97417x.

U.S. 3,438,747 to W. Hertler (DuPont). Salts from decaborane (14) and sulfurldiimines. Benzyl and cyclohexyl isocyanides were co-polymerized with $B_{10}H_{12}S$ which was prepared from decarborane (14); *Chem. Abstr.* **71**, 12525d.

S. African 69 01,922 to I. Hammann, P. Hoffmann, G. Unterstenhoefer, H. Kleimann, D. Marquarding, K. Offermann, and I. Ugi (Farbenfabriken Bayer). Isocyanodiphenyl ethers and thioethers. The title compounds, prepared by dehydration of the corresponding formamides, were found to be powerful acaricides; *Chem. Abstr.* **72**, 111020n.

German 1,943,415 to H. L. Yale (E. R. Squibb and Sons, Inc.). Bacteriostatic and fungistatic *o*-(benzyloxy)- and *o*-(benzylthio)phenyl isocyanides. The preparation of the compounds named above is given and they are claimed to have bacteriostatic and fungistatic properties; *Chem. Abstr.* **72**, 111113v.

French 2,002,568 to Farbenfabriken Bayer. Isonitriles. Procedures for the preparation of a large number of isonitriles particularly by acylation with phosgene, phosphorus oxychloride, sulfonyl chlorides, etc.; *Chem. Abstr.* **72**, 89826f.

# Author Index

Numbers in parentheses are reference numbers and indicate that an author's work is referred to although his name is not cited in the text. Numbers in italics show the page on which the complete reference is listed.

## A

Alagna, B., 74(27), *91*, 97(24), *107*
Allerhand, A., 5(39, 59, 60), *7*, 109, *129*, *131*
Anderson, G. W., *35*
Abraham, N. A., 27(1), *35*
Achenbach, H., *36*
Anderson, J. C., 202(1), *213*
Arlt, D., *35*
Aronovic, P. M., 17(4), 24(4), *35*
Asinger, F., 183(1), *197*
Aumann, R., 80(1), *90*, 226(1), *232*
Aumuller, W., 81(2), *90*

## B

Badger, R. M., 3, *6*
Badin, E. J., 163(2), *197*
Badley, E. M., 223(2), *232*
Bailey, R. A., 24(100, 101), *37*, 191(32), *198*
Bajusz, S., 202(2, 3), *213*
Bak, B., 4, 5(2), *6*
Baker, R. H., 133(2), 134(1, 3), 135(1, 2, 3), 136, *143*
Baiocchi, F., *107*
Bamford, C. H., 220(3), *232*
Banti, G., 18(132), 28(132), *38*
Bardamova, M. I., 19(158), *39*
Barrow, G. M., 5(3), *6*
Barsch, H., 186(6), *197*
Barton, M. A., 202(1), *213*
Bauer, S. H., 3, *6*, 58(12), 61(12), *64*
Beachell, H. C., 24(103), *38*
Beck, F., 22(161), 24(159), 25(159), *39*
Becker, E. I., 19(93), *37*
Behnisch, R., 186(55), *199*
Beichl, G. J., *35*
Bellman, R. E., 211(4), *214*
Bel'skij, N. K., 17(4), 24(4), *35*
Bentler, R., 140(13), *143*

Bergmann, M., 202, *214*
Bergs, H., 186, *197*
Bergstrom, F. W., 17(104), *38*
Berson, J. H., 23(152), 24(152), *39*
Berwin, H. J., 16(138), *38*
Bestmann, H. J., 10(6), 24(6), 25(6), 26(6), *35*
Betz, W., 9(8), 10(7), 19(7), 25(162), *35*, *39*, 158(57), 26(162), 27(162), 29(162), 88(56), *92*, *199*
Bhagwat, W. V., 3(4), *6*
Biddle, H. C., 18(9), *35*
Bigørgne, M., 24(10), *35*, 221(4), 222(4), *232*
Bither, T. A., 24(11), 26(11), *35*
Bittner, G., 111(2, 26), 114(26), *129*, *130*
Bodąnszky, M., 201(6), 202(6, 7), 203(6), *·*211(6), *214*
Bode, K. D., 76(60), *92*
Bodesheim, F., 15(169), *39*, 149(4), 150(4), 153(4, 75), 159(4, 63), 160(4, 63), 161(63), 196(64), 197(64), 80, *92*, *197*, *199*
Böttner, E., 186(65), 187(65), *199*
Bonati, F., 11(96), 19(96), *37*, 217(27), *232*
Bonciani, T., *107*
Booth, M. R., 24(12), *36*
Bose, S., 18(13), 25(13), *36*
Bossert, H., 202(14), *214*
Boulton, A. H., 83(5), *90*, 113(13), 115(13), 123(13), *129*
Bouquet, A., 24(10), *35*, 221(4), 222(4), *232*
Boyd, D. R., 161(5), *197*
Brackmann, W., 20(14), *36*
Brandt, W., 186(7), *198*
Bredereck, H., 14(15, 16, 17, 18), 24(15, 17), 26(17), 27(17), *36*, 29(15, 17), 74(3), *90*
Bremer, H., 202(57), *215*

Bresadola, S., 112(3, 4), 117, 123, *129*
Bricas, E., 203(8), *214*
Bright, J. H., *130*
Brockway, L. O., 3, *6*
Brois, S. J., 127, *129*
Brown, H. C., 126(8, 60), 127(7), 128(71), *129, 131*
Brügel, W., 17(19), *36*
Bruylants, P., 17(105), *38*
Bubnov, Y. N., 116, *129*
Bucherer, H. T., 186, *197, 198*
Burkhardt, G. N., *36*
Burkhardt, S., 205(18), *214*
Bussert, F., 31(21), *36*

**C**

Cahn, R. S., 162(11), *198*
Cairns, T. L., 69(4), 70(4), *90*
Callahan, F. M., *35*
Cappellina, F., 220(11), *232*
Carassiti, V., 222(5), *232*
Carraro, G., 112(3), 123(3), *129*
Carroll, H. F., 58(12), 61(12), *64*
Carter, J. C., 113(64), 117, *131*
Casanova, J., Jr., 15(22), 24(22, 23), 25 (23), 27(23), *36*, 42(1, 2), 43, 60, *64*, 83(5), *90*, 110(14), 111(14), 113(12, 13), 115, 117(11), 122(12), 123, *129, 130*
Casini, V., 75(32), *91*
Cetini, G., 222(6), *232*
Chan, S. C., 59(3), *64*
Chatt, J., 223(2), *232*
Cheng, C. C., *107*
Chien, S. W., 147(23, 24), *198*, 203(21), 205(21), *214*
Ching-Yun, Ch., *106*
Ciurlo, A., 17(33), 24(33), *36*
Coates, G. E., 125(15), *130*
Colwell, I. E., *35*
Condorelli, G., 222(5), *232*
Condorelli-Costanzo, L. L., 222(5), *232*
Corey, E. J., 15(62), *37*, 181(52), *199*
Costain, C. C., 3(6), *6*
Costopanagiotis, A., 202(58), *215*
Cotton, F. A., 18(24), 24(24), 25(24), 27 (24), *36*, 109(16), *130*, 218(8, 10), 220(8, 10), 221(9, 10), 222(7, 10), *232*

Crabtree, E. V., 24(25), 25(25), 27(25), *36, 38*
Craig, A. D., 25(26), *36*
Cresswell, I., 5(68), *7*
Cruse, R., 162(12), *198*
Cuvaeva, T. P., 205(46), *215*

**D**

Dadieu, A., 3, *6*, 17(27), *36*
Dahlmans, J., 202(58), *215*
Dallacker, F., 15(92), 24(92), 25(92), 27 (92), 30(92), 31(92), *37*
Damrauer, R., 19(150), 24(150), *39*
Danto, W., 202(58), *215*
Daughhetes, P. H., Jr., 79(14), *90*
Daumiller, G., 17(19), *36*
Davis, T. L., 18(28), 24(28), *36*, 68(6), *90*
Denkewalter, R. G., 202, *214*
Dessoulavy, E., *106*
Detscheff, T., 22(173), 25(173), *39*
Dewar, M. J. S., 136, *143*
Dietl, A., 72(55), *92*
Domann, H., 231(29), *232*
Dorokhov, V. A., 120, *130*
Dreyfus, H., 19(29), 24(29), 25(29), 27 (29), *36*
Dunath, D., 146(16), 157(16), 182(16), 185(16), 194(16), *198*
Duncan, J. L., 4, *6*
Dunne, T. G., 218(8), 220(8), *232*

**E**

Eastmond, G. C., 220(3), *232*
Eckmann, M., 18(66, 67), *37*
Eholzer, U., 11(164), 12(164), 14(164), 15(47, 49, 50, 51), 20(89), 24(164), 25(164), 26(164), 27(164), 28(50), 29 (49, 50, 164), 30(164), 31(164), 32 (164), 33(164), *36, 37, 39*, 103(8), *106*, 137(7), 141(6), *143*, 204(49), *215*
Eliel, E. L., 161(13), *198*
Emsley, J. W., 5(9), *6*
Engels, E., 202(58), *215*
England, D. C., 101(20), *106*, 135(12), *143*
Etling, H., 15(52, 53), 28(50, 52, 54), 29 (50, 52, 54), *36*, 103(8), *106*

## Author Index 259

Euglin, M. A., 20(94), *37*
Eyring, H., 49(5), *64,* 109(14), *198*

### F

Fabbri, G., 220(11), *232*
Farrar, W. V., 23(30), 25(30), *36*
Farrow, M. D., *36*
Feeney, J., 5(9), *6*
Ferstandig, L. L., 5(10), *6,* 109, *130*
Fetzer, U., 9(168), 11(162, 164), 12(164), 14(164), 18(168), 22(161, 163), 24(164), 25(162, 164, 168), 26(162, 164), 27 (162, 164), 28(164), 29(162, 164), 30(164), 31(164), 32(164), 33 (164), 76(58), 81(59), *92,* 134(5), 135 (5), 136(5), *143,* 145(67), *199*
Feuer, H., 19(32), 24(32), 26(32), 27(32), *36,* 79(7), *90*
Filatov, A. S., 97(17), *106*
Fischbeck, H., 186(8), *198*
Fischer, E., 202, *214*
Fischer, E. O., 80(1), *90,* 218(12), 224 (13), 226(1), *232*
Fischer, H., 218(12), *232*
Fletcher, F. J., 58(4), 59(4), *64*
Föhlisch, B., 14(15, 16, 17, 18), 24(15, 17), 26(17), 27(17), 29(15, 17), *36,* 74 (3), *90*
Francesconi, L., 17(33), 24(33), *36*
Frankiss, S. G., 24(12), *36*
Franklin, J. L., 4(11), *6*
Freund, M., *232*
Fridkin, M., 202(10), *214*
Fujii, H., 93(29a, b), 95(29a), 106(29b), *107,* 137(28), *143*
Fukuda, K., 202(20), *214*

### G

Gambaryan, N. P., 95(1, 3), 101(2, 3, *106,* 145(15), *198*
Gambino, O., 222(6), *232*
Gautier, A., 1, 2, *6,* 9, 10, 17, 24(35, 37, 39, 40), *36,* 79(8), *90*
Geisel, M., 42(1), *64*
Gerhart, F., 83(50), *91,* 104, *106, 107*
Ghosh, S. N., 4(21), *6*
Gilderson, P. W., 41(19), 42(19), 43(19), 47(19), 54(19), *64*

Gillis, R. G., 5(22), 6(23), *6*
Gilman, H., 81(9), *90*
Ginsburg, D., 18(146), 19(146), 25(146), 26(146), *38*
Ginsburg, V. A., 97(17), *106*
Glasstone, S., 3(25), *6,* 49(5), *64*
Glenz, K., 27(143), *38*
Gloede, J., 146(16), 157(16), 182(16), 185 (16), 194(16), *198*
Glushenkova, V. R., 68(21), *91*
Goerdeler, J., 103(5), *106*
Gokel, G., 206(11), 211(11), *214*
Golovaneva, A. F., 97(17), *106*
Gompper, R., 17(42), *36*
Gordy, W., 3, 5, *6, 7*
Grandmougin, E., *106*
Graybill, B. M., 124, *130*
Green, M. L. H., 125(15), *130*
Greenstein, J. P., 203(12), *214*
Gregory, R. A., 202(1), *213*
Griesinger, A., 196(43), *198*
Griffith, W. P., 17(43), *36*
Grigat, E., 21(44), *36*
Griseback, H., *36*
Gross, H., 146(16), 157(16), 182(16), 185 (16), 194(16), *198*
Grundmann, C., 100, *106*
Guillemard, M. H., 17(46), 18(46), 24 (46), 25(46), 27(46), *36, 64*
Gulden, W., 102(10), *106,* 111(27), 117 (28), 118(28), 119(27), *130*
Gutte, B., 202(13), 203(13), *214*
Guttmann, S., 202(14), *214*

### H

Hagedorn, I., 15(47, 48, 49, 50, 52, 53), 24(57), 28(50, 52, 54), 29(47, 49, 50, 52, 54), *36, 37,* 73(10), 74(10), 80 (10), *90,* 103(8), 105(9), *106,* 137, 141(6), *143*
Hagemann, H., 35
Hagihara, N., 79(54), 225(40), 226(39), *233*
Hajela, N., 27(1), *35*
Halleux, A., 80(11), *90*
Hammick, D. L., 3, *6*
Hansen-Nygaard, L., 4(2), 5(2), *6*
Hardy, P. M., 202(1), *213*
Hargreaves, K., 220(3), *232*

Harris, M. M., 161(58), *199*
Hartley, E. G., 18(58, 59), *37*, 231(15), *232*
Hata, T., 21(111), 25(111), *38*
Hauser, W., 18(66, 67), *37*
Haussleiter, H., 113(25), *130*
Havlik, A., 78(12), *91*
Hawthorne, M. F., 124, *130*
Heckert, L. C., 81(9), *90*
Heinle, R., 109(42), *130*
Heitzer, H., 206(27), *214*
Heldt, W. Z., 18(60), 24(60), 29(60), *37*, 227(19, 21), 228(17), 229(18, 20), 231(16, 21), *232*
Hellmann, H., 222(35), *233*
Herlinger, H., 156(71), 162(17,73), 163 (17, 18), 164(71, 72, 73), 165(73), *198, 199*
Hertler, W. R., 15(62), 28(62), *37, 130*
Herzberg, G., 4(30), *7*, 17(63), *37*
Hesse, G., 102, *106*, 110(22), 111(2, 22, 24, 26, 27), 112(24), 113(25), 114(2), 117, 118, 119(27), 123, 124(23), *129, 130*
Hesse, H., 189(36), *198*
Hieber, W., 222(22), *232*
Higashino, T., 22(64), 26(64), *37*
Hillman, M. E. D., 126, *130*
Hine, J., 15(65), *37*
Hine, M., 15(65), *37*
Hirota, K., 67(37, 38, 39, 40, 42, 45), 68 (39, 40, 45), 69(41), 70(45), 71(38), *90*, 223(34), *233*
Hirschmann, R., 202, *214*
Hölzl, F., 18(66, 67, 68, 69), *37*
Hoffmann, P., 9(8), 10(70), 13(71), 14(71), *35, 37*, 145(19), 158(20), 163(20), 164(30), *198*, 204(16, 28), 205(16), 206(11, 16, 27), 209(29), 210(27), 211(11, 51), *214, 215*
Hoffmann, R., 42(28), *64*, 116, *130*
Hofmann, A. W., 1, *7*, 9, 18(72, 73, 74, 76), 19(76), 24(74), 25(73), 27(73), *37*
Holliday, R. E., 42(10), *64*
Hollis, D. P., 5(58), *7*
Holly, F. W., 202(15), *214*
Hooz, J., 129, *130*
Hopf, H., 17(107), *38*
Horn, E. M., 125(37), *130*
Horobin, R. W., *37*

Horrocks, W. D., 5, *7,* 221(24), 222(23), *232*
Hoy, D. J., 16(79), 24(25), 25(25, 79), 27 (25, 79), *36, 37*
Humbel, R. E., 202(23), *214*
Hutley, B. G., *37*
Hyatt, D. E., 124, *130*

**I**

Inatome, M., 127, *130*
Ingold, C. K., 15(78), 24(78), *36, 37,* 42 (7), *64,* 162(11), *198*
Inoue, K., 21(114), 25(114), *38*
Insole, J. M., 42(11), *64*
Ito, Y., 31(144), *38,* 67(36, 37, 38, 39, 40, 42, 45), 68(36, 39, 40, 45), 69(41), 70 (45), 71(38), 72(26), 75(35), 77(13), 78(26, 44), 79(52), 81(43), 85(47, 49), 86(46), 87(52), 89(48), *91,* 223(34), *233*

**J**

Jackson, H. L., 18(80), 24(80), *37*
Jacob, T. A., 202(15), *214*
Jacobson, P., 19(108), *38*
Jaeger, E., 202(56), 203(56), *215*
Jahnke, D., 124, *131*
Jakubke, H. D., 205(17, 18), *214*
Jaquenoud, P. A., 202(14), *214*
Jeffery, G. H., 5(68), *7*
Jennings, J. R., 120, 124(40), *130*
Johnson, H. W., Jr., 79(14), *90*
Jones, K., 83(15), *90*
Joop, N., 96(13), *106*
Joshi, K. K., 219(25), 220(25), *232*
Jungermann, E., 19(81), 31(81), *37*

**K**

Kabbe, H. J., 89(16), *90,* 93(11), 94(11), 95(11, 12), 96(12, 13), *106*
Kabitzke, K. H., 19(148), *39*
Kästner, P., 14(110), *38,* 140(13), *143*
Kagen, H., 23(82), *37,* 137, *143*
Kahlen, N., 24(151), *39*
Kalenda, N. W., 18(157), 26(157), 27 (157), *39*
Kassel, L. S., 44(8), 51(8), *64*

Katchalski, E., 202(10), *214*
Katsoyannis, P. G., 202(19, 20), *214*
Kaufhold, G., 148(66), 164(66), 169(66), 170(66), 171(66), 179(66), 180(66), *199,* 206(50), 211(51), *215*
Kaufler, F., 17(83), 24(83), *37*
Keitel, I., 146(16), 157(16), 182(16), 185 (16), 194(16), *198*
Kelso, A. G., *37*
Kelvin, Lord, 161(21), *198*
Kemp, D. S., 147(22, 23, 24), 149(24), *198,* 203(21), 205(21), *214*
Kemp, M. K., 5(36), *7*
Kenner, G. W., 202(1), *213*
Kessler, M., 3(37), *7*
Khan, N. R., *37*
Kiefer, H. R., 83(5), *90,* 113(12, 13), 115 (13), 122(12), 123(13), *129*
Kimura, M., 79(52), 87(52), *91*
King, R. B., 227(26), *232*
Kinoshita, H., 75(35), *91*
Klabunowski, J. J., 161(25), *198*
Klages, F., 17(85), *37,* 109, *130*
Kleigel, W., 192(85), *199*
Kleimann, H., 162(17), 163(17, 18), *198,* 206(22), 208(22), 209(22), 211(51), *214, 215*
Klostermeyer, H., 202(23), *214*
Klusacek, H., 206(11), 211(11), *214*
Knorr, R., 88(17), 99, *90*
Knoth, W. H., 24(11), 26(11), *35*
Knunyants, I. L., 95(3), 101(3), *106*
Knupfer, H., 11(164), 12(164), 14(164), 24(164), 25(164), 26(164), 27(164), 30(164), 31(164), 32(164), 33(164), *39,* 204(49), *215*
Kobayashi, S., 31(144), *38,* 67(36, 37, 39, 40, 42, 45), 68(36, 39, 40, 45), 69 (41), 70(45), 71(38), 85(47, 49), 86 (46), *90,* 223(34), *233*
Kodaira, Y., 19(112), 24(112), *38*
Kohlmaier, G., 41(9), 43(9), 59(9), *64*
Koller, G., 188(26), *198*
Kornfeld, E. C., 188(82), *199*
Korte, F., 18(136), 15(136), 19(136), *38*
Krakora, J., 18(68), *37*
Krapcho, A. P., 19(86), 25(86), 27(86), *37*
Krespan, C. G., 102(20), *106,* 135(12), *143*
Kreutzkamp, N., 32(87), 33(87), *37,* 141 (9), 142(9), *143,* 149(27), 191(27), 193(27), *198*
Krishnamachari, S. L. N. G., 4(38), *7*
Kühle, E., 20(97), *37*
Kühlein, K., 83(25), *91*
Kuhn, L. P., 127, *130*
Kuivila, H. G., 120, *130*
Kuntz, I. D., Jr., 5(39), *7*
Kuwada, D., 83(5), *90,* 113(3), 115(13), 123(13), *129*

**L**

Lacey, A. B., *37*
Laidler, K. J., 49(5), *64*
LaMar, G. N., 221(24), *232*
Lämmerhirt, K., 32(87), 33(87), *37,* 141 (9), 142(9), *143,* 149(27), 191(27), 193(27), *198*
Lang, Z., 202(2, 3), *213*
Langmuir, I., 3, *7*
Lappert, M. F., 83(15), *90,* 116, 120, *130*
Lapworth, A., *36*
Larchar, A. W., 69(4), 70(4), *90*
Lautenschläger, F., 15(88), 28(88), 29(88), *37*
Lechner, F., 3, 4(4), *7*
Lee, L. T. C., 127(67), *131*
Leicester, J., 5(68), *7*
Lengyel, I., 23(153, 154), 24(153, 154), *39,* 93(32), *107*
Levine, S. D., 202(7), *214*
Lewis, D., 3(25), *6*
Lewis, E. S., 42(10, 11), *64*
Ley, K., 20(89), *37,* 99(15, 16), *106*
Lieb, V. A., 186(9), *198*
Lieke, W., 1, *7,* 17, 24(90), *37*
Lienert, J., 10(6), 24(6), 25(6), 26(6) *35*
Lifshitz, A., 58, 61(12), *64*
Lillien, I., 23(82), *37,* 137, *143*
Lindemann, H., 3, *7,* 18(91), 24(91), 25 (91), 27(91), *37*
Lindsey, R. V., 24(11), 26(11), *35*
Linke, S., 129, *130*
Linn, L. E., 134(1), 135(1), 136(1), *143*
Linnett, J. W., 4(44), *7, 39*
Lipinski, M., 15(169), *39*
Lipp, M., 15(92), 24(92), 26(92), 27(92), 28(92), 29(92), 30(92), 31(92), *37*
Lippincott, E. R., 5(45)

Lloyd, J. E., 124(40), *130*
Locker, D. J., 58(4), 59(4), *64*
Loevy, H., 81(34), *91*
Löw, M., 202(3), *213*
Long, L. M., 155, *198*
Lorenz, D. H., 19(93), *37*
Losco, G., 75(18), *90*
Lowenstein, A., 5(46), *7*
Lübke, K., 201(43), 202(43), 203(43), *215*
Lüdke, G., 206(24), 207(25), *214*
Lüttringhaus, A., 15(51), *36*

**M**

Maasböl, A., 224(13), *232*
McBride, J. J., 24(103), 28(102), 32(102), *38*
McCrosky, C. R., 17(104), *38*
McCusker, P. A., 126(50), *130*
McDiarmid, A. G., 25(26), *36*
McFarland, J. W., 95(18), *106,* 134(11), 141(11), *143,* 190(33), 191, *198*
McFarlane, M. W., 5(47), *7*
McKenna, J., *37*
McKenzie, A., 161(34), *198*
McKervey, M. H., 161(5), *197*
McKusick, B. C., 18(80), 24(80), *37,* 69(4), 70(4), *90*
MacLeod, J. K., 202(1), *213*
Majumdar, M. K., 116, *130*
Makarov, S. P., 20(94), *37,* 97(17), *106*
Malatesta, L., 11(96), 18(95), 19(96), 24(95), *37,* 25(95), 27(95), 217(27), *232*
Malone, L. J., 126(48, 49) *130*
Maloney, K. M., 43(14), 44(14), 45, 47(14), 48(14), 45(14), 53(14), 54(13), 55(13), 56(14), 57(14), 58(13), *64*
Malz, H., 20(97), *37*
Manley, M. R., 126(49), *130*
Mann, R. H., 5, *7*
Manuel, T. A., 19(98), *37,* 230(28), *232* (28), *232*
Marcus, R. A., 47(15, 16), *64*
Margalit, Y., 5 (46), *7*
Marglin, A., 202(26), *214*
Marquarding, D., 13(71), 14(71), *37,* 138(10), *143,* 145(19), 158(20, 29), 162(73), 163(20), 164(30, 73), 165(73), *198, 199,* 204(16, 28), 205(16), 206(11, 16, 27), 209(29), 210(27, 30), 211 (11, 51), *214, 215*
Martin, D., 19(99), 24(99), *37,* 79(19), *90*
Martynova, G. A., 97(17), *106,* 205(46), *215*
Mathieu, J., 161(31), *198*
Matsumura, Y., 87(53), *91*
Matteson, D. S., 24(100, 101), *37* 191 (32), *198*
Mayer, H., *199*
Medzihradszky, K., 202(2, 3), *213*
Meienhofer, J., 202(57), *215*
Meier, I., 15(92), 24(92), 26(92), 27(92), 30(92), 31(92), *37*
Meier-Mohar, T., 18(69), *37*
Meier zu Köcker, T., 15(92), 24(92), 26(92), 27(92), 28(92), 29(92), 30(92), *37*
Menzie, G. K., 71(33), *91*
Merchx, R., 17(105), *38*
Merrifield, R. B., 202(13, 26), 203(13, 31), *214*
Merz, W., 95(25), *107,* 190(45), 195, *198*
Meyer, E., 1, *7, 17, 38*
Meyer, J., 231(29), *232*
Meyer, K. H., 17(107), *38*
Meyer, R., 5, *7,* 9(168), 10(166), 15(165, 169), 18(168), 24(165, 166), 25(166), 26(169), 27(165, 166, 167), 28(166), 29(165, 166), 30(166), *39,* 97(35), *107,* 138(29), 140(29), *143,* 145(67), *198*
Meyer, V., 19(108), *38*
Mickhailov, B. M., 17(4), 24(4), *35,* 116, *129*
Middleton, W. J., 95(19), 101(20), *106,* 135(12), *143*
Miller, C. A., 155(28), *198*
Miller, J. G., *35*
Miller, N. E., 119(51), *130*
Millich, F., 68, *90*
Mills, O. S., 219(25), 220(25), *232*
Misfaludy, L., 202(3), *213*
Mislow, K., 161(35), *198*
Mitin, J. V., 68(21), *91*
Mönkemeyer, K., 17(85), *37,* 109(42), *130*
Mori, K., 18(128), 24(128), 25(128), 26(128), *38,* 28(128), 29(128), 227(30), *232*
Morley, J. S., 202(32), *214*
Morris, R. N., 42(1), *64*
Mott, L., 10(6), 24(6), 25(6), 26(6), *35*

Author Index   263

Müller, E., 14(110), 20(109), 25(109), 28 (109), 30(109), 33(109), *38*, 95(21), 100(21), *106*, 140(13), 141(14, 31), *143*
Mueller, W., 231(29), *232*
Muetterties, E. L., 120, *130*
Mukaiyama, T., 19(112, 113), 21(111, 114), 24(112, 113, 115), 25(111, 114, 115), 26(113), 27(113, 115), *38*
Mumm., O., 189(36), *198*
Musker, W. K., *130*

**N**

Nagarajan, G., 5(45), *7*
Nagasawa, K., 123, *130*
Nakamura, A., 79(28), *91*, 225(32), 231 (31), *232*
Nambu, H., 19(113), 24(113), 26(113), 27 (113), *38*
Narayan, V. K., 202(7), *214*
Narr, B., 20(109), 25(109), 28(109), 30 (109), 33(109), *38*
Nast, R., *106*
Nef, I. U., 2, *7*, 17(118), 19(119), 22(116, 117, 118), 24(121), 27(116), *38*, 69 (22), 72(22,23), 73(22), 76, 78(22), 79(23), 80(22), *91*
Neidlein, R., 24(122, 124), 25(124), 26 (123, 124), 27(123, 124), *38*, 74(24), *91*, 103, *106*, 135(16), 142(15), *143*, 157(39, 40), 182(39), 195, *198*
Nejmyseva, A. A., 20(135), *38*
Neri, A., *38*
Neumann, W. P., 83(25), *91*
New, R. G. A., 3(29), *6*, *7*, 25(126), 26 (126), *38*
Niedenzu, K., 122(54), *130*
Nielsen, A. T., 19(32), 24(32), 26(32), 27 (32), *36*, 79(7), *90*
Nikolajeva, T. V., 20(94), *37*
Nin, C., 202(33), *214*
Nishida, K., 79(52), 87(52), *91*
Nutt, R. F., 202(15), *214*

**O**

O'Brien, D. E., *107*
Occolowitz, J. L., 5(24), 6(23), *6*
Oda, R., 77(13), 78(26), 79(52), *38*, 72

(26), 87(52), *91*, 145(41), *198*
Offermann, K., 11(164), 12(164), 14(164), 24(164), 25(164), 26(162, 164), 27 (162, 164), 28(162, 164), 30(164), 31 (164), 32(164), 33(164), *39*, 88(56), *92*, 149(42), 150(42, 69), 155(42), 156(71), 162(17, 73), 163(17, 70), 164 (68, 71, 72, 73), 165(73), 183(42), *198*, *199*, 204(49), 207(52), 208(52), *215*
Ogg, R. A., Jr., 41(17), 43(17), *64*
Okamoto, M., 19(113), 24(113), 26(113), 27(113), *38*
Okano, M., 72(26), 77(13), 78(26), *91*
Okumura, Y., 67(45), 68(45), 70(45), *91*
Oliveri-Mandala, E., 74(27), *91*, 97(24), *107*
Olofson, R. A., 147(83, 84), *199*
Ondetti, M. A., 201(6), 202(6, 7), 203(6), 211(6), *214*
Opitz, G., 95(25), *107*, 190(45), 195, 196, *198*
Otsuka, S., 18(128), 24(128), 25(128), 26 (128), *38*, 28(128), 29(128), 79(28), *91*, 225(32), 227(30), 231(31), *232*
Owen, D. A., 124(38), *130*

**P**

Pacsu, E., 163(2), *197*
Paetzold, von P. I., 116, *130*
Palazzo, F. C., 17(134), 24(134), *38*
Parish, R. V., 221(9), *232*
Parsons, T. G., 16(138), *38*
Passerini, M., 18(129, 132), 28(132), 31 (129), *38*, 75(29, 30, 31, 32), *91*, 98 (26), 102(27), *107*, 133, 134(18, 22, 25), 135(17, 18, 21, 22), 139(27), 140 (23, 24), 141(19, 22, 25), 87(31), 104 (28), *143*
Pasto, D. J., 129, *130*
Pasynkiewicz, S., 124, *130*
Patchornik, A., 202(10), *214*
Pattison, I., 120(41), *130*, *131*
Paulay, Z., 202(2, 3), *213*
Pauling, L., 3, *6*, *7*
Pauson, P. L., 219(25), 220(25), *232*
Pavlou, S. P., 54(13), 55(13), 58(13), *64*
Pavlovskaya, 97(17), *106*
Pecile, C., 112(3), 123(3), *129*

Peratoner, A., 17(134), 24(134), *38*
Petrov, K. A., 20(135), *38*
Piette, L. H., 5(58), *7*
Pilgram, K., 15(136), 18(136), 19(136), *38*
Piquet, A., 21(174), 25(174), *39*
Placzek, D. W., 48(18), *64*
Pless, J., 202(14), *214*
Ploquin, J., 23(137), *38*
Pomeranz, C., 17(83), 24(83), *37*
Porosiu, K. T., 205(46), *215*
Powers, J. C., 16(138), *38*
Poziomek, E. J., 16(79), 24(25), 25(25, 79), 27(25, 79), *36, 37, 38*
Pracejus, H., 161(46), *198,* 207(34, 35, 36), *214*
Prelog, V., 161(47), 162(11), *198*
Preston, J., 202(1), *213*
Pritchard, H. O., 5, 7, 41(24, 25), 61(25), 84(51), *64, 91*
Pütter, R., 21(44), *36*
Puosi, G., 112(4), 117(4), *129*
Purcell, K. F., 5, 7

**R**

Rabinovitch, B. S., 41(9, 20, 22, 23), 42(19, 22, 23), 43(9, 14, 19, 22, 23), 44(14, 22), 45, 46(22), 47(14, 19, 22, 23), 48(14, 18, 22, 29), 49(14, 22), 50, 52(14, 22), 54(13, 19, 20, 23), 55 13, 23), 56(14), 57(14), 58(13, 22), 59(3, 4, 9), *64*
Ragnarsson, U., 204(54), *215*
Ragni, G., 139(27), *143*
Rathke, M. W., 126(8, 59, 60), *129, 131*
Ray, J. D., 5(58), *7*
Reichel, L., 17(140), *38*
Reinheckel, H., 124, *131*
Restrup-Andersen, J., 4(2), 5(2), *6*
Rice, O. K., 47(16), 52, *64*
Richards, R. L., 223(2), *232*
Ring, H., *7*
Robinson, E. B., *36*
Rommel, O., 17(19), *36*
Rokhlin, E. M., 95(3), 101(2, 3), *106*
Rosenblum, M., 209(37), *214*
Rosendahl, F. K., 15(169), 16(170), 25 (170), *39,* 105, *107,* 142(30), *143,* 153 (48, 75), 157(48, 74), *199*
Rosetto, F., 112(4), 117(4), *129*

Ross, D., 26(141), *38*
Rothe, W., 15, *38*
Rowland, F. S., 41(27), 61, 63(27), *64*
Rubinstein, H., 19(32), 24(32), 26(32), 27 (32), *36,* 79(7), *90*
Ruch, E., 161(50), 162(49, 50), 164(49), *199,* 206(11, 38, 39), 207 (39), 211 (11), *214,* 220(33), *233*
Rücker, D., 162(17, 51), 163(17, 18, 51), *198, 199*
Rundel, W., 14(110), *38,* 140(13), *143*
Rupe, H., 19(143), 27(143), *38*
Ruppersberg, H., 188(26), *198*
Ryan, J. W., 71(33), *91*
Rzepkowska, Z., 124(57), *130*

**S**

Sachs, F., 81(34), *91*
Saegusa, T., 31(144), *38,* 67(36, 38, 39, 40), 68(36, 39, 40, 45), 69(41), 70 (45), 71(38), 75(35), 78(44), 81(43), 85(47, 49), 86(46), 89, *91,* 93(29a, b), 95(29a), 106(29b), *107,* 131(144), 137, *143,* 223, *232*
Sakiyama, F., 14(145), 29(145), *38,* 204 (40), *214*
Samuel, D., 18(146), 19(146), 25(146), 26 (146), *38*
Sandria, E., 202(14), *214*
Sarges, R., 202(41), *214*
Sayigh, A. A. R., 27(147), *38,* 230(38), 232(38), *233*
Schaeffer, R., 119, 123, *131*
Scharf, R., 202(56), 203(56), *215*
Schenk, H., 103(5, 30), *106, 107*
Schlesinger, A. H., 133(2), 135(2), 136 (2), *143*
Schleyer, P. von R., 5(39, 59, 60), *7,* 109, *129, 131*
Schlögl, K., 210(42), *215*
Schmidt, G., 202(58), *215*
Schmidt, U., 19(148), *38*
Schnabel, E., 202(57), *215*
Schneider, F. W., 41(19, 22, 23), 42(19, 22, 23), 43(19, 22, 23), 44(22), 46 (22), 47(19, 22, 23), 48(22), 49(22), 50, 52(22), 54(19, 20, 23), 55(23), 58 (22), *64*
Schneidewind, W., 19(149), 27(149), *39*

## Author Index 265

Schöllkopf, U., 83(50), *91,* 104(4,31), *106, 107*
Schröder, E., 201(43), 202(43), 203(43), *215*
Schubert, H. W., 196(43), *198*
Schuster, R. E., 15(22), 24(22, 23), 25 (23), 27(23), *36,* 110(14), 111(14), 42 (2), 43(2), 60(2), *64, 130*
Schwyzer, R., 202(44, 45), *215*
Seidner, R., 16(138), *38*
Setser, D. W., 54(20), *64*
Seyferth, D., 19(150), 24(150, 151), *39*
Sharkey, W. A., 24(11), 26(11), *35*
Shaw, B. W., 219(25), 220(25), *232*
Shaw, D. H., 6, *7,* 41(24, 25), 61(25), 84 (51), *91*
Shchekotichin, A. I., 97(17), *106*
Sheehan, J. C., 23(152, 153, 154), 24(152, 153, 154), *39,* 93(32), *107,* 181(52), *199*
Sheppard, R. C., 202(1), *213*
Shimizu, T., 81(43), *91*
Shingaki, T., 18(155), *39*
Shono, T., *38,* 72(26), 78(26), 79(52), 87 (52, 53), *91,* 145(41), *198*
Shpansky, V. A., 97(17), *106*
Shukla, R. P.
Sibnev, V. A., 205(46), *215*
Sidgwick, N. V., 3(29), *6, 7*
Sieber, P., 202(45), *215*
Sim, G. A., 223(2), *232*
Simonyan, L. A., 95(3), 101(3), *106*
Sims, L. B., 43(30), 47(30), 55(30), *64*
Sinclair, R., 68, *90*
Sjöberg, K., 95(33), *107,* 145(53), 183(54), 185, *199*
Slater, N. B., 44, 51(26), *64*
Slotta, K., 186(55), *199*
Smit, P. J., 20(14), *36*
Smith, F. W., 19(81), 31(81), *37*
Smith, P. A. S., 20(156), 26(157), 27(157), *39*
Smyth, C. P., 3(63), *7*
Sonogashira, K., 79(54), *91*
Speier, J. L., 71(33), *91*
Spicer, L., 59(3), *64*
Spiesecke, H., 5, *7*
Stanonis, D., 134(3), 135(3), 136, *143*
Starowieski, K., 124(57), *130*
Steglich, W., 102, *107*

Steinbrückner, C., 9(168), 18(168), *39,* 145(76), 156(78), 157(78), 158(57), 182(56), 185(78), 192(56, 76), 193 (77), 195(77), *199,* 206(53), *215*
Steiner, W., 186(10), *198*
Stenson, J. P., 224(36), *233*
Stevens, R. R., *130*
Strang, E., 188(26), *198*
Strasser, O., 17(140), *38*
Streib, W., 115, *131*
Strohmeier, W., 222(35), *233*
Stubbs, W. H., 219(25), 220(25), *232*
Stutman, J. M., 5(45), *7*
Suhr, H., 140(13), *143*
Summerford, C., 120(41), *130*
Suttcliffe, L. H., 5(9), *6*
Sutton, L. E., 3(29, 55), *6, 7,* 25(126), 27 (126) *38*
Szwarc, M., 4(65), *7*
Szyszka, G., 186(55), *199*

### T

Taft, R. W., Jr., 60, *64*
Takahashi, S., 79(54), *91*
Taka-ishi, N., 89(48), *90,* 93(29a, b), 95 (29a), 106(29b), *107,* 137(28), *143*
Takebayashi, M., 18(155), *39*
Takeda, N., 67(42), *91*
Tanaka, J., 113(64), 117, *131*
Tatsuno, Y., 79(28), *91,* 231(31), *232*
Taylor, J. W., 4(65), *7*
Taylor, R. C., 221(24), 222(23), *232*
Teearu, P., 229(37), *233*
Thiel, M., 183(1), *197*
Thompson, H. W., 5, *7*
Tille, A., 207(36), *214*
Ting, C. T., 41(27), 61, 63(27), *64*
Todd, L. J., 119, 123, 124(38), *130, 131*
Tönjes, H., 15, *36, 37,* 105(9), *106*
Tometsko, A. M., 202(20), *214*
Tomita, S., 75(35), 78(44), *91*
Tonooka, K., 21(114), 25(114), *38*
Trambarulo, R., *7*
Treibs, A., 72(55), *92*
Treichel, P. M., 224(36), *233*
Tronov, B. V., 19(158), *39*
Troutman, H. D., 155(28), *198*
Tsai, C., 115, *131*
Tschugaeff, L., 229(37), *233*

Tschuikow-Roux, E., 48(18), *64*
Tucker, B., 230(38), 232(38), *233*
Tufariello, J. J., 127(66, 67), *131*
Turco, A., 112(3), 123(3), *129*
Turner, E. E., 161(58), *199*

## U

Ugi, I., 5, 7, 9(8), 10(166), 11(7, 162, 164), 12(164), 13(71), 14(71, 164), 15(165, 169), 16(170), 18(168), 19(7), 22(161, 163), 23(160), 24(159, 164, 165, 166, 168), 25(159, 162, 164, 166, 170), 26(159, 162, 164, 169) 27(159, 162, 164, 165, 166, 167, 169), 28 (159, 164, 166), 29(162, 164, 165, 166), 30(159), 31(159, 164, 166), 32 (159, 164), 33(159, 164), *35, 39*, 76, 80, 81(59), 88(56), *92*, 95(34), 97(35, 36), 105, *107*, 134(5), 135(5), 136 (5), 138(29), 140(29), 142(30), *143*, 145(19, 59, 60, 62, 67, 76), 148(60, 66), 150(69), 153(73), 156(71, 78), 157(74, 78), 158(20, 57), 159(63), 160(63), 161(50, 63), 162 (17, 49, 50, 59,.61, 62, 73), 163(20, 70), 164(30, 49, 66, 68, 71, 72, 73), 165(73), 169 (66), 170(66), 171(66), 179(66), 180 (66), 183(79), 185(78), 186(65), 187 (65), 192(76), 193(77), 195(77), 196 (64), 197(64), *198, 199*, 204(16, 28, 49), 205(16), 206(11, 16, 22, 24, 27, 38, 39, 47, 48, 50, 53), 207(25, 39, 48, 52), 208(22, 52), 209(22, 29), 210 27, 30), 211(11, 51), *214, 215*
Ulrich, H., 27(147), *38*, 230(38), 232(38), *233*
Urenovitch, J. V., 25(26), *36*

## V

Van Dine, G. W., 42(28), 43(28), *64*
Varga, S. L., 202(15), *214*
Veber, D. F., 202(15), *214*
Verhulst, J., 17(105), *38*
Videijko, A. F., 20(94), *37*
Viditz, F., 18(69), *37*
Vitali, R. A., 202(15), *214*
Vlasov, G. P., 68(21), *91*
Vogel, A. I., 5(68), *7*

Voigt, A., 205(17, 18), *214*
Volquartz, H., 189(36), *198*
von Saltza, M. H., 202(7), *214*

## W

Waage, E. V., 48(29) *64*
Wade, J., 17(171), 24(171), 25(171), *39*
Wade, K., 120(41), 125(5), *130, 131*
Waitkins, G., 17(104), *38*
Wald, M., 78(12), *91*
Walter, W., 76(60), *92*
Walz, K., 14(15, 16, 17, 18), 24(15, 17), 26(17), 27(17), 29(15, 17), *36*, 74(3), *90*
Wang, Y., 202(33), *214*
Ware, E., 186(80), *199*
Watkins, K. W., 58(4), 59(4), *64*
Weill-Raynal, J., *198*
Weinraub, B., 18(146), 19(146), 25(146), 26(146), *38*
Weise, A., 19(99), 24(99), *37,* 79(19), *90*
Weiss, E., 222(22), *232*
Weith, W., 19(172), *39,* 79(61), *92*
Wendleberger, G., 202(56), 203(56), *215*
Werner, A., 21(174), 22(173), 25(173, 174) *39*
Werner, N. D., 15(22), 24(22, 23), 25(23), 27(23), *36,* 42(2), 43(2), 60(2), *64*
Wettau, J. F., 43(30), 47(30), 55(30), *64*
Weygand, F., 102, *107,* 203(54), *215*
Whitfield, R. C., 19(175), *39*
Whitten, G. Z., 48(18), *64*
Wiegrebe, L., 3, *7,* 18(91), 24(91), 25(91), 27(91), *37*
Wilkinson, G., 109(16), *130*
Willems, H., 202(14), *214*
Williams, R. L., 5, *6, 7*
Winitz, M., 203(12), *214*
Winkelmann, H.-D., 24(57), *37,* 137(7), *143*
Winterfeldt, E., 98(39), *107*
Wischhöfer, E., 183(79, 81), *199*
Witkop, B., 14(145), 29(145), *38,* 202(41), 204(40), *214*
Witte, H., 102(10), *106,* 110(22), 111(2, 22, 24, 26, 27), 112(24), 113(25, 69), 114(2, 26), 116(69), 117(28), 118(28), 119(27), 123, 124(23), *129, 130, 131*
Wittmann, R., 10(176), 24(176), 26(176), 27(176), *39*

Wojtkowski, P., *131*
Wojtkowski, P. W., 129, *130*
Wood, J. S., 218(8), 220(8), *232*
Woodward, R. B., 147(24, 83, 84), 149(24), 188(82), *198, 199*
Wright, G. F., 15(88), 28(88), 29(88), *37*
Wünsch, E., 202(55, 56), 203, *215*
Wyatt, B. K., 120(41), *130*

**Y**

Yakubovich, A. Y., 97(17), *106*
Yamagami, K., 18(128), 24(128), 25(128), 26(128), 28(128), 29(128), *38*, 227(30), *232*
Yamamoto, Y., 225(40), 226(39), *233*
Yamazaki, H., 225(40), *233*
Yaroslavsky, S., 99, *107*
Yasuda, N., 31(144), *38*, 85(47, 49), *91*
Yelland, W. E., 18(28), 24(28), *36*, 68(6), *90*
Yoshida, Z., 24(115), 25(115), 27(115), *38,* 127, *131,* 225(32), *232*
Yoshika, H., 223(34), *233*
Yoshioka, H., 67(39, 40), 68(39, 40), 71(38), *91*

**Z**

Zahn, H., 202(57, 58), *215*
Zalut, C., 202(20), *214*
Zeek, B., 6(70), *7*, 95(21), 100(21, 41), *106, 107*, 140(13), 141(14, 31), *143*
Zeifman, Y. V., 95(3), 101(2, 3), *106*
Zervas, L., 202, *214*
Zimmermann, J. E., *35*
Zingales, F., 18(24), 24(24), 25(24), 27(24), *34*, 218(10), 220(10), 221(10), 222(10), *232*
Zinner, G., 192(85), *199*
Zobel, R., 202(57), *215*
Zweifel, G., 128(71), *131*

# Subject Index

## A

Acetylene dicarboxylate diesters, 243, 245
Acid chlorides, see α-Addition, halide
Acid component, see Four-component condensation, acid component
Activating group, see Peptide synthesis, activating group
Active methylene compounds, see α-Additions
α-Acylaminocarbonamide (table), 156–157
Acyl hydrazones, see Four-component condensation, acyl hydrazones
Acylimine, 101–102
Acyl isocyanate, 103, 142
Acyl isothiocyanate, 103
α-Acyloxycarbonamide, see Passerini reaction
α-Addition, 2, 65–90
    acetylenes, 75
    alcohols, 69–70
    carbenes, 80–81
    carboxylic acids, 73
        formic, 73
    catalysis by group IB and IIB element compounds, 67–71
    copper catalysis, 67–71, see also individual compounds
    dialkylphosphines, 85
    formic acid, 73
    halides, 76–78
        acid chlorides, 76
            α-ketoimidoyl chlorides, 76
            α-ketoamides, 76
            α-ketothiocarbonamides, 76
        amide chloride, 77
            α-ketoamide from, 77
        $N,N$-dialkylamide chloride, 76–77
        dimethylformamide chloride, 76
            α-dimethylaminomalonamide from, 77
        nitrogen–halogen compound, 77
            $N$-bromoamide, 77

oxygen halogen compounds, 77–78
    alkyl hypochlorite, 77
    $t$-butyl hypochlorite, 77–78
    phosgene, 76
sulfur halogen compounds, 78
    sulfonyl chloride, 78
    Vilsmeier reagent, 76
halogen, 78
hydrazoic acid, 74
hydrogen, see isonitriles, reduction
hydrogen halides, 73
hydroxylamine, 72–73
immonium ions and anions, see Four component condensation, immonium ion intermediate
mercaptans, see α-Addition, thiols
methylene compounds, CH acidic, 75
naphthols, 75
nitrenes, 80–81
    from azides, 81
    from chloramine-T, 81
nonmetal hydrides, 67–71
olefins, 75
organolithium compounds, 241
organometallic compounds, 81–84
    aminometallation, 83
    aminoplumbane derivatives, 83
    aminostannane derivatives, 83
    Grignard reagent, 81
oxygen, see Isonitriles, oxidation
Passerini reaction, 138
phenol, 75
phosphine, 68, 85
pyrrole derivatives, 72
silanes, 71
stannanes, 85
sulfur, 79
thiols, 70, 86
tropylium ion, 88–89
water, acid-catalyzed, 73
zinc chloride catalysis, 72
α-Adduct, Passerini reaction, see Passerini reaction

## Subject Index

### A

β-Alanine derivatives, cleavable, see Peptide synthesis, model reactions
Alcohols, see α-Addition, alcohols
Aldehyde, synthesis, 241
Alkali metals, see Isonitriles, reduction
Alkane thiols, see α-Addition, thiols
1-Alkyl-3-acyl-1,4-dihydroquinoline carbonamides, 186–188
Alkylation of cyanide complexes, 17–18
Alkylboranes, see Boranes, alkyl-
Alkyl hypochlorite, see α-Addition, oxygen halogen compounds
Aluminum
 organic compounds of, 124–125
 triphenyl-, 124
Aluminum azide, 139
Amide chlorides, see α-Addition, halides
Amidines, see Four-component condensation, amidines
Aminals, see Four-component condensation, aminals
Amine component, see Four-component condensation, amine component
Amino acid, see specific amino acids
β-Amino acids, see Four-component condensation, β-lactam
α-Aminoalkylation of isonitriles and acids, see Four-component condensation
α-Aminoalkyl azlactone, see Four-component condensation, α-aminoalkyl azlactones
α-Amino amidines, see Four-component condensation, amidines
α-Amino carbonamides, see Four-component condensation, α-amino carbonamides
Aminometallation, 83
α-Aminoselenoamides, see Four-component condensation, selenoamides
α-Aminothioamides, see Four-component condensation, thioamides
Aryl boranes, see Boranes, aryl-
N-Aryl-N', N'-diacyloxamides, 142
1-Azabenz[b]azulene, 248
Azides, see α-Addition, nitrenes
α-Azidoamines, see Four-component condensation, α-azidoamines
Azines, see Four-component condensation, azines

### B

Back donation, in mixed complexes, 221–222, see also Isonitrile complexes, back donation
Beckmann rearrangement, abnormal, 20–22
Benzene diazonium-2-carboxylate, 88, 99, see also Isonitriles, reaction with benzyne
Benzocoumarone derivatives, 98
Benzoyl peroxide, 87, see also Isonitriles, radical reactions of,
Benzyl isocyanide, base-catalyzed dimerization, 252
Benzyne, see Isonitriles, reaction with benzyne
Bis(diethyl sulfide)decaborane, 124
Bonding orbitals, irreducible representations, 220
Boranes,
 alkyl-, 110
 aryl-, 110
 polyhedral, 124
 tri-$n$-butyl-, 102, 110, 118–119
 triethyl-, 83, 110, 123
 trimethyl-, 110
 triphenyl-, 114
Boron halides, 109
Bucherer–Bergs reaction, 186
$n$-Butyl borane, see Borane, tri-$n$-butyl-
$t$-Butyl diphenylketenimine, 239, see also Cyclohexyldiphenylketenimine
$t$-Butyl hypochlorite, see α-Addition, oxygen halogen compounds
$t$-Butyl isocyanide
 dimer, 89
 trimer, 89
$t$-Butyl isocyanide dimerization, 246
Butyllithium, see α-Metallation of isonitriles

### C

Carbene complexes, catalytic intermediates, 223–224
Carbenes, 239, see also α-Addition, carbenes
Carbethoxynitrene, 239
Carbodiimide derivatives, 239

## Subject Index

Carbomethoxyphenylacetylene, 244
Carbonyl component, see Four-component condensation, carbonyl component
Carboxylic acids, 73, 133–140, 155–185, 189–191, see also α-Additions, carboxylic acids
Carbylamine reaction, see Hofmann's method
Catalysis, see α-Addition, catalysis by group IB and IIB element compounds
Catalytic hydrogenation, see Isonitriles, reduction
Chirality product, 161–162
Chloral, 94, 97, 140
Chloramine-T, 81
Cholest-4-en-3-one, 245
Chugaev's salt, revised structure, 254
Copper catalysis, see α-Additions, copper catalysis
Copper isonitrile complexes, see Isonitrile complexes, copper(II)
Corresponding reactions, see Four-component condensations, pairs of corresponding reactions
Curtin–Hammett principle, 169
Cyanates, synthesis of isonitriles from, 19
Cyanic acid, see Four-component condensations, hydantoin-4-imides
Cyanide complexes, alkylation of, see Alkylation of cyanide complexes
Cyano group
 charge distribution, 4
 electron density, 4
Cyclization reactions, 93–106
 dipolar intermediates, 103
 five-membered rings, 97–104
 four-membered rings, 95–97
  formation from chloral, 94
  from nitrosotrifluoromethane, 97
 secondary reactions, 96
Cyclization reactions
 diiminooxetanes, 95
 α-lactams, 93
 oxetanes, 93–95
 seven-membered rings, 106
 six-membered rings, 104–106
Cycloadditions, see Cyclization reactions
Cyclohexyldiphenylketenimine, 238, 239

see also t-Butyl diphenylketenimine
N-Cyclohexyliminodioxolane, 101
Cyclopentane trione, derivatives, 97
Cysteine derivatives, see also Four-component condensation, cysteine derivatives
 from ethyl isocyanoacetate, 251

## D

Decaborane-14, 124
Depsipeptides, see Passerini reaction
α-$N_\alpha$,$N_\beta$-Diacylhydrazinocarbonamides, 160, see also Four-component condensation, α-$N_\alpha$,$N_\beta$-diacylhydrazinocarbonamides
Diacyl imides, see Four-component condensation, diacyl imides
N,N-Dialkylamide chlorides, see α-Addition, halides
Dialkyl phosphines, see α-Addition, phosphines
Diazadiboretidine, 115–116
Diazadiborolidine, 115, 116
Diaziridones, 242
Diboradihydropyrazine derivatives, see Pyrazines
Dibroane, 110, 117
Diboradihydropyrazine, see Pyrazines
Dicarbomethoxyacetylene, 98
Dichloro carbene, carbylamine reaction intermediate, 19
Diene amines, see Four-component condensation, diene amines
Diethyl phosphine, see α-Addition, phosphine
Dihydro-1,3-oxazines, 249
2,3-Diiminooxetanes, see Cyclization reactions
α,γ-Diketoamides, see Passerini reaction, related reactions
Dimerization of t-butyl isocyanide, see t-Butyl isocyanide, dimerization
Dimethylacetylene dicarboxylate, see Dicarbomethoxyacetylene
α-Dimethylaminomalonamide, see α-Addition, halides
Dimethyl sulfoxide, see Isonitriles, oxidation
Diphenylcyclopropenone, 246, 247

Diphenyldiazomethane, 238, 239
Diphenyl ketene, 142
Diphenyl ketene p-tolyl imine, 23
Diphenylphosphinylmethyl isocyanide, four-component condensations of, 253
1,3-Dipolar additions, 119
1,3-Dipole, 114
1,4-Dipole, 99
Dithiobiurets, 105
1,1-Ditrifluoromethyl-2,2-dicyanoethylene, 95
Divalent carbon, in isonitrile complexes, see Isonitrile complexes, divalent carbon
$\sigma$-Donors, 109
Dynamic programming, see Peptide synthesis, optimum tactics

# E

Einhorn–Woodward acid, 188
Electrolysis, see Isonitriles, electrolysis
$\alpha$-Eliminations, 10–16
Enamines, see Four-component condensation, enamines
Ethylborane, see Borane, triethyl
Ethyl isocyanide, vibration and rotation spectra, 235, 236
Ethyl isocyanoacetate
    Claisen condensation, 237
    condensation with carbonyl compounds, 250, 251

# F

$\alpha$-Ferrocenylalkyl amines, see Four-component condensation, $\alpha$-ferrocenylalkyl amines
$\alpha$-Ferrocenylalkyl carbonium ions, 209–211
$\alpha$-Ferrocenylethyl amine, see also Four-component condensation
    absolute configuration, 210
    stereoselectivity, 161–181, 201–211
Formally divalent carbon, 133
Formamide chlorides, see $\alpha$-Addition, halides
Formic acid, see $\alpha$-Addition, formic acid
$\alpha$-Formylaminoacrylic esters, 251, see also Ethyl isocyanoacetate

Four-component condensations, 95, 145–213, see aslo individual headings
    acid component, 146–147
    $\alpha$-acylaminocarbonamides, 155–185, see also Peptide synthesis
    acyl hydrazones, 150–153, 159–161, 198
    $\alpha$-adduct, 146–148, 155, 158, 169, 182–184, 189, 191
    reactions with nucleophiles, 158, 189, 191
    amidines, 191
    aminals, 193
    amine component, 146–147
    $\alpha$-amino acids, see Four component condensation, $\beta$-lactam
    $\alpha$-aminoalkyl azlactones, 158–159
    $\alpha$-aminocarbonamides, 190–191
    asymmetric induction, see Four-component condensations, stereoselectivity
    $\alpha$-azido amines, 196
    azines, 161, 197
    carbonyl component, 146
    carboxylic acids, see Four-component condensation, $\alpha$-acylaminocarbonamides
    cysteine derivatives, 183
    $\alpha$-$N_\alpha$,$N_\beta$-Diacylhydrazinocarbonamides, 159–161
    diacyl imides, 189–190
    diene amines, 190–195
    enamines, 189–193
    $\alpha$-ferrocenylalkyl amines, 209–211
    hydantoin-4-imides, 149–151
    hydrazine derivatives, 151–153, 159–161, 197
    hydrazoic acid, see Four-component condensation, tetrazoles
    hydrogen cyanate, see Four-component condensation, hydantoin-4-imides
    hydrogen selenide, see Four-component condensation, selenoamides
    hydrogen thiocyanate, see Four component condensation, 2-thiohydantoin-4-imides
    $\alpha$-hydroxy amides by-product, 191
    hydroxylamine derivatives, 253
    immonium betaines, 182
    immonium ion intermediates, 148–169

intermediates, 148, 161–181
internal stereoselectivity, 170–177
isonitrile component, 146
β-lactam, 181–183
mechanism, see Four-component condensation, reaction mechanism
monomethyl carbonate, see Four-component condensation, urethanes
p-nitrophenyl cyanamide, 155
optically active primary amine component cleavability of, 206–211
pairs of corresponding reactions, 164–181
Passerini reaction, competition with, 191
penicillamine derivatives, 183
penicillanic acid derivatives, 183–185
reaction coordinate, monotonic, 175
reaction mechanism, 146, 161–181
reactive intermediates, see Four-component condensation, intermediates
Schiff bases, 149–150, 156, 159, 164–169, 194–195
secondary reaction of α-adducts, 145–149
selenoamides, 192
semicarbazones, 160
stereoselectivity, 161–181, see also Peptide synthesis
  coefficients, 181
  kinetic control, 169
stoichiometric matrix, 171
stoichiometric relations, independent, 171
sulfonyl hydrazones, 153, 161, 197
tetrazoles, 193–197, see also Tetrazoles
$\Delta^3$-thiazolines, 152, 183
thioamides, 192
2-thiohydantoin-4-imides, 149–155
thiosulfate, see Four-component condensation, thioamides
transannular reacylation, 182, 184
urethanes, 185
yields, 149
Fragment strategy, see Peptide synthesis, fragment strategy

**G**

Gautier's method, 17–18

Glyoxylic acid derivatives, 100
Grignard reagent, see α-Addition, organometallic compounds

**H**

Halogen, see α-Addition, halogen
Hexafluoroacetone, 101
Hexafluoroacetone acylimines, as 1,4-dipoles, 101
Hexafluorobutyne-2, reaction with isonitriles, 249
Hofmann degradation, anomalous, 23
Hofmann's method, 17–18
Homogeneous catalysts, see Isonitriles, oxidation
Homopolymerization, see Isonitriles, homopolymerization
Hydantoin-4-imides, see Four-component condensation, hydantoin-4-imides
Hydantoins, see Bucherer–Bergs reaction
Hydrazine derivatives, see Four-component condensation, hydrazine derivatives
Hydrazoic acid, 74, 97, 139
Hydrides, nonmetal, see α-Additions, nonmetal hydrides
Hydrogen, see Isonitriles, reduction
Hydrogen bonds, 109
Hydrogen ferrocyanide, alkylation, see Isonitrile complexes, synthesis by alkylation of hydrogen ferrocyanide
Hydrogen halides, see α-Addition, hydrogen halides
Hydrogen thiocyanate, 105, 149–155
Hydroxyacrylamides, 96
α-Hydroxyamides, see Passerini reaction
α-Hydroxyisocyanides, 249, see also β-Hydroxyisocyanides
β-Hydroxyisocyanides, 249
γ-Hydroxyisocyanides, see Dihydro-1,3-oxazines
Hydroxylamine, see Four-component condensations, hydroxylamine; α-Addition, hydroxylamine

**I**

Imidazole derivatives, 114
  from α-metallated isonitriles, 250, 252
Imines, see Schiff bases

Iminoepoxides, *see* Passerini reaction
Iminooxiranes, *see* Passerini reaction, mechanism
Immonium ions, *see* Four-component condensation, immonium ion intermediate
Indigodianil, *see* Phenyl isocyanide, tetramer
Indole carbonamides, 100
Indolenins, 100
Insulin, *see* Peptide synthesis, optimum tactics
Internal stereoselectivity, *see* Four-component condensation, internal stereoselectivity
Inverse yield function, *see* Peptide synthesis, optimum tactics
Ion pair, oriented, *see* Four-component condensation, reaction mechanism
Isobutyraldehyde-(S)-α-phenylethyl imine, 164–181
Isocyanamides, 15, 140
Isocyanates, *see also* Isonitriles, oxidation reaction with enamines and isonitriles, 99
 reduction, 19
α-Isocyano acids, 14, 158–159
α-Isocyanoacyl peptide esters, 201
1-Isocyanocyclohexyl carboxylic acid, 138
α-Isocyano esters, 13–14, 102, 201
Isocyano group
 bond refraction, 5
 charge distribution, 4
 electron density, 4
 force constants, 4
 fundamental vibrations, 5
 heat of formation calculations, 5
 infrared spectra, 5
 polarizability, 5
 thermodynamic data, 4
Isocyano peptides, 14, 204, 205, 211
Isodiazomethane, *see* Isocyanamide
Isomerization
 cationic, *see* Isonitriles, cationic isomerization
 of isonitriles, *see* Isonitriles, thermal rearrangement
Isonitrile complexes, 217–232
 antibonding orbitals, 218
 back donation, 218–222

 bending of CNR axis, *see* Isonitrile complexes, geometry
 bond linearity, *see* Isonitriles, coordinated geometry
 catalytic activity in oxygen transfer, 231
 catalytic intermediates, 230
 charge distribution, 218–222
 copper (II), 254
 "divalent carbon," 218
 donor bonds
  π-type, 218
  σ-type, 218
 electrophilic substitution, 227
 geometry, 218–219
 IR absorption, 218
 photochemistry, 222
 synthesis, 231–232
  by alkylation of hydrogen ferrocyanide, 231
  of silver ferrocyanide, 231
  via carbonium ions, 227
  by ligand substitution, 232
 transalkylation, 229
Isonitrile component, *see* Four-component condensation, isonitrile component
Isonitriles
 acceptor strength, 221–222
 analytical procedures, 9
 anodic methoxylation, 241
 base-catalyzed dimerization, 249
 cationic oligomerization, 89
 coordinated, *see* Isonitrile complexes
 cyclization reactions, *see* Cyclization reactions
 dipole moment, 3
 electrolysis, 87–88
 electron diffraction, 3
 formal divalency of carbon, 2, 66
 history, 1
 homopolymerization, 68
  via isonitrile complexes, 226
 hydrogen bonding, 5
 insertion reactions into nickel–alkyl bonds, 225–226
 isomerization
  cationic, 89
  radical catalysis, 61, 84
 Lewis base character, 5, 109

mass spectra, 6
α-metallated, 104, see also
    α-Additions, organometallic compounds
methods of preparation, see Isonitriles, syntheses
microwave data, 3
moments of inertia, 53
multicomponent reactions of, 99, 103, 115, 119, 133–143, 145–197, 201–213
nomenclature, 1
nuclear quadrupole coupling in, 5
odor, 9
organoborane reactions, 109–124
oxidation, 79
    dimethyl sulfoxide, 79
    homogeneous catalysts, 79
    ozone, 79
    metal oxides, 79
    by nitrogen oxides, 242
    peracids, 79
    pyridine $N$-oxide, 79
perturbation by coordination, 217
photoisomerization, see Methyl isocyanide, photoisomerization
photoreaction, 248
polarographic reduction, see Isonitriles, reduction
polyfunctional, 14
polymerization, see Isonitriles, homopolymerization
preparation of, 12–13
protective group for, 227
purification of, 12
quadrupole coupling constants, 236
reaction
    with benzyne, 88, 99
    radical, 84–88
reduction, 80
    alkali metals, 80
    catalytic hydrogenation, 80
    hydrogen, 80
    polarographic, 241
structure, 1
syntheses, 9–35
    by abnormal Beckmann rearrangement, Beckmann rearrangement, abnormal
    from cyanides, 17–18
    from isonitrile dichlorides, 20
    from olefins, 18
    from onium dicyanoargentates, 237
    by redox reactions, 19–20
    table of, 24–35
    thermal rearrangement
        dependence of activation energy on pressure, 49
        fall-off shape parameters, 46, 56
        gas phase, 43–45
        kinetics, 41–64
        pressure dependence of unimolecular rate constants, 44–45
        radical catalysis, 61, 84
        radical mechanism, 237
        rate constant, fall-off plot, 43, 51
        secondary isotope effect, 54
        shock tube, see Methyl isocyanide, rearrangement, shock tube
        solution, 59–60
        theoretical calculation of rate constants, 45
    toxicity to mammals, 2
    trimerization, see $t$-Butyl isocyanide trimer
    triple additions, tetrazole formation, 139
UV spectra, 5
vibrational spectra, 3
Isothiocyanates, see also α-Addition, sulfur, thiols
    reaction with enamines and isonitriles, 99
    synthesis of isonitriles from, 19

### K

Ketenes, 97
α-Ketoacids, 241
α-Ketoamides, see
    α-Addition, halides; Cyclization reactions, four-membered rings
α-Ketoimidoyl chlorides, see
    α-Addition, halides
α-Ketothiocarbonamides, see
    α-Addition, halides

### L

α-Lactams, 23, see also Cyclization reactions

## Subject Index

β-Lactam, 96, see also Four-component condensation, β-lactam
Lactones, see Passerini reaction
Lewis base character of isonitriles, see Isonitriles, Lewis base character
Ligand substitution, see Isonitrile complexes, synthesis by ligand substitution
α-Lithiated isonitriles, 83

### M

Mercaptans, see α-Additions, thiols
Merrifield synthesis, see Peptide synthesis, Merrifield's method
Metal-isonitrile bond, dissociation, 222
α-Metallation of isonitriles, 83, 249
Metal oxides, see Isonitriles, oxidation
4-Methoxyphenyl oxazole, 103
Methyl acrylate, see α-Additions, olefins
Methyl isocyanate, 97
Methyl isocyanide
    molecular data, 3, 236
    photoisomerization, 62–64
    rearrangement
        in shock tube, 58–59
        by transfer of vibrational energy, 238
    thermal rearrangement, see Isonitriles, thermal rearrangement
Methylphenyldiazoacetate, 239
2-Methylpropane-2-thiol, see α-Addition, 25
N-Monosubstituted formamides
    dehydration, 10–16, 73
    by phosgene, see Phosgene method
Multicomponent reactions, see Isonitriles, multicomponent reactions of
Mumm rearrangement, 189

### N

α-Naphthol, 98
Naphthols, see α-Addition, naphthols
Nitrenes, 239, see also α-Addition, nitrenes
Nitrilium ion, see Four-component condensation, reaction mechanism
Nitrogen halogen compounds, see α-Addition, halide

Nitrones, 240
Nitroso alkanes, see Diaziridones
Nitrosobenzene, 104
α-Nitroso-β-naphthol, 102
Nitrosotrifluoromethane, see Cyclization reactions, four-membered rings
Nomenclature, see Isonitriles, nomenclature
Nomenclature of diastereomers, 162
p-n-Nomenclature, see Nomenclature of diastereomers

### O

Olefins
    from oxazoline decomposition, 104
    synthesis of, 83
Oligomerization, see Isonitriles, cationic oligomerization
    of phenyl isocyanide, see Phenyl isocyanide, oligomerization
Organoaluminum compounds, see Aluminum, organic compounds of
Organoborane-isonitrile reactions, see Isonitriles, organoborane reactions
Organolead-amino compounds, see α-Additions, organometallic compounds
Organometallic compounds, see α-Addition, organometallic compounds
Organotinamino compounds, see α-Addition, organometallic compounds
Oxazolidine, synthesis of, 102
Oxazoline(s), 101, 104
$\Delta^2$Oxazolines, 248, see also α-Metallated isonitriles
Oxazolinium betaine, 114
Oxetane(s), 100, see also Cyclization reactions
Oxidation of isonitriles, see Isonitrile, oxidation
Oxygen, see Isonitriles, oxidation
Oxygen, halogen compounds, see α-Addition, halogen compounds
Ozone, see Isonitriles, oxidation

### P

Passerini reaction and related reactions, 97, 133–136

α-acyloxycarbonamides from, 136–139 (table), 134–135
α-adducts in, 138
asymmetric induction in, 136
depsipeptides from, 136
α,γ-diketoamides from, 142–143
  table, 142
α-hydroxyamides, 140–142
  table, 141
iminooxiranes from, 138
lactones from, 136
mechanism of, 136–139
sterically hindered ketones in, 133
tetrazoles formed by, 139–140
α,β-unsaturated ketones in, 133
Penicillamine derivatives, see Four-component condensation, penicillamine derivatives
Penicillanic acid derivatives, see Four-component condensation, penicillanic acid derivatives
Peptide synthesis
  activating groups, 203
  classical concept, 202–204
  fragment strategy, 203, 211
  Merrifield's method, 203
  model reactions, 155–159, 161–181, 204–211
    cleavable β-alanine derivatives, 208–209
    resonance stabilized vinyl amines, 207
    stereoselective cleavage, 208–209
  optimum tactics, 211–213
  protecting groups, 203
  tactics, 211
Peracids, see Isonitriles, oxidation
Perfluorobutyne, see Hexafluorobutyne-2, reaction with isonitriles
Pernitrosocamphor, 140
Pernitrosomenthol, 141
Phenols, see α-Addition, phenols
Phenylacetylene, 109
N-Phenylbenzaldimine, 118
Phenyl isocyanide, 98
  oligomerization, 104
  p-substituted, thermal rearrangement, see Isonitriles, thermal rearrangement in solution
Phenyl isocyanide tetramer, 100

Phenyl methoxycarbonylketene-$N$-$t$-butylimine, 239
Phosgene, see α-Addition, halides
  method for preparing isonitriles, 11–14
Phosphines, 240, see also α-Additions, phosphines
Photoisomerization, see Methyl isocyanide, photoisomerization
Phthalimide, 99
Polyhedral boranes, see Boranes, polyhedral
Polymerization, see Isonitriles, homopolymerization
Protecting groups, see Peptide synthesis, protecting groups
Pyrazines,
  diboradihydro-, 252
  2,5-dibora-2,5-dihydro, 110, 123
  2,5-dibora-3,6-dihydro, 110, 115
  by hydrogenation of diisocyanides, 105
  table, 111–113
Pyrazolone derivatives, see α-Addition, active methylene compounds
Pyridine $N$-oxide, see Isonitriles, oxidation
Pyrrole derivatives, electrophilic substitution, see α-Additions, pyrrole derivatives

**Q**

Quadrupole moments, see Isonitriles, nuclear quadrupole coupling
Quinazoline-3-oxide, 22
Quinoline quinone derivative, 104
Quinolinium ions, 186–188

**R**

Radical reactions
  of isonitriles, see Isonitriles, radical reactions of,
  stannanes, see α-Addition, stannanes
Radicals, see Isonitriles, thermal rearrangement
Reduction of isonitriles, see Isonitriles, reduction
Ring closures, see Cyclization reactions

## S

Saegusa reaction, 228
Schiff bases, 248, see also Four-component condensation, Schiff bases cyclization reactions with isonitriles, 243
Selenoamides, see Four-component condensation, selenoamides
Semicarbazones, see Four-component condensations, semicarbazones
Silanes, see α-Additions, silanes
Silver ferrocyanide, alkylation, see Isonitrile complexes, synthesis by alkylation of silver ferrocyanide
Sodium isocyanotrihydroborate, 236
Stannanes, see α-Addition, stannanes
Stereochemical linear free energy relationship, 207
Stereoselective cleavage, see Peptide synthesis, model reactions
Stereoselective four component condensations, see Four component condensations, stereoselectivity
Stereoselectivity coefficient, see Four-component condensation, stereoselectivity coefficients
Sterically hindered ketones, see Passerini reaction
Succinimide derivatives, 99
Sulfenyl chlorides, see α-Addition, sulfur halides
Sulfonyl hydrazones, see Four-component condensations, sulfonyl hydrazones
Sulfur, α-addition, see α-Addition, Sulfur; Isonitriles, oxidation
Sulfur halogen compounds, see α-Addition, halides

## T

Tetrazole derivatives, see Tetrazoles
Tetrazoles, 74, 97, 139–140, 193–197, see also α-Addition, hydrazoic acid
improved synthesis, 242
table of, 140
Thermal rearrangement of isonitriles, see Isonitriles, thermal rearrangement
Thiazole diones, 103

$\Delta^3$-Thiazolines, see Four-component condensations, $\Delta^3$-thiazolines
Thioacyl isocyanates, 103
from thiazole diones, 103
Thioamides, see Four-component condensation, thioamides
Thiocyanic acid, 105, 146, 149–153, see also Hydrogen thiocyanate
Thiohydantoin-4-imides
oxidative desulfurization, 154
reductive desulfurization, 154
2-Thiohydantoin-4-imides, see Four-component condensations, 2-thiohydantoin-4-imides
Thiols, see α-Additions, thiols
Transannular acyl migration, see Four-component condensation, transannular reacylation
Transannular reacylation, see Four-component condensation, transannular reacylation
Trialkyltin hydrides, see α-Addition, stannanes
Triazines, from hydrogen thiocyanate and isonitriles, 105
Tri-n-butylborane, see Borane, tri-n-butyl-
Triethylborane, see Borane, triethyl-
Trifluoroalanine, 102
N-Trifluoromethyl-N'-methylcarbodiimide, 97
Tripeptides, 102
synthesis, 206
Triphenylaluminum, see Aluminum, triphenyl-
Trimethylborane, see Borane, trimethyl-
Triphenylborane, see Borane, triphenyl-
Triphenyl phosphine fluoborate, 240
Triple additions of isonitriles, see Isonitriles, multicomponent reactions; Passerini reaction
Tropylium ion, see α-Addition, tropylium ion

## U

Ugi reaction, 95, 97, see also Four-component condensations
Ugi reaction
hydrazine derivatives, 252
hydroxylamine derivatives, 253

Unsaturated iminoamides, 95
α,β-Unsaturated ketones, *see* Passerini reaction
Urethanes, *see* Four-component condensation, urethanes

## V

Vilsmeier reagents, *see* α-Addition, halides
Vinyl amines, resonance stabilized, *see* Peptide synthesis, Model reactions

## W

Water, *see* α-Addition, water; Four-component condensation, α-Aminocarbonamides

## X

Xanthocillin, 15
 biosynthesis, 15
 $O,O'$-dimethyl ether, 15
Xanthocillin dimethyl ether, hydrogenation, 105

## Y

Yield, inverse, *see* Inverse yield function

## Z

Zinc chloride catalysis, *see* α-Addition, zinc chloride catalysis